Geomorphological Hazards and Disaster Prevention

Human activities, especially in the last two centuries, have had a huge impact on the environment and landscape through industrialisation and land-use change, leading to climate change, deforestation, desertification, land degradation, and air and water pollution. These impacts are strongly linked to the occurrence of geomorphological hazards, such as floods, landslides, snow avalanches, soil erosion, and others. The work undertaken by geomorphologists includes not only the understanding but also the mapping and modelling of Earth's surface processes, and many of these processes directly affect human activities and societies. In addition, geomorphologists are now becoming increasingly involved with the dimensions of societal problem solving, which can be expressed through vulnerability analysis, along with hazard and risk assessment and management. The work of geomorphologists is therefore of prime importance for disaster prevention.

This volume, with chapters written by an international team of geomorphologists:

- provides state-of-the-art knowledge about the contribution of geomorphology to the comprehension of hazards;
- links the work undertaken by geomorphologists to the framework of the likely impacts of climatic change and global environmental change;
- shows the significance of technology (remote sensing and Geographical Information Systems) for hazard and risk assessment and management;
- demonstrates the role of geomorphology in vulnerability and risk analysis, disaster prevention and sustainability.

The language is scientifically rigorous but accessible to a wide audience of geomorphologists and other Earth scientists, including those involved in environmental science, hazard and risk assessment, management and policy.

IRASEMA ALCÁNTARA-AYALA was born in Mexico City in November, 1970. She received her first degree in Geography at the National Autonomous University of Mexico (UNAM). In 1997, she obtained her Ph.D. in Geography with speciality in Geomorphology from King's College London, University of London. Afterwards, she carried out a postdoctoral stay in the Department of Civil and Environmental Engineering at the Massachusetts Institute of Technology, Boston. She is currently Director and Professor of the Institute of Geography at the National Autonomous University of Mexico (UNAM). Her research is concentrated on mass movement processes, natural hazards, risks, vulnerability, and prevention of disasters. She has published numerous peer reviewed papers and book chapters, and has presented her work at numerous international meetings. In 2005 she was awarded the 3rd Evelyn Pruitt Lecture by the Department of Geography and Anthropology, Louisiana State University, Baton Rouge, USA. She is on the Editorial and Advisory Editorial Boards of renowned scientific journals including *Earth Surface Processes and Landforms, Journal of Mountain Science, Singapore Journal of Tropical Geography*, among others. From 2002 to 2005, she acted as a member of the International Association of Geomorphologists (IAG) Executive Committee and was appointed as the International Geographical Union (IGU) Representative within the Earthquakes and Megacities Initiative. In 2007, she represented IGU at the conference "Global Scientific Challenges: Perspectives from Young Scientists" (an international conference celebrating 75 years of ICSU). She was IGU theme leader of the "International Year of Planet Earth" (IYPE) related to the topic of Deep Earth – from crust to core. She is President of the Mexican Society of Geomorphology and Chair of the Geomorphological Hazards Working Group of the International Association of Geomorphologists (IAG). She is a Member of the International Council for Science (ICSU) Committee on Scientific Planning and Review (CSPR), and of the International Consortium on Landslides. Recently she was elected as TWAS Young Affiliate Fellow, and at present she is Vice-President of the International Geographical Union (IGU).

ANDREW GOUDIE was Professor and Head of Department of Geography at Oxford University. A distinguished physical geographer, he was awarded a DSc by Oxford University, received a Royal Medal from the Royal Geographical Society, the Prize of the Royal Belgian Academy, the British Society for Geomorphology's David Linton Award (2009) and the Geological Society of America's Farouk El-Baz Award for Desert Research. He has been President of the Oxford University Development Programme, Pro-Vice-Chancellor of the University, and Delegate of Oxford University Press. He has recently been President of the Geographical Association, President of Section E of the British Association and Chairman of the British Geomorphological Research Group. Professor Goudie became Master of St Cross College, Oxford, in October 2003, and continues to lecture at the Oxford University Centre for the Environment.

Since 2005 he has been President of the International Association of Geomorphologists. In addition to being author of nearly 200 scientific papers, he is the author or co-author of the following books (amongst others): *The Human Impact, The Nature of the Environment, Environmental Change, The Encyclopedia of Global Change, Geomorphology of Deserts, Geomorphological Techniques, Chemical Sediments and Geomorphology, The Geomorphology of England and Wales, The Warm Desert Environment, Discovering Landscape in England and Wales, Landshapes, The Encyclopedic Dictionary of Physical Geography, Desert Geomorphology, The Student's Companion to Geography, The Earth Transformed, Aeolian Environments, Sediments and Landforms, The Encyclopedia of Geomorphology, The Oxford Companion to Global Change* and *Wheels Across the Desert*.

Geomorphological Hazards and Disaster Prevention

Irasema Alcántara-Ayala

Universidad Nacional Autonóma de México, Mexico City

Andrew S. Goudie

St Cross College, Oxford

CAMBRIDGE
UNIVERSITY PRESS

CAMBRIDGE UNIVERSITY PRESS

Cambridge, New York, Melbourne, Madrid, Cape Town, Singapore,
São Paulo, Delhi, Dubai, Tokyo

Cambridge University Press
The Edinburgh Building, Cambridge CB2 8RU, UK

Published in the United States of America by Cambridge University Press, New York

www.cambridge.org
Information on this title: www.cambridge.org/9780521769259

First published 2010

Printed in the United Kingdom at the University Press, Cambridge

A catalogue record for this publication is available from the British Library

ISBN 978-0-521-76925-9 Hardback

Contents

Contributors

Irasema Alcántara-Ayala
Instituto de Geografía, UNAM
Circuito Exterior, Ciudad Universitaria
04510, Coyoacán, México, D.F.
México

Gilles Arnaud-Fassetta
Université Paris-Diderot
Case Postale 7001
75205 Paris Cedex 13
France

Perry Bartelt
Avalanches, Debris Flows and Rockfall Research Unit
WSL Institute for Snow and Avalanche Research SLF
Flüelastrasse 11
CH-7260 Davos Dorf
Switzerland

Gerardo Benito
Laboratorio de Geomorfología e Hidrología
Centro de Ciencias Medioambientales, CSIC
Serrano 115 dup.
28006, Madrid
Spain

John Boardman
Environmental Change Institute,
School of Geography and the Environment
South Parks Road
Oxford, OX1 3QY
UK

Lisa Borgatti
Dipartimento di Ingegneria delle Strutture, dei Trasporti,
delle Acque, del Rilevamento, del Territorio
Alma Mater Studiorum Università di Bologna
Viale Risorgimento, 2
41036 Bologna
Italy

William B. Bull
6550 N. Camino Katrina
Tucson, AZ, 85718-2022
USA

Michael Bründl
Warning and Prevention Research Unit
WSL Institute for Snow and Avalanche Research SLF
Flüelastrasse 11
CH-7260 Davos Dorf
Switzerland

Etienne Cossart
UMR Prodig 8586 – CNRS
Université Panthéon-Sorbonne (Paris 1)
2 rue Valette,
75005 Paris
France

Michael Crozier
Victoria University of Wellington
School of Geography, Environment and Earth Sciences
PO Box 600
Wellington
New Zealand

Monique Fort
UMR Prodig 8586 – CNRS
Université Paris Diderot (Paris7)
Département de Géographie
UFR GHSS, Case 7001
75205 Paris Cedex 13
France

Thomas Glade
Department of Geography and Regional Research
University of Vienna
Universitaetsstr. 7
A-1010 Vienna
Austria

Andrew S. Goudie
St Cross College
St Giles
Oxford, OX1 3LZ
UK

Avijit Gupta
School of Geography
University of Leeds
Leeds, LS2 9JT
UK

Francisco Gutiérrez
Universidad de Zaragoza
Departamento de Ciencias de la Tierra
Edificio Geológicas
C/. Pedro Cerbuna, 12
50009 Zaragoza
Spain

David Higgitt
Department of Geography
National University of Singapore
1 Arts Link, Kent Ridge
117570 Singapore

Paul F. Hudson
Department of Geography and the Environment
University of Texas at Austin
Austin, TX 78712
USA

Gabi Hufschmidt
School of Geography
Victoria University of Wellington
P.O. Box 600
Wellington
New Zealand

Margreth Keiler
Department of Geography and Regional Research
University of Vienna
Universitaetsstr. 7
A-1010 Vienna
Austria

Molly McGraw
Southeastern Louisiana University
SLU-10686
Hammond, LA 70402
USA

David Petley
Durham University
Department of Geography, Science Laboratories
South Road
Durham, DH1 3LE
UK

Jürg Schweizer
Snow and Permafrost Research Unit
WSL Institute for Snow and Avalanche Research SLF
Flüelastrasse 11
CH-7260 Davos Dorf
Switzerland

Olav Slaymaker
University of British Columbia
Department of Geography
1984 West Mall
Vancouver, V6T 1Z2
Canada

Mauro Soldati
Dipartimento di Scienze della Terra
Università di Modena e Reggio Emilia
Largo S. Eufemia, 19
41121 Modena
Italy

Jean-Claude Thouret
Laboratoire Magmas et Volcans UMR 6524 CNRS et
OPGC
Université Blaise Pascal Clermont II
5 rue Kessler
63038 Clermont-Ferrand Cedex
France

Cees J. van Westen
International Institute for Geo-Information Science and
Earth Observation (ITC)
Hengelosestraat 99
PO Box 6
7500 AA Enschede
The Netherlands

Heather Viles
School of Geography and the Environment
South Parks Road
Oxford, OX1 3QY
UK

Harley J. Walker
Department of Geography and Anthropology
Louisiana State University
Baton Rouge, LA 70803
USA

YANG Xiaoping
Institute of Geology and Geophysics
Chinese Academy of Sciences
P.O. Box 9825
Beijing 100029
China

1 Introduction

Andrew S. Goudie

As Rosenfeld (2004, p. 423) wrote, 'A significant practical contribution of geomorphology is the identification of stable landforms and sites with a low probability of catastrophic or progressive involvement with natural or man-induced processes adverse to human occupance or use. Hazards exist when landscape developing processes conflict with human activity, often with catastrophic results.' Geomorphic events can kill people and damage property. Although high-magnitude, low-frequency catastrophic events, such as hurricanes or earthquakes, gain attention because of the immediacy of large numbers of casualties and great financial losses, there are many more pervasive geomorphological changes that are also of great significance for human welfare. These may have a slower speed of onset, a longer duration, a wider spatial extent and a greater frequency of occurrence. Examples include weathering phenomena and soil erosion. In this volume we discuss both types of geomorphological hazard: the catastrophic and the pervasive.

Indeed, there is a great diversity of geomorphological hazards. One major category is mass movements, such as rockfalls, debris flows, landslides and avalanches. There are also various fluvial hazards, such as floods and river channel changes (e.g. avulsion). In volcanic areas there are disasters caused by eruptions, lava flows, ash falls and lahars. Seismic activity is another type of hazard associated with tectonic activity. In coastal environments one has inundation and erosion caused by storm surges, rapid coastal erosion and siltation, sand and dune encroachment, shoreline retreat and sea-level rise. In glacial areas hazards may be posed by such phenomena as glacial surges, outwash floods and damming of drainage. Permafrost regions may be hazardous because of ground heave, thermokarst development, icings and other such phenomena. There is also a wide range of subsidence hazards caused by solution of limestone, dolomites and evaporites (e.g. gypsum or halite), degradation of organic soils, hydrocompaction of sediments and anthropogenic removal of groundwater and hydrocarbons. In desert regions hazards are posed by wind erosion and deflation of susceptible surfaces, dust storm generation, and by dune migration. More generally, water erosion causes soil loss and gully or badland formation, while weathering can be a threat to a wide range of engineering structures.

The incidence of such hazards can be increased or triggered by human activities, and in particular by land use and land cover changes. There is also an increasing concern that the incidence of hazards will be changed in a warmer world. However, another important consideration is the extent to which human societies are placing themselves at an increased risk as population levels increase and new areas are exploited or settled. Potentially hazardous areas, such as floodplains or steep, deeply weathered slopes, may become occupied, placing human groups at risk. Large urban populations may be especially at risk (see, for example Cooke, 1984). There is evidence that for these sorts of reasons, damage to property and loss of life caused by geomorphological hazards are increasing (Alcántara-Ayala, 2002).

The roles of the geomorphologist in hazard research are many. Of great importance are: the mapping of hazard-prone areas (Griffiths, 2001); constructing the history of occurrence of past hazardous events; establishing their frequency and magnitude; predicting the occurrence and location of future events; monitoring geomorphological change; and using knowledge of the dynamics of geomorphological processes to advise on appropriate mitigation strategies.

In recent years the capabilities of geomorphologists in these roles have increased and the application of

Geomorphological Hazards and Disaster Prevention, eds. Irasema Alcántara-Ayala and Andrew S. Goudie. Published by Cambridge University Press. © Cambridge University Press 2010.

2 Regional seismic shaking hazards in mountains

William B. Bull

2.1 Introduction

The steep, crumbly and wet mountains of New Zealand have frequent mass movements and floods, and landscape instability related to glaciers. The Southern Alps sit astride the Australian–Pacific plate boundary (Figure 2.1) with its highly active dextral-oblique thrust faults, active volcanoes, and subduction zones. This chapter summarizes a hazardous geomorphological consequence of widespread seismic shaking in the South Island of New Zealand.

Rocks tumble downhill during earthquakes and the times of arrival on a scree slope can be dated by measuring the largest lichen of *Rhizocarpon* subgenus *Rhizocarpon* on many blocks. This discussion focuses on how to use lichenometry to better understand the intensity and history of hazardous seismic shaking so as to minimize risk. A typical study site is described, the precision of dating is noted, and maps revealing areal intensities of seismic shaking are examined to better understand sizes and sequences of earthquakes.

2.2 Lichenometry site characteristics

The Mt. John site summarizes the desirable features needed for seismic-hazard evaluations. The valley side here was trimmed by a late Pleistocene glacier, creating cliffy outcrops of crumbly greywacke sandstone that have shed blocks to be stored in talus accumulation areas. Hazards such as snow avalanches and water floods are absent. I would have chosen an active talus cone if I had wanted to study the frequency and sizes of avalanches. *Rhizocarpon* subgenus *Rhizocarpon* grows well at this altitude and climate, but locally is impinged upon by other lichens and by mosses.

Each tongue of scree between the bands of dark bushes (Figure 2.2) was inspected for blocks with isolated, circular to elliptical lichens (Figure 2.3) with longest axes clearly demarcated with abrupt margins suitable for measurement with digital calipers. Only the largest *Rhizocarpon* subgenus *Rhizocarpon* was measured on each of the blocks, which range in size from 0.2 to 2 m. The number of lichens to evaluate and choose from generally increased with block size.

As a quality control measure, I made a subjective evaluation of the relative quality of each lichen-size measurement. The main questions to be answered were:

1. Is this really a single thallus, or have several lichens grown together?
2. Is the lichen sufficiently well preserved to reveal the endpoints of the longest axis of growth?
3. Are the margins at the two measurement points sharp and well defined?

Assigned quality control numbers range from 1 to 4:

1. A superb lichen – nearly circular single thallus with abrupt prothallus rims at the two measuring points – that has you reaching for your camera to take a picture of it.
2. Close to ideal for a reliable lichen-size measurement.
3. Nothing special but we feel quite comfortable in including it in the dataset.
4. We hesitate about including this lichen size in the dataset; it has borderline characteristics.

The lichens shown in Figure 2.3 are only of quality class 3, but the rating was raised to 2 because of the presence of a second lichen on this block of about the same size as the largest lichen.

A dataset of 546 lichen-size measurements was collected in 2008. The longest axes range from 6 to 137 mm (age range is AD 1150 to 1980). I have been

Geomorphological Hazards and Disaster Prevention, eds. Irasema Alcántara-Ayala and Andrew S. Goudie. Published by Cambridge University Press. © Cambridge University Press 2010.

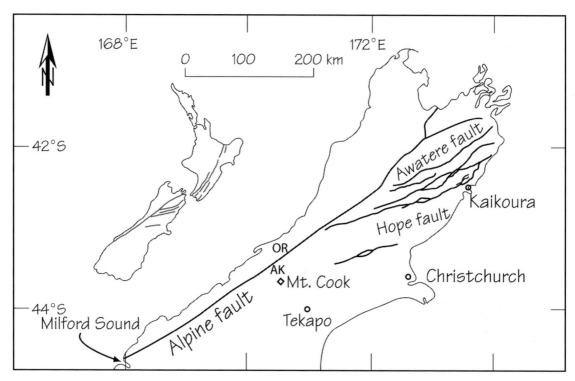

FIGURE 2.1. New Zealand location map. The plate-bounding Alpine fault splits into dextral-oblique thrust faults in the northeast corner of the South Island (the Marlborough district). OR and AK are the Oroko Swamp and Alex Knob tree-ring analysis sites.

FIGURE 2.2. Mt. John scree lichenometry site, east of the Southern Alps of New Zealand. The largest lichen was measured on 546 blocks in the long tongues of talus below rock outcrops of highly fractured, slightly metamorphosed, greywacke sandstone. Buildings at summit are part of an astronomical observatory. Site is at 43° 59′ 19″ S, 170° 28′ 6.67″ E and at an altitude of 815–930 m.

measuring lichens in New Zealand since 1989, so direct comparison of this dataset with those of previous years is not possible unless the data are normalized, in this case to my base year of 1992.

Some Mt. John rockfalls are caused by frequent processes such as frost wedging, but the Figure 2.4 histogram consists of many distinct peaks indicative of pulses of outcrop collapse. These data are meaningful only when compared to my South Island dataset of 37 000 lichen sizes measured at 101 sites. All of the Figure 2.4 prominent lichen-size peaks occur at other sites. Peak B dates to the time of a tree-ring dated Alpine fault earthquake

TABLE 2.1. *Comparison of the precision of lichenometry and tree-ring methods of dating regional seismic shaking events in the South Island of New Zealand*

Dating method	1715 event	1615 event	1580 event
Lichenometry	$n=10$	$n=9$	$n=8$
Mean age, two standard deviations	AD 1716 ± 2.80 years	AD 1613 ± 1.78 years	AD 1579 ± 2.08 years
Tree-ring analyses	$n=12$	$n=9$	$n=9$
Mean age, two standard deviations	AD 1716 ± 2.78 years	AD 1615 ± 2.58 years	AD 1578 ± 3.12 years

FIGURE 2.3. The longest axes of these two elliptical *Rhizocarpon* subgenus *Rhizocarpon* on this rockfall block both are 74.89 mm, including the black prothallus fungal rim. These date to *c.* AD 1550.

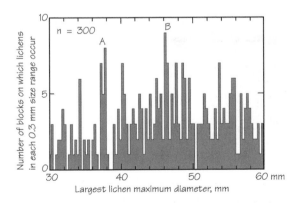

FIGURE 2.4. History of seismic shaking at Mt. John as revealed by histogram of lichen sizes on numerous coseismic rockfalls. Class interval is 0.3 mm. Peak A records two regional seismic shaking events (AD 1788 and 1792 ± 2 years as dated in Figure 6.28 of Bull (2007)). Peak B records the Alpine fault earthquake of AD 1717.

(Table 2.1). Mt. John is 76 km from the Alpine fault. Peak A would be regarded as a single rockfall event if only the Mt. John site is considered. But a twin peak is a defining characteristic for two earthquakes, closely spaced in time, at 67 other sites. Of course some of the smallest peaks are merely dataset noise, but many record weak or distant seismic shaking. A lichen-size peak is indicative of regional seismic shaking if it also occurs at three or more other sites.

Event ages are estimated using the modal value lichen of a lichen-size peak on a standard histogram or by decomposing a composite probability density plot into its component peaks (Bull and Brandon, 1998). Careful selection of the most appropriate class interval or Gaussian kernel size is essential. Unwanted noise is included if too small, and combining of real lichen-size peaks into meaningless peaks results if too large. Comparison of local data with sizes from other sites has led to the realization that using closely spaced lichen-size peaks is warranted. Fortunately, the rate of lichen growth here is sufficiently fast to separate events only 4–6 years apart, but slow enough to date 1,000-year-old events.

The growth rate of several sections of *Rhizocarpon* subgenus *Rhizocarpon* is the same and was determined at sites of known age (Bull and Brandon, 1998):

$$D = 315.31 - 0.1552t,$$

where D is the mean size of a lichen-size peak and t is the substrate-exposure age in years.

Dating precision and accuracy are excellent. The times of three recent prehistorical Alpine fault earthquakes are summarized in Table 2.1. The tree-ring analyses included both simple counting of annual growth rings and dendrochronologic cross dating. The lichenometry age estimates for these three earthquake events are virtually the same as estimated by tree-ring analyses. Accuracy of lichenometric dating is also ±~2 years in both New Zealand and California (Bull, 2003, 2004, 2007, Figures 6.8 and 6.47). Assuming that age estimates presented here are within ±5 years seems reasonable. This large, precise database allows studies of regional variations in the seismic shaking for a particular earthquake.

2.3 Regional seismic shaking

The size of the 84 mm lichen-size peak varies from site to site, and contouring of these data creates a pattern that is parallel to, and close to, the Alpine fault. The resulting peak-size (seismic-shaking index) map (Figure 2.5) suggests rupture of all of the central section of the Alpine fault at about AD 1490. Tree-ring analyses provide strong support for a major surface rupture at this time. Marked suppressions in annual growth of cedar trees at this time were noted in three Oroko Swamp trees and five Alex Knob trees. The event in these eight trees dates as AD 1487 ± 2.56 years. Decomposition of a lichen probability density plot (Bull, 1996, Figure 14) indicates an age of AD 1488.9 ± 4 years (2σ again). The Figure 2.5 map does not cross the Alpine fault because the climate on the northwest side of the Southern Alps is too wet for *Rhizocarpon* subgenus *Rhizocarpon*. The map would be similar if the Alpine fault were offshore. I next continue this type of analysis for the faults of the Marlborough district.

In Figure 2.6, decomposition of a single Gaussian probability density plot for the times of regional rockfall events at 47 lichenometry sites suggests that three coseismic rockfall events occurred during a 6-year time span in the Marlborough district. Estimated calendric ages for these three regional rockfall events are about AD 1842, 1838,

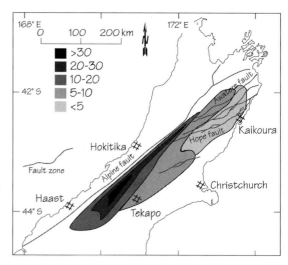

FIGURE 2.5. A seismic-shaking index map for the c. AD 1490 Alpine fault earthquake. The peak-size index is the area of a modeled subpopulation relative to the total area of a Gaussian composite probability density plot in the 6 mm range of lichen sizes that bracket the event time.

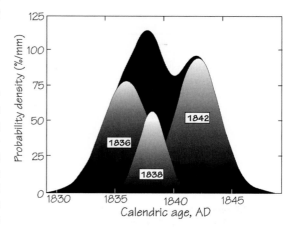

FIGURE 2.6. Decomposition of a Gaussian probability density plot, in black, of the times of regional rockfall events at 47 lichenometry sites. Gaussian kernel size is 1.0 year. (From Figure 22 of Bull and Brandon, 1998.)

and 1836. Can we really separate seismic-shaking events only a few years apart? Seismic-shaking index maps that describe regional patterns of sizes of lichen-size peaks provide an unequivocal answer.

The peak-size pattern for the two older events (Figures 2.6, 2.7A) is suggestive of a composite pattern of seismic shaking resulting from two earthquakes. The smaller event occurred in about AD 1838 near the western edge of the study area. The AD 1836 event occurred along the Conway segment of the Hope fault. The linear area of

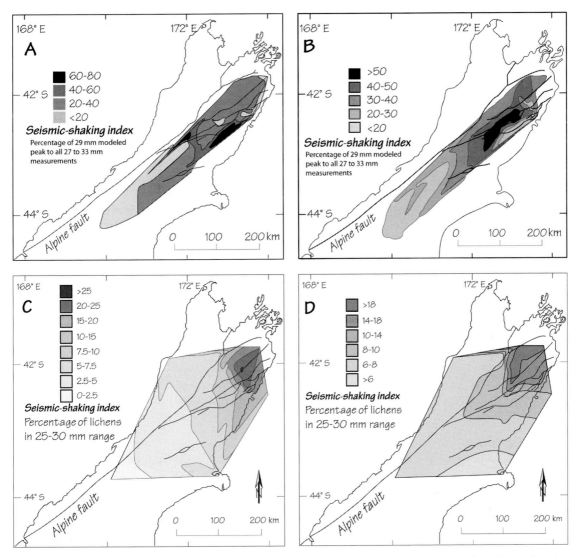

FIGURE 2.7. Seismic-shaking index maps based on regional variations in rockfall abundance. A. Two events at edges of the study region. (From Figure 23A of Bull and Brandon, 1998.) B. Event emanating from center of study region. (Figure 23B of Bull and Brandon, 1998.) C. AD 1848 historical earthquake seismic shaking. (From Figure 6.32C of Bull, 2007.) D. AD 1855 historical earthquake seismic shaking. (From Figure 6.32D of Bull, 2007.)

most intense seismic shaking parallels the fault trace, but continues on to the southwest for another 40 km. This apparent extension of seismic shaking may have resulted as a directivity effect of a fault rupture that had its propagation energy absorbed at the southwest terminus.

The simpler pattern associated with the next oldest seismic-shaking index event (Figure 2.7B) clearly suggests seismic shaking associated with a single large earthquake in about AD 1842 on the Clarence and Elliott faults (between the Hope and Awatere faults on Figure 2.1). The earthquake epicenter, if near the north end, was only about 30 km southwest of the historic AD 1848 surface rupture on the

Awatere fault. Peak sizes for the AD ~1842 event are anomalously low near the Conway segment of the Hope fault, most likely because the AD ~1836 earthquake had already dislodged most of the unstable blocks on those hillslopes.

The internal consistency of the two seismic-shaking index maps and the three peaks of Figure 2.6 support the hypothesis of three earthquakes in 6 years. The AD ~1842 Clarence–Elliott earthquake is not included in lists of historical earthquakes (post AD 1840 European settlement) because the first attempt to find a route up the remote Awatere valley was in AD 1850.

The historic Mw magnitude ~7.4 earthquake of 16 October 1848 damaged many cob houses of early settlers. It occurred on the eastern half of the Awatere fault (Grapes et al., 1998; Benson et al., 2001) A seismic-shaking index map of this event (Figure 2.7C) suggests the same location as the geologic field studies.

A very large Mw magnitude ~8.2 historic earthquake in the southern part of the North Island occurred on 23 January 1855. Beaches were raised as much as 6 m (Hull and McSaveney, 1993) and as much as 12 m of horizontal displacement was noted for the surface rupture (Grapes and Downes, 1997). The intensity of seismic shaking in north-eastern Marlborough (Fig. 2.7D) was about the same as the local 1848 event even though the 1855 earthquake epicenter was ~85 km away. Strong seismic shaking extended far to the southwest.

The 1848 and 1855 seismic-shaking index maps both use the percentage of lichens in their respective lichen-size peaks relative to the total number of rockfalls in the 25–30 mm lichen-size range. This allows construction of a map showing the relative seismic shaking for these two events (Figure 2.8). Marlborough lichenometry site responses to seismic shaking range greatly; many new rockfall blocks at some sites, few at others. The southwestern two-thirds of the map clearly shows the progressively greater relative seismic-shaking intensity of the 1855 event. Although the 1848 earthquake was large, comparatively it was quite local. The 1855 earthquake was enormous; it maintained hillslope disruptive power as it rumbled much farther southwest. The hazards potential is much higher for Mw 8 than for Mw 7 earthquakes.

Perhaps these four recent earthquakes in the transpressional plate boundary are related and occurred as a sequence at 6-year intervals. Their temporal spacing and northward spatial progression support this speculation. Figure 2.9 shows the locations of the primary faults, but the secondary cross faults were involved too. Event A was in 1836 on the Conway segment of the Hope fault. Event B was the 1842 earthquake on the Clarence–Elliott faults. Right-lateral displacement of about 7 m (Nicol and Van Dissen, 2002) indicates a Mw magnitude >7.0 earthquake for this event. The magnitude ~7.4 Marlborough earthquake in 1848 (Event C) had a surface rupture whose southwestern end at Barefells Pass (Benson et al., 2001) was only 10 km north of the Clarence fault. This 105 km long rupture extends towards the northeast. The M ~ 8.2 Wairarapa earthquake of 1855 (Event D), described by Van Dissen and Berryman (1996), occurred 6 years after the Marlborough earthquake and an additional 85 km farther northeast. Benson et al. (2001, p. 1,090) note, "The dextral

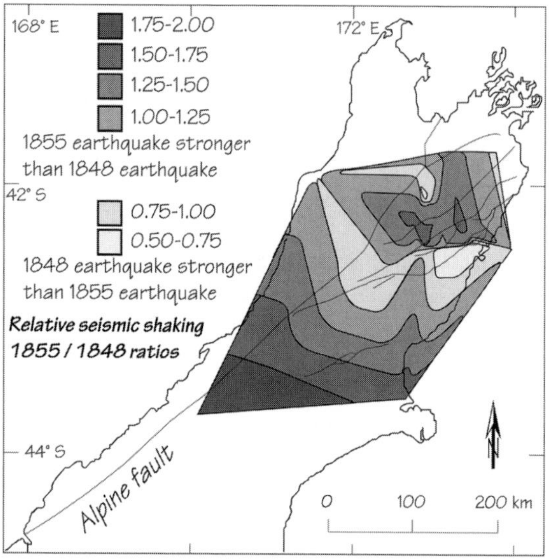

FIGURE 2.8. Relative intensity of seismic shaking (rockfall block abundance) for the Mw magnitude 8.2 Wairarapa earthquake of 1855 (earthquake epicenter in the North Island), compared to the Mw magnitude 7.4 Marlborough earthquake of 1848 (earthquake epicenter on eastern Awatere fault). Either event can be dominant near the Awatere fault, but seismic shaking during 1855 is progressively more important to the southwest. (From Figure 6.33 of Bull, 2007.)

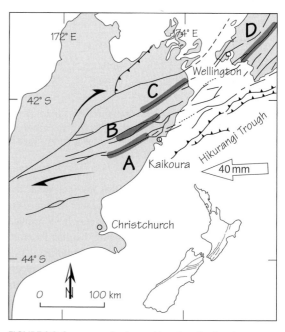

FIGURE 2.9. Sequence of epicentral locations for four large earthquakes in the Marlborough–Wellington plate transpressional zone. (From Figure 6.34 of Bull, 2007.)

strike slip Wairarapa fault can be traced offshore from lower North Island into Cook Strait and to within 20 km of the northeastern end of the Awatere fault (Carter *et al.*, 1988). Although there may be no direct fault connection between the two faults, it might be speculated that the 1848 rupture of the Awatere fault precipitated failure of the Wairarapa fault during the 1855 earthquake." This sequence – four earthquakes at 6-year intervals on different fault zones – is unlikely to be repeated. The earthquake-prone mountains near the town of Kaikoura have not experienced hazardous seismic events since this sequence of four earthquakes relieved much of the accumulated stresses on this part of the plate boundary.

2.4 Conclusions

The frequency and intensity of landslide hazards in mountainous landscapes is a function of rock type, climate, and earthquakes. Use of lichens to study geomorphic processes has progressed to where individual rockfall events can be dated with a precision of ±5 years. Spatial variation in abundance of rockfall blocks of a certain age can be used to make seismic-shaking intensity maps as good as Mercalli intensity maps. These maps identify the most likely fault(s) for a given time of earthquake(s), and the approximate position of the earthquake epicenter. Series of maps may depict earthquake sequences as stress is transferred from fault to fault.

References

Benson, A. M., Little, T. A., Van Dissen, R. J., Hill, N. and Townsend, D. B. (2001). Late Quaternary paleoseismic history and surface rupture characteristics of the eastern Awatere strike-slip fault, New Zealand. *Geological Society of America Bulletin*, **113**, 1079–1091.

Bull, W. B. (1996). Prehistorical earthquakes on the Alpine fault, New Zealand. *Journal of Geophysical Research, Solid Earth, Special Section "Paleoseismology"*, **101**, 6037–6050.

Bull, W. B. (2003). Lichenometry dating of coseismic changes to a New Zealand landslide complex. *Annals of Geophysics*, **46**, 1155–1167.

Bull, W. B. (2004). Sierra Nevada earthquake history from lichens on rockfall blocks. Sierra Nature Notes, http://www.yosemite.org/naturenotes/LichenIntro.htm.

Bull, W. B. (2007). *Tectonic Geomorphology of Mountains: A New Approach to Paleoseismology*. Oxford: Blackwell Publishing.

Bull, W. B. and Brandon, M. T. (1998). Lichen dating of earthquake-generated regional rockfall events, Southern Alps, New Zealand. *Geological Society of America Bulletin*, **110**, 60–84.

Carter, L., Lewis, K. B. and Davey, F. (1988). Faults in Cook Strait and their bearing on the structure of central New Zealand. *New Zealand Journal of Geology and Geophysics*, **31**, 431–446.

Grapes, R. and Downes, G. (1997). The 1855 Wairarapa, New Zealand, earthquake-analysis of historical data. *Royal Society of New Zealand Bulletin*, **30**, 271–368.

Grapes, R., Little, T. A. and Downes G. (1998). Rupturing of the Awatere Fault during the 1848 October 16 Marlborough earthquake, New Zealand: historical and present day evidence. *New Zealand Journal of Geology and Geophysics*, **41**, 387–399.

Hull A. G. and McSaveney M. J. (1993). *A 7000-year Record of Great Earthquakes at Turakirae Head, Wellington, New Zealand*. New Zealand Earthquake Commission Research Paper 011, Project 93/139.

Nicol, A. and Van Dissen, R. (2002). Up-dip partitioning of displacement components on the oblique-slip Clarence Fault, New Zealand. *Journal of Structural Geology*, **24**, 1521–1535.

Van Dissen, R. J. and Berryman, K. R. (1996). Surface rupture earthquakes over the last 100 years in the Wellington region, New Zealand, and implications for ground shaking hazard. *Journal of Geophysical Research*, **101**, 5999–6019.

3 Volcanic hazards and risks: a geomorphological perspective

Jean-Claude Thouret

3.1 Introduction

The burial of Pompeii and Herculaneum by the AD 79 eruption of Vesuvius, the devastation produced by the tsunami generated during the eruption of Krakatoa in 1883 and the city of St Pierre laid waste by Mount Pelée's 'nuées ardentes' are but a few awesome examples of disasters caused by powerful volcanic eruptions. Since AD 1783 eruption-related deaths have totalled 220,000 (Tanguy *et al.*, 1998). Most resulted from post-eruption famine and epidemic disease (30%), pyroclastic flows and surges (27%), lahars (17%) and volcanogenic tsunamis (17%). Volcanic fatalities are small compared to those of floods and earthquakes but the potential threat from a massive eruption is greater today than ever before, because of the large concentrations of populations living around volcanoes. More than 500 million people live in active volcanic areas (Tilling, 2005) and at least 200 million now live in cities within 200 km of an active or potentially active volcano (Chester *et al.*, 2001). A volcanic eruption of 'modest' size (VEI 3), such as the Nevado del Ruiz event in 1985, results in about 7.7 billion US$ in loss. This figure may look small when compared to the aftermath of the Kobe earthquake in Japan but the modest 1985 eruption affected an estimated 20% of Colombia's domestic growth.

The sense of volcanic risk can be defined by Fournier d'Albe's qualitative 'risk equation' (1979): *volcanic risk = hazard × vulnerability × value*. There are, however, subtle differences in the terms hazard and risk, as well as in related terms.

Hazard refers to a potentially dangerous event that occurs at a particular site within a given period of time. *Risk* is the probability of loss or the degree of harm caused when the dangerous event occurs. *Vulnerability* is the proportion of the elements exposed to a hazard (e.g. humans, housing, lifelines and staple economic activity) and likely to be lost if the dangerous event occurs. *Value* may include the number of people at a particular site, the monetary value of property, as well as the potential replacement value in case of destruction.

A *volcanic hazard* is produced by eruptive phenomena or eruption-related processes on and around active volcanoes, among them PDCs (pyroclastic density currents) and debris avalanches, as well as widespread ash fallout and aerosols dispersed from high eruption columns. The magnitude and extent of these phenomena can be driven by magma composition and tectonic environment, in particular by the relationships between the edifice and its substratum (e.g. volcano spreading on ductile bedrock, sector collapse and fault system). Interactions between ascending magma, vesiculation and fragmentation mechanisms into the conduit also lead to changes in the size and internal composition and behaviour of eruption columns. A special note is needed for extinct volcanoes: although there is apparently no hazard linked to the eruptive activity, recent catastrophes such as the Casita debris avalanche and subsequent debris flow in Nicaragua should remind us that impending risk to nearby dwellings can be linked to the hazardous combination of unconsolidated, hydrothermally altered volcanic rocks, heavy rainfall and seismic shaking or failure triggering. The principal factors that transform volcanic hazard into a volcanic risk in a particular zone and at a particular time are: population density, properties and resources; presence of large cities; intense cropping or grazing activities on volcano's flanks; and infrastructure in radial valleys draining the edifice's ring plain. More critical factors lie on the social side: the vulnerable subgroups of population; the level of education on natural hazards and risk perception bearing on people's behaviour in case of an alert; the state of emergency procedures; and the capacity of a community to cope with a crisis.

Geomorphological Hazards and Disaster Prevention, eds. Irasema Alcántara-Ayala and Andrew S. Goudie. Published by Cambridge University Press. © Cambridge University Press 2010.

A *geomorphic hazard* results from any landform change that adversely affects the geomorphic stability of a site and that intersects the human use system with adverse socio-economic impacts (Slaymaker, 1996). Geomorphic hazards are characterized by magnitude, frequency and areal extent. On volcanoes, geomorphic hazards can be endogenous (i.e. caused by eruptive and/or seismic processes), and exogenous (i.e. caused by meteorological and climatic processes leading to weathering and erosion). They are therefore manifold and more numerous than on any other mountain type (Thouret, 2004) and investigations into geomorphic hazards indirectly related to volcanic eruptions are increasing nowadays (e.g. floods triggered by breakout of lakes, dammed by ignimbrites, long after the eruptive episode).

A *volcanic disaster* or *catastrophe* strikes when the scale or particular nature of a volcanic event exceeds the capacity of a community to cope with the unusual consequences. The scale of the phenomenon need not be large (e.g. the Nevado del Ruiz 1985 event) to unleash an unprecedented catastrophe in cities where the threatened population is unaware of or unprepared for the particular characteristics of the volcanic flows or lahars of unusual magnitude.

The *capacity of a community to cope* (Schmincke, 2004) depends on the economic and cultural potential of a community or country and the strength of private and public institutions capable of quickly and effectively responding to a volcanic crisis (e.g. timely warning system, organized evacuation roads, designated shelters, 'preparedness' programme etc.). Disaster mitigation, either passive or active, includes all activities toward reducing risk, either the hazards (by means of engineering works) or vulnerability (by developing hazard-zone and contingency maps, land-use planning, and education). It also includes preparation of responsible administrative bodies and civil protection authorities as well as full information for the public and eventually of evacuation measures.

At active or potentially active volcanoes, geomorphic hazards are superimposed on volcanic hazards. The extent and rate of geomorphic processes of denudation are exacerbated on volcanic edifices due to the thickness of unconsolidated materials, to earthquakes, and to hydrothermal alteration. Volcanoes are also subjected to specific processes, such as ash falls, pyroclastic flows and surges, and volcanic mudflows or lahars. This is the rationale for distinguishing direct (eruption-induced) hazards at active volcanoes, and indirect, often more widespread, and delayed geomorphic hazards around dormant or extinct volcanoes. By indirect we mean lahars triggered by rainfall or snowmelt, and landslide- and/or earthquake-induced debris avalanches and tsunamis. These hazards also act altogether at

distinct ranges in space and distinct rates in time (see Figure 11.2 after Chester in Thouret, 2004). Usually volcanic hazardous phenomena threaten and affect a modest in size (<100 km^2) area on a volcano and narrow areas along the drainage network that crosses the ring plain surrounding the edifice. Except for ash fallout and aerosols, and for exceptionally long runout lahars, the effects are usually restricted to within 10 km to 40 km around the vent. However, the geomorphic perspective offers a broader view of the volcanically induced hazards: geomorphic effects (erosion, sedimentation) can become widespread following a large-scale eruption (>VEI 4) and can be delayed as much as 10 years or more (e.g. lahars around Pinatubo after 1991). Geomorphic hazards can also trigger erosion processes on slopes and in valley channels during the periods of quiescence very long after the end of the last eruption.

In this chapter we will distinguish (1) the direct volcanic hazards induced by the eruptive activity on active volcanoes; (2) the geomorphic hazards following eruptions over short- and long-time periods; and (3) the long-term hazards induced by the remobilization of debris produced by volcanoes that are no longer active. In the second section of the chapter we will focus on methods of hazard assessment (hazard-zone maps, eruption scenarios), and on quantitative methods of risk assessment (the probabilistic approach) based on recent case studies.

3.2 Direct volcanic hazards around active volcanoes

Direct volcanic hazards are ranked according to their lethal and destructive impact, as well as to their geomorphic effect (see Table 3.1). We distinguish hazardous processes due to flowage (pyroclastic density currents, debris avalanches), followed by hazardous processes linked to fall processes. We emphasize that more than one hazardous process can occur during the same eruption, each process having its own temporal and spatial scales; this fact renders volcanoes rather distinctive from any other type of mountain. A combination of volcanic hazards and geomorphic hazards can be triggered at the same time and repeatedly during an eruption or an eruptive episode.

3.2.1 Pyroclastic density currents: flows, surges and laterally directed blasts

Many of the largest volcanic disasters in history have involved pyroclastic flows and surges (e.g. 29,000 killed at Mount Pelée, Martinique, in May 1902). Pyroclastic

TABLE 3.1. *Principal types of hazardous volcanic processes, characteristics pertinent to risk and selected examples*

Type	Characteristics pertinent to risk	Examples
Direct hazards		
Fall processes: Tephra fall VF; Ballistic projectiles C	Can extend 1000+ km downwind; can produce impenetrable darkness; surface crusting encourages runoff; high impact energies; fresh bombs above ignition temperatures of many materials	Vesuvius, 1631, 1906; Soufrière Saint-Vincent, 1812
Lava flowage: Lava flow F; Domes R	Bury or crush objects in their path; noxious haze from sustained eruptions	Kilauea, 1960, 1983–…; Merapi, 1994; Soufrière Hills, Monserrat, 1995–…
Pyroclastic flowage: Pyroclastic flows VF; Pyroclastic surges F; Laterally directed blast VR	Small flows travel 5–10 km down topographic lows, but large flows travel 50–100 km; large flows climb topographic obstructions; knock down all constructions	Pinatubo, 1991; Unzen, 1991–5; Mount Pelée, 1902; Taal, 1960; Bezymianny, 1956; Mount St. Helens, 1980
Debris flowage: Primary (eruption-triggered) debris flows VF; Jökulhlaups C	Velocities may exceed 10 m s^{-1}; rapid aggradation, incision or lateral migration; hazard may continue for months or years after eruption; can occur with little or no warning	Nevado del Ruiz, 1985; Kelud, 1919; Katla, 1918; Grimsvötn, 1996
Sector collapse and flank failure: Debris avalanche C; Magmatic origin R; Phreatic origin R; No eruption; Seismogenic C	Emplacement velocities up to 100 m s^{-1}; create topography, pond lakes; can produce tsunamis in coastal areas	Mount St. Helens, 1980; Bezymianny, 1956; Bandaï-san Ontake, 1984; Shimabara, 1792
Other eruptive processes: Phreatic explosions VF; Volcanic gases and acid rains VF	Damage limited to proximal areas but can be lethal; corrosive, reactive; low pH in water; CO_2 in areas of low ground	Soufrière Guadeloupe, 1976; Dieng plateau, 1979
Indirect hazards		
Earthquakes and ground deformation F	Limited damage; subsidence may affect hundreds of km^2	Sakurajima, 1914; Usu, 2000
Tsunami R	Can travel great distances; exceptionally, waves to 30+ m	Krakatoa, 1883
Secondary debris flows VF	Can continue for years	Santa Maria, 1902–24
Posteruption erosion and sedimentation F	Can affect extensive areas for years after eruption	Irazu, 1963–4; Pinatubo, 1991–2000
Atmospheric effects C (Air shocks, lightning)	Limited effects	Mayon, 1814; Agung, 1960
Posteruption famine and disease R	Limited effects at present	Lakagigar, 1783

Processes having a direct impact on geomorphic hazards are italicized. Historical frequency for adverse effect, damage and/or death is as follows: VF, very frequent; F, frequent; C, common; R, rare; VR, very rare.
After Tilling, 1989, modified; Blong, 1984, 2000, modified.

flows contain ground-hugging mixtures of hot and dry fragments and magmatic gas, descending slopes at very high speeds (Nakada, 2000). They follow topographic lows and valleys as they move. Pyroclastic surges are more diluted flows, containing a low concentration of particles but more gas. Surges can be generated directly by column collapse, and also by separation from the moving, relatively dense, main body of the pyroclastic flow. Surges rarely have sufficient force to destroy buildings in their path and are short lived compared with pyroclastic flows because of their lower momentum and density. The area impacted by surges is limited to a few kilometres from the source, although they can run over topographic barriers. 'Blasts' or laterally directed explosions are also short-lived diluted turbulent currents, but they may travel large distances, such as 35 km from the Mount St Helens' summit on 18 May 1980.

Owing to their mass, high temperature (300–800 °C) and gas content, high velocity and great mobility, pyroclastic flowage processes are amongst the most deadly of volcano hazards, causing asphyxiation, burial, incineration and physical impact. Pyroclastic density currents (PDCs) are so fast that they exert high pressures on obstacles such as buildings that they encounter. The pressure exerted on a wall at right angles to a PDC with a bulk density of $\rho = 1,000$ kg m^{-3} and a speed of $u = 100$ m s^{-1} will be $0.5\rho u^2 = 5$ MPa, which is hundreds of times greater than typical load-bearing strengths of the strongest building materials. It is therefore not surprising that even more dilute pyroclastic surges can cause structural damage (Parfitt and Wilson, 2008).

Pyroclastic flows can be derived from gravitational collapse of Plinian eruption columns. For example, at Pinatubo in 1991, more than 4 km^3 of ash and pumice flowed from such a collapse within a few hours. Pyroclastic flows can also originate directly from the vent, as produced by the August 1980 eruptions of Mount St Helens. Lateral explosions directly from a summit dome may produce small but devastating pyroclastic flows like those at Mount Pelée, Martinique, on 8 and 20 May 1902. Finally, pyroclastic flows may originate from the gravitational collapse of lava domes or lava flows: they were frequently observed during the 1991–5 eruption at Unzen (Japan) and the 1995–2008 eruption at Soufrière Hills, Montserrat. At Unzen, partial collapse of the lava dome on 3 June 1991 produced pyroclastic flows that reached a distance of 5 km, and the associated ash-cloud surge killed 43 people in the evacuated zone (Nakada, 2000).

The relief of volcanoes can influence some physical properties of flows. Pyroclastic flows and surges are hot and very mobile, and their speed is generally over 10 m s^{-1} and sometimes over 100 m s^{-1}. Their travel distance depends on the mass of tephra thrown out and the height at which the pyroclastic flow starts to descend. The ratio of the height (H) dropped over the travel distance (L) is as small as 0.2–0.29 for large-scale pyroclastic flows and as high as 0.33–0.39 for small block-and-ash flows. This shows that larger flows are more mobile than smaller flows. High-mobility pyroclastic flows may be fluidized by volcanic gases released from the interior of lava particles or by expansion of heated air trapped in the moving front. From the relationship between volume erupted and the H/L ratio, one can predict the travel distance from the starting

point of pyroclastic flows, assuming a constant apparent friction (H/L) for the flow. Since dome eruptions commonly last a few years, the pyroclastic flow hazards change, as shown by the ongoing eruption of Soufrière Hills in Montserrat. As a lava dome increases in volume and height with time, pyroclastic flows tend to travel larger distances due to the increase in H, leading to larger-scale dome collapse. As the valleys around a dome become filled, later pyroclastic flows are likely to descend further and to spread out more. Observations of recent eruptions suggest that rugged topography and high relief may affect transport and deposition of PDCs (Nakada, 2000). The interactions of pyroclastic currents with topography include blocking, downslope drainage, formation of secondary pyroclastic flows in valleys, and development of dividing streamlines, as indicated by the flow-surge laid down by the 18 May 1980 blast at Mount St Helens.

3.2.2 Sector collapses and debris avalanches

Active and potentially active volcanoes are unstable structures, because they are built of weak or unconsolidated rocks, have steep slopes, are highly fractured and faulted, and may be undergoing deformation related to magma movement and/or to hydrothermal pressurization (Tilling, 2005). Thus, structural collapse at restless volcanoes can precede, accompany or follow eruptive activity, triggering rockfalls, rockslides and debris avalanches, which can move rapidly downslope and may pose significant hazards. Debris-avalanche deposits form a large talus apron of hummocky terrain with water-filled depressions, steep flow margins, and thick deposits of unsorted, unstratified angular-to-subangular debris (Siebert, 1996; Ui, 2000). Debris avalanches are highly destructive, burying and destroying everything in their paths. Subsequently, the 'dewatering' of a debris avalanche can generate lahars and floods downvalley, as observed at Mount St Helens in 1980 (Janda et al., 1981).

A relationship exists between the distance travelled by an avalanche and the failure volume. Large-volume debris avalanches can extend as far as 85 km beyond their sources and can deposit volcanic debris tens of metres thick over areas of 100 to 1,000 km^2 (e.g. 700 km^2 and 45 km^3 at Mount Shasta in California : Tilling, 2005). The ratio of vertical drop H to travel length L ranges from 0.09 to 0.18 for Quaternary volcanic avalanches <1 km^3 in volume and from 0.04 to 0.13 for avalanches >1 km^3 (Siebert, 1996). Volcanic debris avalanches are more mobile than their non-volcanic counterparts, and, for a given volume and vertical drop, volcanic debris avalanches travel farther (Siebert, 1996), suggesting that low-rigidity, partially fluidized avalanches are capable of travelling tens of kilometres. With sufficient momentum, debris avalanches can run up slopes and cross topographic barriers up to several hundred metres high.

Widespread slope failure in a variety of tectonic settings suggests that it may be the dominant catastrophic process modifying volcanoes. Not only are composite cones susceptible to slope failure, but also less voluminous and steep-sided lava-dome complexes. The summit of Mount Augustine, Alaska, has repeatedly collapsed and regenerated, averaging 150–200 years per cycle, during the past 2,000 years (Beget and Kienle, 1992). The unprecedented frequency of summit failure was made possible by sustained lava effusion rates over ten times greater than is typical of arc volcanoes. Regardless of how extensively a volcano is eviscerated, the conical shape is reconstructed very rapidly. At Parinacota in northern Chile, reconstruction since a massive collapse 13,000 years ago has virtually rebuilt the entire edifice.

The causes of edifice collapse are uncertain. Structural factors such as steep dip slopes, zones of weakness and local extension promoted by parallel dike swarms can contribute to flank failure. Siebert (1996) emphasized differences in processes triggering flank failures: magmatic eruption of Bezymianny type, non-magmatic explosions of Bandai type, and cold avalanches of Ontake type. But van Wyk de Vries and Francis (1997) argued that the volcanic edifice itself (e.g. Mombacho, Nicaragua) contributes to the weakness of its bedrock. In contrast to radial spreading, preferential spreading in one direction is critical to collapse development. Radial spreading induces inward-dipping faults, which inhibit collapse, whereas sector spreading generates failure-prone outward-dipping structures. Spreading in a preferential direction may be caused by buttresses, the regional slope of basement beds, regional stress, weak basement or by high fluid pressures under one side. There are many sector collapses with similar wedge-shaped scars that could have involved such processes.

One of the most remarkable pieces of evidence of topographic control on the emplacement mechanisms of debris avalanches has been obtained by Kelfoun et al. (2008) from the study of the exceptionally well preserved 26 km^3 Socompa debris-avalanche deposit in Northern Chile. Using a composite 3D ortho-image of the Socompa long-runout debris-avalanche deposit, along with field observations, Kelfoun et al. have reconstructed the sequence of events during avalanche emplacement. The avalanche spread 40 km across a pre-existing basin and as it impinged on the western and northern margins it was reflected back, forming a secondary flow that continued to travel 15 km down a gentle slope at an oblique angle to the primary flow, the front of the return wave being preserved frozen on the surface of the deposit as a prominent escarpment. They

have shown evidence for large-scale topographic reflection and secondary flow. Evidence for avalanche reflection includes clearly recognizable secondary-slide masses, sub-parallel sets of curvilinear shear zones, headwall scarps separating the primary levee from the secondary terranes, extensional jigsaw breakup of the surface lithologies during return flow, and cross cutting, or deflection, of primary flow fabrics by secondary terranes. The secondary flow occurred as a wave that swept obliquely across the primary avalanche direction, remobilizing the primary material, which was first compressed, then stretched, as it passed over and rearward of the wave front. Secondary flow took place on slopes of only a few degrees, and the distal lobe flowed 8 km on a slope of ~1°. Overall the avalanche is inferred to have slid into place as a fast-moving sheet of fragmental rock debris, with a leading edge and crust with near-normal friction and an almost frictionless fluidal interior and base.

More intriguing is the air of permanence of large, low-angle shield volcanoes that belies their inherent instability: particularly noteworthy are 70 landslides on the Hawaiian ridge, which have removed volcano-flank sectors that exceed 1000 km^3 in volume (Moore et al., 1994). Instability and collapse are a major process and hazard in the evolution of the basaltic oceanic islands. The Canarian and Hawaiian volcanoes share common constructional and structural features, such as rift zones, progressive volcano instability and multiple gravitational collapses (Carracedo, 1999). Gravitational stresses and dyke injections progressively increase the mechanical instability of these edifices, especially in the most active shield-stage phases of growth.

What processes trigger slope instability on low-angle shield volcanoes? Keating and McGuire (2000) have identified 23 endogenetic and exogenetic processes that contribute to edifice collapse on volcanic ocean islands. Endogenetic causes of failure dominate during periods of active volcanism whereas exogenetic sources may cause failure at any time. Endogenetic instability includes repeated dike injection along rift zones; swelling due to intrusion of a new magma batch with or without eruption; unstable foundations, detachment surfaces linked to volcanic intrusions, thermal alteration and faulting. Other endogenetic factors are: increased interstitial pore pressure related to intrusives; and hydrothermal alteration of the interior of the volcano making rocks susceptible to failure, either catastrophically or by creep of the flanks of the volcano. Exogenetic sources of instability and failure are: the heterogeneous combination of rock types having poor cohesion and low strength; and oversteepening of one flank of a volcano due to long-term sustained erosion processes. The steepness of an edifice appears to determine how quickly it becomes unstable and susceptible to lateral failure.

3.2.3 Tephra falls and ballistic projectiles

Tephra triggers the widest-ranging direct hazard from volcanic eruptions. Tephra fall includes all airborne products of explosive eruptions except gas. Volcanic bombs, blocks and lapilli that leave the vent with ballistic trajectories are called projectiles. Particles that fall from the eruption column or the plume are termed tephra fall. Tephra fall consists of ash and lapilli, particularly pumiceous lapilli with low specific density, which can be transported more than 100 km downwind. As the tephra disperses downwind the concentration of particles declines, so that the thickness deposited also declines exponentially with distance. The areas experiencing tephra fall in a major eruption may be very large: the ash fall around Krakatoa (Java) in 1883 covered about 827,000 km^2, but much of this area received less than 1 cm thickness.

It is impact energy rather than size or density that determines projectile hazardousness (Blong, 1984). Distances travelled by projectiles vary between 0.3 km for sizes in tons and 3–6 km for sizes in kilograms. The temperature on impact is significant because projectiles can set fire to vegetation and man-made objects several kilometres from the vent. The hazards produced by tephra fall result principally from the accumulation of the deposits and extreme darkness. The weight of ash deposited can bring down roofs and cause serious damage, as well as injury to people. Most of the 300 deaths due to the eruption of Mount Pinatubo in 1991 occurred outside the 30 to 40 km radius of evacuation as a result of roofs collapsing under the weight of ash. The ash on the roofs had been made denser by the effect of rain: a wet ash deposit of 15 cm may have a load of 200 kg m^{-2}. After eruptions, fine-grained tephra deposits may represent a greater hazard than coarse-grained tephra because they will be more easily eroded by wind and water. In the well-documented Irazu eruption in Costa Rica in 1963 (Waldron, 1967), the ash destroyed the vegetation cover and then hardened to form an impervious crust, which led to increased runoff and flash floods.

3.2.4 Lava flows

Most lava flows move slowly enough for people to avoid them, but some are fast enough to cause loss of life. Two eruptions of high-discharge, low-viscosity lava flows have occurred on Hawaii and one in Congo during historical times, with velocities of 40–100 km h^{-1}. As many as 600 people died in the extremely fluid, fast-moving lava flows in 1977, when the lava lake at Mount Nyiragongo, Congo, rapidly emptied. Again in January 2002, a fissure-fed lava poured down the slopes of Nyiragongo through the eastern Congolese city of Goma, 18 km south of the crater, destroying the city, displacing hundreds of thousands of people, and killing about 50 people. Other examples of devastating lava flows in recent decades include those of Kilauea (Hawaii), Etna (Italy) and Paricutín (Mexico). The land covered by lava is generally rendered uninhabitable and unproductive, perhaps for centuries. However, lava flows are efficient geomorphic agents as they build up land on coasts, particularly on oceanic islands. The best recent example is provided by the current effusive period at Kilauea volcano in Hawaii where the Pu'uŌ'ō and Kupaianaha lava flow field, with a DRE volume of 3.3. km^3, has covered 119 km^2 on the southeast flank and has added 169 ha to the coast of the Big Island since 1983 (USGS Hawaii Volcano Observatory website).

Large-scale (tens of km^3) lava-flow-forming eruptions have a greater geomorphic impact on land as well as an indirect impact on climate. They are less frequent (once every 100,000 years on average) but their magnitude is such that they can create plateaus termed traps in both continental and marine settings. The Deccan plateau (512,000 km^2; 67–65 Ma) and the Colombia River Basalt plateau (164,000 km^2; 17–15.5 Ma) are examples of the first case and the Ontong-Java plateau illustrates the second, oceanic in nature. Flood basalt eruptions have the potential to release immense amounts of sulphur dioxide (~1,000 Mt yr^{-1}) by similar mechanisms to the Laki event in 1783–4 in Iceland (Rampino and Self, 2000). The high discharge of lavas during a relatively short geologic time period (less than 2 million years) is likely to have a severe impact on life: at least 7 out of 11 mass extinctions of species in the biological record are apparently correlated to the timing of large flood basalt eruptions (Parfitt and Wilson, 2008).

3.2.5 Volcanic gases and phreatic explosions

Every volcanic volatile is directly dangerous to most living creatures. The commonest volcanic gas is water vapour. Although water has no direct adverse effects on plants and animals on the ground, it may have very adverse effects when high eruption columns carry it into the stratosphere. A complex chain of chemical reactions produces the layer of ozone that absorbs many wavelengths of ultraviolet light that, if they reach the ground, can cause skin cancers. Excessive amounts of water vapour can modify the reactions that maintain the ozone (Parfitt and Wilson, 2008). The next commonest volcanic gas is carbon dioxide: denser than air, it collects in topographic hollows, and people and animals there rapidly fall unconscious. Carbon dioxide emission, accompanying a phreatic eruption from the southwest side of the Dieng volcano, killed 149 people on 20 February 1979. Carbon dioxide is a corrosive and poisonous acid gas that is released from shallow magma bodies and seeps upward to collect as a dissolved gas in the water in lakes. If an event such as a landslide disturbs the density stratification of the lake, the water may overturn: water from the bottom, which is saturated in gas, rises to the top where the pressure is much less. Here it becomes supersaturated, and explosively exsolves a dense cloud of carbon dioxide. The cloud of gas, denser than air, will hug the ground like a PDC and travel downhill. During the catastrophes of the Monoun and Nyos crater lakes in Cameroon in 1984 and 1986, more than 1,700 people were killed by a descending carbon dioxide cloud. Other gases such as fluorine, even denser than carbon dioxide, can be equally lethal in various ways. Sheep and cattle have been poisoned in large numbers by accumulation of this gas. Almost equally poisonous are the high concentrations of sulphur dioxide and hydrogen sulphide that basaltic volcanoes sometimes release. Sulphur dioxide can dissolve in atmospheric water drops and react with oxygen to form sulphuric acid. Tiny droplets or aerosols can alter the way the atmosphere reflects and absorbs sunlight, thus changing the climate. The hazard of gas release for buildings is small, but economic losses such as the death of cattle can be significant. The early warning of an impending gas eruption is extremely short. Even the emission of gas (CO_2, SO_2) from fumaroles in active volcanoes can cause major damage to crops and is a significant health hazard.

3.2.6 Large-scale, caldera- and ignimbrite-forming explosive eruptions

Large explosive eruptions are likely to induce regional and even global impacts. Large-scale explosive eruptions (more than 50 km^3 in volume) and very large volume (>300 km^3) 'super-eruptions', which produce widespread (hundreds to thousands km^2) ignimbrites, are historically rare (300 km^3 once in 20,000 years and 1,000 km^3 once in 100,000 years: S. Self, personal communication). These powerful eruptions are responsible for the highest death toll in recent history: Krakatoa in 1883 caused 36,000 deaths, mainly from tsunamis associated with the caldera formation. 'Super-eruptions' probably represent the most catastrophic geologic events that have ever affected the Earth's surface, and occur more frequently than large meteorite impacts.

In recent years, intense seismic unrest has occurred at three calderas: Long Valley in the United States (1980 to present), Campi Flegrei in Italy (1983 to present) and Rabaul in Papua New Guinea (1971–94). Both Campi

Flegrei and Rabaul have already experienced multiple caldera-subsidence events associated with large eruptions, so additional caldera-forming events are a significant long-term hazard. The unrest at Rabaul culminated in 1994, when two volcanoes erupted simultaneously on opposite sides of an old caldera, but no large eruption occurred. One of the future tasks for volcanologists will be to recognize precursors and determine the time intervals between precursors and future, potentially devastating, caldera-forming eruptions. If an eruption comparable in size to that of Toba in Sumatra (2,000 km^3, 73,000 yr ago) should occur in a heavily populated area such as Naples, the effects would be devastating (Lipman, 2000).

The super-eruption impact will not only be destructive on life but will also induce continental or global climate change and indirectly exert a climatic control on geomorphic processes. The examination of historical records of volcanic activity and climate variation, and from satellite monitoring, shows that volcanic eruptions can cause short-term climate change. In general the effect of a volcanic eruption is to cause global cooling. The effect of historic eruptions has been to cause small but significant cooling (typically ≤0.5 °C) for 2–3 years after the eruption (e.g. Chichón in 1982, Pinatubo in 1991; Parfitt and Wilson, 2008). A number of factors determine the effect of a given eruption on climate. These include the height reached by the eruption plume, the erupted volume, the geographical location of the volcano, the composition of the erupted magma, and the eruption duration. Although the general effect of an eruption is to cause overall surface cooling, when examined in detail the effects are more complex because cooling can disrupt normal weather systems: some areas may actually experience net warming rather than cooling. Changes in rainfall patterns may also occur in the aftermath of a large eruption. Volcanic eruptions affect climate by injecting ash and volcanic gas into the atmosphere to form sulphuric acid aerosols. The ash and aerosols intercept some of the incoming solar radiation and scatter it back out to space. This causes a reduction in the amount of sunlight reaching the surface and hence surface cooling. The aerosols have the most significant effect because they scatter more light and they have a longer residence time in the atmosphere than the ash particles. The effects of the largest eruptions seen in the geological record are expected to be much more significant than those of historical eruptions. Current evidence suggests that flood basalt eruptions are likely to be more important than the largest rhyolitic eruptions in causing climate change because of the higher sulphur content of basaltic eruptions and their extended durations (e.g. the 8-months-long Laki event in 1783). There is a strong correlation between the occurrence of flood basalt eruptions and mass extinction events, which suggests that environmental stress caused by these events may act as a trigger for the extinctions.

3.3 Indirect volcanic hazards and geomorphic impact

Volcanic hazards are termed indirect when they affect people, housing, transport links and infrastructure as a consequence of an eruption in conjunction with other elements (sea, lake, snow and ice, rainfall, atmosphere, earthquake, crust deformation). The indirect hazardous processes affect a widespread area beyond the usual perimeter of the direct hazards of eruptive phenomena. Most of them do not cause casualties but their aftermath can induce severe impacts on human or animal health (and therefore deaths in historic times) and can alter significantly the climate or the ground. Indirect volcanic hazards, ranked on the basis of impacted surface area and geomorphic effects, are the following: volcano landslide and volcanogenic tsunami, sediment–water flows, effects of ash in the atmosphere, and post-eruption famine and disease (see Table 3.1). Other short-term indirect volcanic hazards may have a local geomorphic impact, such as ground-shaking and movements caused by volcanogenic earthquakes and crust deformation (e.g. in Toya city, Hokkaido, during the Usu volcano eruption in 2000).

3.3.1 Volcanogenic landslides and tsunamis

Volcano landslide or flank failure can occur before or during an eruption (Mount St Helens or Bezymianny type) but most take place without an eruption: they are called Bandaï-San debris avalanches (Siebert *et al.*, 1996) after the name of a Japanese volcano, where the debris avalanche apparently followed a phreatic event. Landslides and tsunamis cause a wide range of impacts on the environment. Historical records of volcanic eruptions show that about 17% of volcano victims have died as a result of tsunamis. The lethal effect of volcanogenic tsunamis is in the unexpected transfer of energy from isolated volcanoes to sea waves, which travel rapidly to densely populated shorelines. In the historical record, 92 examples of tsunamis of volcanic origin can be attributed to eight causal mechanisms. Earthquakes accompanying eruptions, pyroclastic flows and submarine explosions each account for 20% of all volcanogenic tsunamis. Caldera collapse is responsible for about 10%, avalanches of cold and hot materials about 14%, with lahars, air waves and lava avalanching of minor extent.

In 1792, the collapse of a parasitic lava dome of Unzen volcano (Japan) caused a tsunami resulting in about 15,000

fatalities. The most calamitous volcanogenic tsunamis in recorded history occurred during the eruption of Krakatoa on 27 August 1883. Two large waves swept along the Sunda Strait shorelines to be followed by a gigantic wave >15 m that inundated the entire coastal strip on the northern coast of Java and southern coast of Sumatra. About 36,000 lives were lost, mostly from drowning. The tsunamis probably originated by one of three mechanisms (Schmincke, 2004). The first is large-scale collapse of the northern part of Krakatoa Island as part of a caldera-forming process. The second is the discharge of about 12 km^3 of subaerially generated pyroclastic flows which were violently emplaced into the sea. The third is a major debris avalanche into the sea north of Krakatoa, as shown by a hummocky submarine morphology.

3.3.2 Geomorphic hazards from volcanogenic sediment–water flows

Although some lahars are syn-eruptive (i.e. produced by instantaneous transformation of pyroclastic flows, surges and avalanches into debris flows, e.g. Mount St Helens, 1980), most lahars are secondary, i.e. produced by remobilization of pyroclastic material by rainfall on slopes and in valleys. This can occur on active volcanoes between eruption episodes and on dormant volcanoes as well. Lahar is an Indonesian term that describes a flowing mixture of rock debris and water, other than normal stream flow, from a volcano, which encompasses a continuum from debris flows (sediment concentration more than 60% per volume) to hyperconcentrated flows (sediment concentration from 20 to 60% per volume). People in distal areas commonly neither expect the event nor anticipate the destructive power of lahars (Vallance, 2000). Hydrological disasters are known around volcanoes that are ice clad and also in wet areas where annual rainfall exceeds 2 m, such as the equatorial and monsoon belt. Lahars are more deadly and devastating than pyroclastic flows for several reasons. They occur more frequently and over longer periods of time than pyroclastic flows. They flow farther down slopes to the more heavily populated plains. The rock fragments carried by lahars make them especially destructive, while abundant liquid allows them to flow over gentle gradients and inundate areas far away from their source. Requiring only the sudden mixture of large amounts of water with abundant, loose and easily eroded debris on a volcano slope, they can be formed in a variety of ways. We further distinguish sediment-water flows and jökülhlaups.

Sediment-water flows encompass three categories (Pierson and Costa, 1987): (1) debris flows, involving saturated slurries of rock debris and water; (2) hypercon-centrated flows occur as the more dilute downstream runout flows of some large debris flows; (3) unsaturated, predominantly granular flows of snow and admixed rock debris, termed 'volcanic mixed avalanches', can occur as a consequence of explosive events, such as the 13 November 1985 eruption of Nevado del Ruiz, Colombia (Pierson and Janda, 1994; Thouret et al., 1995). Unusual 'ice diamicts', comprising clasts of glacier ice and subordinate rock debris in a matrix of ice, snow and coarse ash, were emplaced during the 15 December 1989 eruption of Redoubt volcano in Alaska (Waitt et al., 1994). Transient, mixed avalanches transformed to initial 'snow slurry' lahars, then either to dilute or concentrated lahars on Ruapehu, New Zealand, in 1995 (Cronin et al., 1996).

Emplacement of hot pyroclastic rock debris on snow packs on volcanoes can trigger hazardous rapid flows of sediment and water, which can extend far beyond the flanks of the volcano, producing catastrophic consequences more than 100 km downstream. Historical eruptions at five snow-clad volcanoes (Tokachidake, Japan, in 1926; Nevado del Ruiz, Colombia, in 1985; Cotopaxi, Ecuador, in 1877; Mount St Helens, USA, in 1982–4; Ruapehu, New Zealand, in 1995–6) have demonstrated that snowmelt-generated debris flows can have peak discharges as large as $10^5 \, \mathrm{m^3 \, s^{-1}}$, attain velocities as high as 20–40 m s^{-1}, mobilize as much as $10^8 \, \mathrm{m^3}$ of debris, and travel long distances as debris flows in valleys draining the edifices (Pierson, 1999). The risk to human life from such events was tragically demonstrated on 13 November 1985, at Nevado del Ruiz, where snowmelt-triggered lahars rushed down adjoining canyons to nearby villages and inundated the town of Armero and 23,000 of its inhabitants (Pierson et al., 1990; Thouret, 1990, 2004; Thouret et al., 2007). This was the second worst volcanic disaster in the twentieth century and the fourth most disastrous eruption in recorded history. The total financial loss was over 7 billion US dollars (Voight, 1990). The lahars with a volume of $9 \times 10^7 \mathrm{m^3}$ were initiated within minutes of the onset of the small eruption (<VEI 3) and incorporated unconsolidated materials, increasing the total flow and peak flow rate, a process termed 'bulking'. Over the 104 km distance travelled by lahars, net flow volume increases by factors of 2 to 4 occurred. This finding has a major social implication because lahars may far exceed their initial volume during their passage and thus inundate inhabited regions very far from the mountain.

Jökulhlaups are impressive but much less frequent worldwide because interaction between ice and volcanic activity is limited to Iceland, Antarctica and arctic North America. Here, many volcanoes lie beneath ice caps and, due to prolonged geothermal activity or to sudden explosive activity, large quantities of water escape from subglacial reservoirs. These glacial outbursts, named jökulhlaups in Icelandic, have been responsible for creating the extensive sandur of the south Icelandic coastline. Jökulhlaups include water floods as well as hyperconcentrated flows that carry more sediments than normal floods. Their immense size is illustrated by the 1918 jökulhlaup generated by the Katla subglacial volcano and by the November 1996 and 2004 jökulhlaups resulting from Gjálp and Grímsvötn, two volcanoes under the Vatnajökull ice cap. Given the intense scientific and media activity before the Gjálp event, no casualties were incurred but that was not the case in historical subglacial eruptions (Björnsson, 2004).

3.3.3 Ash in the atmosphere

There are two indirect hazards linked to the presence of ash and finer particles in the atmosphere. The first hazard that led to catastrophes in the recent past was the encounter of aircraft with high eruption columns that rise into the stratosphere or distant ash clouds that drift from a volcano over a continent for several days or even weeks. During the past two decades, beginning with the unexpected eruption of Galunggung volcano (Indonesia) in 1982, more than 80 modern jets were damaged when inadvertently flying through volcanic ash and almost 10 large aircraft experienced in-flight loss of engine power and barely avoided a crash (Schmincke, 2004). A famous example is the near-crash of a new Boeing 747-400 with more than 300 passengers aboard, more than 200 km northeast of erupting Redoubt volcano in Alaska on 15 December 1989. The main cause of engine thrust loss was the accumulation and solidification of ash on the turbine nozzle guide vanes. Fortunately no passengers were injured, but the cost of replacing all four engines and repairing other damage exceeded US$ 80 million These near-crash events spurred the creation of a new body of scientists, engineers, pilots and people in charge of the civil aviation safety under the IAVCEI umbrella, and in 1991 the aviation industry decided to set up a network of Volcanic Ash Advisory centres (Casadevall, 1994).

The second hazard from ash particles settling to the ground stems from the fact that the smallest particles (in the range of 10–50 micrometres) can be inhaled by people in the fallout area, particularly if free silica is present. A layer of moist ash on the lung surface of living creatures stops the air from reaching the alveoli in the lungs. It behaves like a layer of wet cement and can cause death or at least long-term damage to the lungs (silicosis, asthma and other respiratory problems) comparable to that associated with exposure to asbestos fibres or coal dust.

3.3.4 Post-eruption famine and disease

Volcano hazards can also adversely affect the daily lives of people more importantly in the disruption of food supply due to the immediate loss of livestock and crops, and the longer-term loss of agricultural productivity of farmlands buried by eruptive materials. Before the twentieth century, post-eruption famine and epidemic disease – in the aftermath of the 1783 eruption of Laki (Iceland) and the 1815 Tambora eruption – constituted the deadliest volcano hazard (Tilling, 2005). A recent estimate of the deaths caused by post-eruption starvation after the Tambora eruption is about 49,000 (Tanguy et al., 1998). The estimate of the number of starvation deaths (~10,000) caused by the Laki eruption is more reliable; the deaths represented about a quarter of Iceland's entire population at the time. In the twentieth century, fatalities from post-eruption famine and epidemic disease have been greatly reduced.

3.4 Post-eruption geomorphic impacts and responses

The fact that volcanically induced geomorphic processes keep on acting on the landscape, eroding or constructing landforms and remobilizing deposits, renders volcanoes very distinct. No other hazardous geologic process includes such a long-lasting impact after the onset of the phenomenon. The most obvious geomorphic impacts of eruptions occur on land through volcano landforms construction and destruction (Ollier, 1988; Thouret, 1999): fast construction of an edifice (months to years), instantaneous destruction of an edifice (caldera formation, horseshoe-shaped scar of sector collapse), new land addition through lava flow field or even volcanic island creation. Other more local changes occur in volcano morphology on an edifice: new vent or vent erosion, crust deformation, aprons of new pyroclastic deposits and debris-avalanche deposits, erosion and re-sedimentation in valleys and ring plains, etc. In particular along the drainage network on the volcano flanks, devastating geomorphic impacts are complex, with an initial stage of accelerated erosion followed by an exponential decrease over a few years. Rain-generated runoff, lahars and floods are a serious hazard at any time during the rainy seasons following an eruption. We distinguish two categories of geomorphic responses to small and to large eruptions. In addition, long-term geomorphic hazards include geomorphic processes that occur without a link to eruptions on inactive volcanoes.

3.4.1 Hazardous geomorphic response to small eruptions

Geomorphic hazards following small to moderate eruptions are severe but restricted in time, usually only one year, whereas geomorphic hazards after large-scale eruptions are worse and protracted. Many studies of recent tephra erosion, such as Paricutin 1943–52 in Mexico (Segerstorm, 1950), Irazu 1963–5 in Costa Rica (Waldron, 1967) and Usu 1977–82 in Japan (Chinen and Kadomura, 1986), suggest that the erosion rate peaked soon after the tephra erupted and then declined rapidly, from 25–100 mm yr^{-1} in the first two years to 1–5 mm yr^{-1} within five years of the eruption. Sediment yields calculated at several composite cones following moderate eruptions range between 1.1 and 2.7×10^5 m^3 km^{-2} (e.g. Galunggung in 1982–3, Merapi in 1994–5 and Semeru in 2000, in Java; Unzen, Japan, in 1991–3). The range of sediment yields compares well with data compiled by Major et al. (2000) for catchments affected by eruptions on composite volcanoes in a wide range of humid climatic settings. Persistently active composite cones characterized by small but frequent eruptions deliver a very large amount of sediment worldwide to rivers.

After deposition of tephra, geomorphic hazards tend to be concentrated in valley channels whereas they decline rapidly on hillslopes and divides (Collins and Dunne, 1986). If the 20-year perspective from Mount St Helens (Major et al., 2000) can serve as a guide, yields from basins affected solely by hillslope disturbance will diminish rapidly, probably within one or two years, whereas yields from basins that experience dominantly channel disturbance will likely remain elevated for as much as several decades. In contrast, erosion of tephra-mantled hillslopes typically subsides within several months. This decline is not caused by a recovery of vegetation, but by increased infiltration and decreased erodibility of the tephra layer, by exposure of more permeable and less erodible substrates, and by the development of a stable rill network.

3.4.2 Severe geomorphic hazards in catchments disturbed by large eruptions

The largest sediment yields worldwide (i.e. from 10^3 to 10^6 t km^{-2} yr^{-1}) are delivered in volcano catchments affected by debris flows in the aftermath of explosive eruptions of large magnitude. In contrast, in basaltic regions tephra is commonly coarse, and post-eruptive sedimentary response is muted. After a large (several km^3) eruption, sediment yields can exceed pre-eruption yields by several orders of magnitude. Annual suspended sediment yields following the 1980 Mount St Helens eruption were as much as 500 times greater than typical background levels (Major et al., 2000). After 20 years, the average annual suspended sediment yield from the 1980 debris-avalanche deposit remained 100 times (10^4 t km^{-2}) greater than typical background levels (10^2 t km^{-2}). Within five years of the eruption, annual yields from valleys coated by lahar deposits remained roughly steady, and average yields remain about 10 times greater than background levels. The long-term instability of eruption-generated debris means that effective mitigation measures must remain functional for decades. Prolonged excessive sediment transport after an eruption can cause environmental and socio-economic harm exceeding that caused directly by the eruption, as shown by the protracted lahars around Pinatubo.

The geomorphic response to large eruptions (more than 5 km^3 of tephra) had not been documented prior to the study of the Mount Pinatubo lahars. The explosive eruption of Mount Pinatubo on 15 June 1991 deposited 5 to 6 km^3 of loose pyroclastic-flow deposits in the heads of valleys draining the volcano and about 0.2 km^3 of tephra on the volcano's flanks that would later be the primary source of sediment for lahars. Lahars were still flowing into densely populated areas of central Luzon over the following 10 years, causing deaths, leaving more than 50,000 people homeless, affecting more than 1,350,000 people, and causing enormous property losses and social disruption (Janda et al., 1997). Sediment yields set world records during the first three post-eruption years: 10^6 m^3 km^{-2} yr^{-1} in 1991, i.e. nearly an order of magnitude greater than the maximum sediment yield computed following the 18 May 1980 eruption of Mount St Helens (Pierson et al., 1992). The prodigious sediment yield from Pinatubo has geomorphic after-effects such as watershed disruption, channel piracy and avulsion, and blockage of tributaries. Blockage of tributaries at their confluence with the main channel formed temporary lakes and impoundments. Floods triggered by breaching of these lakes provided an additional hazard. Unlike rain-induced lahars, these floods can occur in the absence of rainfall, and limit the capability to warn threatened areas. Although the areas affected by lahars have expanded, the frequency of events has decreased and the number of impacted river systems had dwindled to four in 1995, as source materials were depleted.

The Pinatubo case study raises at least three important points (Major et al., 1997): (1) heavy rainfall alone was not responsible for generating the lahars in 1991; (2) geomorphic accidents affecting watershed and channels play a significant role in erosion and redistribution of sediments.

These have fostered more lahars in the ensuing years and more flooding beyond the alluvial fans onto the densely populated plains; (3) after the geomorphic impact of the devastating 1991 lahars, the subsequent lahars triggered by the seasonal monsoon rains and other geomorphic accidents had far greater social and economic impacts.

3.4.3 Delayed geomorphic hazards in caldera and ignimbrite contexts

The contrast between hydrographic and sedimentary responses to intracaldera eruptions with respect to composite cones is exemplified by Tarawera (White et al., 1997) and Taupo caldera (Manville et al., 1999) in New Zealand. Temporary storage of water, which is later released in catastrophic floods, commonly takes place in lakes of calderas and dome complexes in temperate to humid environments. The delayed channel disturbance following the 1886 Tarawera event may characterize the response to Plinian eruptions in humid-environment caldera complexes, which have modest relief, weakly integrated drainage systems and numerous lakes. Sediment yield is minimal prior to a breakout flood, but then increases to a maximum only after the breakout flood modifies the trunk stream valley. Sediment yield rises rapidly as gullies sap headwards from the flood-scoured valley, and the increase is reflected in rapid downstream aggradation of stream deposits. Sediment yield declines gradually as streams restabilize but the gradual decline in sediment yield is distinct at calderas in having an initial period of stability.

Delayed and indirect volcanogenic hazards are significant in caldera environments with large ignimbrites that have temporarily dammed rivers around them, even where direct primary consequences of volcanic eruptions are minor or absent. Studies of breakout floods from volcanic areas have generally focused on floods from intracaldera lakes and on those caused by failure of pyroclastic and/or debris-avalanche-dammed riverine lakes (e.g. in Alaska: Waythomas, 2001). Kataoka et al. (2008) described the breakout flood from an ignimbrite-dammed valley after the 5 ka Numazawako eruption in northeastern Japan. The 4 km^3 of valley-confined ignimbrite dammed the Tadami River to a depth of >100 m, temporarily impounding 1.6 km^3 of water. The catastrophic release of the dam lake and the ensuing flood redeposited pyroclastic material tens of metres thick as far as the coastal Niigata Plain >150 km downstream of the volcano. Palaeohydraulic estimation techniques indicate a peak discharge of 30,000–50,000 m^3 s^{-1} at the breach point (Kataoka et al., 2008). Large-scale breakout floods with very high peak discharges and volumes of water released can occur as a result of volcanic activity at small to moderate-sized volcanoes, even though the eruption itself was moderate sized and primary eruption impacts covered a small area. Many large cities in Japan and in comparable volcanic arcs are concentrated on alluvial plains away from volcanoes, reducing the perceived volcanic hazards. This fact points to the need for emergency and land-use planners to reconsider their hazard mitigation concepts to recognize volcano-hydrologic hazards.

3.5 Long-term geomorphic hazards around inactive volcanoes

There are no precise or generally accepted definitions for the terms active, dormant or extinct volcano. However, all volcanoes that have erupted within the past 10,000 years are commonly regarded as active (Simkin and Siebert, 1994). Dormant volcanoes should not be underestimated as sources of hazards and risk, not only because they can reawaken, but because they may be the site of non-eruption-induced mass movements.

The most hazardous volcanoes are commonly those in which eruptive phases are separated by hundreds or thousands of years of quiescence (e.g. Pinatubo had been dormant for 650 years prior to the climactic 1991 eruption). A major cause of vulnerability is also the attitude of people who live in the vicinity of a long-term dormant volcano (Schmincke, 2004). It is thus commonly difficult to evacuate people from potentially affected areas. When volcanoes believed to be extinct erupt, catastrophes often follow, such as at Mount Lamington (Papua New Guinea, 1951) or El Chichón (Mexico, 1982). Volcanoes and volcano fields can be active for millions of years with relatively short periods of volcanic activity alternating with long periods of quiescence. The most important criteria for regarding volcanoes as extinct (i.e. not expected to erupt again) or dormant (i.e. could become active again) are the total lifetime of a volcanic complex and the overall frequency of individual eruptions. This requires very precise analysis to arrive at reasonable estimates of the probability of future eruptions (Schmincke, 2004). The significance of this problem becomes very obvious with respect to long-term storage of nuclear waste, e.g. for all nuclear waste generated in the USA, a huge cave system has been dug into Tertiary ignimbrites at Yucca Mountain in Nevada. The age of the youngest scoria cone, Lathrop Wells, 15 km south of the storage area, is about 0.07–0.1 million years. This has to be compared to a period of 10^4 to 10^5 years until the radioactivity of high-level waste has decayed. The question of whether or not volcanic eruptions may occur in the area of the deposit is therefore crucial to the final decision to house waste at Yucca Mountain (Connor et al., 2000; Hinze et al., 2008).

Geomorphic processes acting on dormant and extinct volcanoes are related to mass movements and debris flows, common on all steep mountains but exacerbated by unconsolidated pyroclastic material. These hazards are unrelated to eruptions and stem from the remobilization of old tephra. On dormant volcanoes, catastrophic erosion processes are uncommon but can reach dramatic proportions. Examples of landslide disasters on dissected volcanoes are Mount Ontake (Japan), triggered by an earthquake in 1984 (Voight and Soussa, 1994), and at Casita volcano (Nicaragua), induced by the heavy rains of Hurricane Mitch in 1998 (Scott et al., 2005). Characteristically, the volume transported changes during movement due to entrainment and deposition along the flow path, and the sequence of movement commonly shows stop-start behaviour as a result of landslide dam formation and breaching.

Unlike other volcanic hazards, lahars and debris avalanches do not require an eruption as they can be triggered by rainfall or by edifice failure long after an eruption. The 1984 event of Mount Cayley volcano in southwest British Columbia illustrates the movement mechanisms typical of rapid landslides on dissected volcanoes, i.e. significant entrainment of debris following initial failure, high velocity, and a reach extended by a distal debris flow (Evans et al., 2001). A disintegrating rock mass of Quaternary pyroclastic rock slid from the western flank mass, entrained more material and then formed a rock avalanche that travelled at 42 m s^{-1} a horizontal distance of 3.45 km from its source over a vertical elevation difference of 1.2 km. The rock avalanche was partially transformed into a distal debris flow that travelled a further 2.6 km down a narrow path to the main river, temporarily blocking it. The landslide is one of seven high-velocity rock avalanches that have occurred in the dormant Garibaldi volcanic belt of southwest British Columbia since 1855.

Secondary lahars occur during volcanic quiescence and can be generated by avalanches, crater lake breakout, heavy post-eruptive rain, and earthquake-induced avalanches. Crater or caldera lakes, and volcanic debris dammed lakes can break out many years after eruptions. The 1998 Casita disaster further emphasizes the hazards of highly mobile debris flows beginning as landslides, their potential to transform, and their ability to amplify during flow. A single wave of debris flow, triggered by hurricane precipitation, killed more than 2,000 residents of two towns near Casita volcano on 30 October 1998 (Scott et al. 2005). The flow wave began as a small flank failure near the summit of the inactive edifice. Peak precipitation from Hurricane Mitch was the flow trigger: almost a metre of rain had fallen between 26 and 30 October. Downstream, the flow volume enlarged by a factor of at least nine, transforming to a catastrophic debris flow over 1 km wide and 3–6 m deep. Only land-use planning based on debris-flow hazard assessment could have prevented the Casita disaster. The towns – established in the 1980s on apparently safe terrain 4 km from the volcano – were found afterwards to have been located in a pathway of prehistoric debris flows.

3.6 Methods and goals of volcano hazard and risk assessment

Lessons from recent eruptions and results acquired during the International Decade for Natural Disaster Reduction (1990–2000) show that considerable progress has been achieved in detecting eruptive phenomena based on observations and monitoring of Decade volcanoes, and on experimental modelling in the laboratory. Retrospective examination of geophysical crises has greatly helped to reduce the effects of major eruptions. New approaches assess damage and loss, and explore the roots of the vulnerability system and all values at stake, including institutional and political factors. The new paradigm has questioned the so-called dominant or radical perspective of behaviour theory and is based on research into the deep-seated roots of vulnerability beyond the hazard domain and on environmental, cultural and political impact analysis. This approach requires a series of tasks aimed at sustainable mitigation of the effects of volcanic catastrophes in the frame of coherent institutional projects.

3.6.1 Coping with volcano hazards

As the number of people and the quantity of economic infrastructure at risk from volcano hazards increases worldwide, what can be done to reduce risk from volcanoes? An effective approach to volcano-hazards mitigation encompasses four important elements (Tilling, 2005):

1. Acquisition of a good understanding of a volcano's past behaviour by deciphering its eruptive history from basic geology, geochronology and other geosciences studies. This will lead to long-term forecasts of potential future activity. Having identified a dangerous volcano, the group of scientists must: (i) determine when and where the eruption will take place on a volcano; (ii) understand how the eruptive products can be emplaced, from which the damage-bearing phenomena can be inferred; (iii) elaborate and rank likely eruption scenarios based on occurrence probabilities, according to the past and present behaviour of the volcano. This stems from knowledge of the activity of the edifice over the past thousands of years, aiming at reconstructing styles and

frequency of eruptions; (iv) identify and delineate areas likely to be affected in case of eruption according to the selected eruption scenarios.
2. Initiation or expansion of volcano monitoring to determine a volcano's current behaviour, to characterize its current baseline level of activity, and to detect, reliably and quickly, any significant departure from it. Instrumental monitoring and satellites dedicated to natural hazards now allow scientists to follow the evolution of a volcano that is remote or difficult to get access to. An instrumental network is essential and can be complemented by modelling of volcanic mass flows. This will lead to a short-term forecast of potential future activity.
3. Equally important is the creation of emergency-response or preparedness plans, as well as the selection of risk reduction strategies (Blong, 2000). This requires a census of threatened populations and of factors of human, cultural, economic and political vulnerability. The aim is to elaborate databases and use GIS, which overlay distinct data layers and attribute tables, and modify the subsequent eruption scenarios at the same time as the recorded activity is evolving. At the same time planning for risk prevention is to be implemented: civil protection and contingency plans for relief organisation and emergency measures; preparedness and education of population; land-use and urban planning.
4. Establishment or maintenance of credible, effective communications from scientists with civil authorities, the media and the general public.

Blong (2000) points out three methods of adjustment to reduce the risk associated with the occurrence of hazards: (1) Prevent the hazard itself, though this is rarely possible with volcanic hazards. The most publicized diversion of a lava flow occurred at Etna in May 1992. (2) Modify vulnerability to the impact of the hazard (see his Table 3, p. 1225). Examples include building roofs that withstand ash loads, protecting crops under plastic sheeting, diverting lahars, land-use planning, and broadcasting volcano forecasts, which allow inhabitants to plan daily activities. (3) Share the loss associated with eruption damage with a wider community through disaster relief and insurance.

We present two sets of operations that form part of a combined approach to hazard and risk assessment of the effects of eruptions, the focus of the most recent research programmes aimed at laboratory volcanoes (e.g. Vesuvius, Teide, Soufrière de Guadeloupe): (1) hazard and risk maps and scenarios based on a geomorphic and modelling approach; (2) long-term assessment based on a quantitative analysis of vulnerability.

3.6.2 The role of geomorphology in hazard and risk zone maps and risk assessment

Geomorphology can contribute to risk assessment through two zonation approaches: geomorphic hazard zones and composite risk zones. Geomorphic hazard domains are established according to the capacity of each hazard to affect geomorphic stability, the perceived vulnerability of people, and the priority of elements at stake such as urban settlements and land-use types. A composite risk zonation, incorporating geomorphic mapping and analysis of risk and of vulnerability, can therefore be achieved.

Geomorphic surveys with the aid of satellite imagery form a logical starting point for natural hazard zoning. Geomorphic hazard zonation aims at recognizing old deposits, mapping flow paths and delineating hazard zones, which are primary inputs in elaborating eruption scenarios. Potential inundation areas can be delineated based on the palaeohydrologic record of flows in specific areas, on relationships between failure volumes and corresponding inundation areas, and with the help of slope stability analysis of a digital elevation model (DEM). Additional risk assessment requires the development of a series of scenarios in which eruption magnitudes, hazard types, composite risk zonation, and the vulnerability of people and infrastructure are adequately considered. Eruption scenarios are useful for the preparation of emergency evacuation plans and long-term land-use planning.

Recently an automatic GIS-based system was proposed for a first-order hazard assessment of the volcano under study in a very short time and using standard computational facilities. The automated system helps to elaborate volcanic hazard and scenario maps (Felpeto *et al.*, 2007). The generation of both maps is based on the use of numerical simulation of eruptive processes. The user can select on a toolbar one hazard and then decide whether to generate a scenario map (with a unique vent) or a hazard map (with a broader source area). The tool has been designed in such a way that the inclusion of new numerical models and functionalities is quite simple. Each numerical model is programmed and implemented as an independent program that is launched from the system and, when it finishes the computation, returns the control to the GIS, where the results are shown. This structure allows for further analyses such as risk analyses to be automated inside the system. This automatic system has near real-time access to different datasets such as monitoring data in case of an evolving eruptive crisis.

An alternative delineation of hazard zones can be obtained with the aid of mathematical models that simulate the evolution of volcanic phenomena and compute the effects at ground level, allowing estimation of the area affected by an event according to a certain scenario. Geomorphic and hydrologic parameters are critical input requirements for the use of DEMs and GIS in long-term planning. The use of DEMs and simulation codes has enabled Iverson *et al.* (1998), Rowland *et al.* (2005), and Sheridan *et al.* (2001), to gauge volcanic flow hazards in densely populated areas around Mount Rainier, Mauna Loa, and Popocateptl volcanoes, respectively.

Iverson *et al.* (1998) proposed a method of delineating lahar hazard zones in valleys that head on volcano flanks, which provides a rapid and reproducible alternative to traditional methods. The rationale derives from scaling analyses of generic lahar paths and statistical analyses of 27 lahar paths documented at 9 edifices. These analyses yield semi-empirical equations that predict inundated valley cross-sectional areas (A) and planimetric areas (B) as functions of lahar volume (V). The predictive equations ($A = 0.05V^{2/3}$ and $B = 200V^{2/3}$) provide all the information necessary to calculate and plot inundation limits on topographic maps. By using a range of prospective lahar volumes to evaluate A and B, a range of inundation limits can be plotted for lahars of increasing volume and decreasing probability. Iverson *et al.* (1998) automated hazard-zone delineation by embedding the predictive equations in a GIS computer program that uses a DEM of topography. The simulation code LAHARZ provides a rapid, automated means of applying predictive equations to regions around edifices and comparison of the results with the hazard-zone boundaries established in the field by mapping lahar deposits. Lahar hazard zones computed for Mount Rainier, USA, mimic those constructed on the basis of intense field investigations. The computed hazard zones illustrate the potentially widespread impact of large lahars, which on average inundate planimetric areas 20 times larger than those inundated by rock avalanches of comparable volume.

Rowland *et al.* (2005) have used the FLOWGO thermorheological model to determine cooling-limited lengths of channel-fed 'a'a lava flows from Mauna Loa (Hawaii). The program starts lava flow every fourth line, and samples on a 30-m spatial resolution SRTM DEM within regions corresponding to the NE and SW rift zones and the N flank of the volcano. The model is not deterministic in that all potential vent sites for lava flows are considered with an equal likelihood to erupt. Each model run represents an effective effusion rate, which for an actual flow coincides with it reaching 90% of its total length. The authors ran the model at effective effusion rates ranging from 1 to 1,000 m^3 s^{-1}, and determined the cooling-limited channel length for each. Lava coverage includes the SW rift zone and all the SW flanks even at low effusion rates, the N flank, the NE rift zone and the city of

Hilo to be reached by lava flows with an effective effusion rate of 400 m^3 s^{-1}, and finally the SE flanks.

Wadge *et al.* (1998) used a physically based simulation of an avalanche based on a three-fold parameterization of flow acceleration for which they chose values using an inverse method. Multiple simulations based on uncertainty of the starting conditions and parameters, specifically location and mass flux, have been used to map hazard zones and successfully model the pyroclastic flows generated on 12 May 1996 by Soufrière Hills, Montserrat. Hooper and Mattioli (2001) used a kinematic modelling of 'dense' flows initiated by gravitational collapse pyroclastic flows and employed a graphical computer model FLOW3D to simulate this type of volcanic flow. The program constructs a digital terrain model based upon a 3D network of triplets, while a synthetic dome was added to the topographic model to improve the accuracy of the simulations. Simulated flow pathways, runout distances and velocities closely approximated observed block-and-ash flows on Soufrière Hills. While the simulations presented do not elucidate the physics of pyroclastic flow, this type of kinematic modelling can be completed easily and without extensive knowledge of specific parameters other than topography. It thus serves as a rapid and inexpensive first approach for initial hazard assessment. However, such computer-based simulations require accurate digital topographic information, because initial topographic boundary conditions critically influence the extent and velocity of pyroclastic flows.

DEM- and GIS-based computer codes for simulating lahars and pyroclastic flows have been used by Sheridan *et al.* (2001) to gauge volcanic hazards at Popocateptl in Mexico, in addition to a detailed survey of the past eruptive history and a close monitoring of the present activity. Popocateptl stands 60 km southeast of Mexico City and 40 km west of Puebla. The combined population of these two metropolitan areas exceeds 30 million, so the 2000–1 activity has received intense public scrutiny. The present eruptive activity, which started in December 1994, has already forced the government to evacuate 20,000 to 50,000 people at a time from towns in the state of Puebla. Several domes have formed in the crater and domes will probably continue to grow until the crater is entirely filled by lava, a process that may take a few more years at the current rate of dome emplacement.

The major hazard associated with the current activity must be considered to be lahars on surrounding villages. Rock avalanches and pyroclastic flows could produce materials for lahars. Assuming likely source areas, inundation zones for lahar volumes of 10^7 and 10^8 m^3 were simulated with ARCINFO using the LAHARZ model. The smaller volume model lahar represents the water

available from glacier ablation by pyroclastic flows and the larger volume represents the water from complete melting of the current glacier. The maximum possible lahar, based on the size of the Ventorillo glacier, is about 10^8 m^3. Simulations of pyroclastic flows and rock avalanches used the FLOW3D model, which provides velocity histories of particle streams along flow paths in three dimensions (Sheridan *et al.*, 2001). Simulations were used to create the hazard map at Popocateptl. Bit-mapped and colour-coded overlays of multiple themes, including the flow paths and velocities, were used to produce a realistic image useful for non-professional observers. The interactive platform of FLOW3D allows the observer to adjust the perspective and distance for the desired view.

3.6.3 A quantitative analysis of vulnerability

The case study of Mount Cameroun, where approximately 450,000 people live or work around the volcano, displays the methodology used to provide the authorities with a first quantitative analysis of the vulnerable context. The method used by Thierry *et al.* (2008) involves several stages: (1) identifying the different geological hazard components; (2) defining each phenomenon's threat matrix by crossing intensity and frequency indices with extension; (3) mapping the hazards; (4) listing and mapping the exposed elements; (5) analyzing their respective values in economic, functional and strategic terms; (6) establishing typologies for the different element-at-risk groups and assessing their vulnerability to the various physical pressures produced by the hazard phenomena; and (7) establishing risk maps for each of the major element-at-risk groups. This has enabled the group to identify the main critical zones and points within the area, and to provide quantified orders of magnitude concerning the dimensions of the risk by producing a plausible eruption scenario.

Because of the devastating potential of volcanoes close to urbanized areas, the EXPLORIS research EU-funded programme has pursued an even more quantitative vulnerability and risk assessment rooted in a previous method based on the event tree (Newhall and Hoblitt, 2002; Marzocchi *et al.*, 2006) scheme to estimate the probability of all the relevant possible outcomes of a volcanic crisis. The quantitative method has thrived and now includes more sophisticated event trees for eruption behaviour, and structured elicitation of expert judgement for probabilistic hazard and risk assessment (Aspinall, 2006), and a new evidence-based approach (Baxter *et al.*, 2008). Scientists have taken the initial information that encompasses three consequential steps (the acquisition of information, the past data to assess long-term volcanic hazard, and the monitoring observations to assess mid- to short-term volcanic hazard). This information has been used in a Bayesian statistical framework to assess long-term volcanic hazard. In particular they use the data and specific probability density function in the prior and posterior statistical distributions to calculate transition probabilities on the event tree. The event tree is a branching graph representation of events in which individual branches are alternative steps from a general prior event, state or condition, and that evolve through time into increasingly specific subsequent events. An event tree attempts to graphically display all relevant possible outcomes of volcanic unrest in progressively higher levels of detail. The advantage lies in the fact that the scheme can take all available information into account and can continuously update probabilities. The results are very useful for cost-benefit analysis of risk mitigation actions, and for appropriate land-use planning.

Event trees have been constructed for European volcanoes such as Vesuvius (Neri *et al.*, 2008), Teide (Marti *et al.*, 2008) and Soufrière de Guadeloupe (Spence *et al.*, 2008). They rely on quantitative assessment and volcanic loss models (e.g. tephra fall and pyroclastic flow impacts on buildings). To achieve coherence, many diverse strands of evidence had to be unified within a formalized structure, and linked together by expert knowledge. For this purpose a Vesuvius event tree was created to summarize in a numerical-graphical form (Figure 3.1), at different levels of detail, all the relative likelihoods relating to the genesis and style of eruption, development and nature of volcanic hazards, and the probabilities of occurrence of different risks in the next eruption crisis. The event tree formulation provides a logical pathway connecting generic probabilistic hazard assessment to quantitative risk evaluation.

The 'evidence-based volcanology' has been applied in the EXPLORIS research programme (Baxter *et al.*, 2008) to contribute to crisis planning and management. The analytical approach enumerates and quantifies all the processes and effects of the eruptive hazards of the volcano known to influence risk. This is a scientific challenge that combines field data on the vulnerability of the built environment and humans in past volcanic disasters with theoretical research on the state of the volcano, and includes evidence from the field on previous eruptions as well as numerical simulation modelling of eruptive processes, formal probabilistic reasoning under uncertainty and a decision analysis approach. The combined operations have provided the basis for the development of an event tree for a future range of eruptions. The most likely future eruption scenarios for emergency planning were derived from the event tree. Modelling the impacts in these scenarios and quantifying the consequences for the threatened area provide realistic assessments for disaster planning and for showing the potential risk-benefit of mitigation measures (timely evacuation, building protection, etc.).

3.7 Concluding remarks and perspectives

Volcanoes possess all the geomorphic processes and hazards of mountains worldwide but they have additional hazards posed by eruptive activity. Geomorphic hazards of volcanic origin are direct, related to explosive activity, and indirect, due to secondary processes such as lahars. Geomorphic hazards often recur many years after an eruption. The geomorphic consequences of large volcanic eruptions are severe, long lasting, and disturb slopes and drainage processes on and around the volcanic mountain for decades. Rates of growth and erosion are very rapid at active volcanoes, particularly at composite cones, thus increasing non-eruptive hazards on slopes, in channels and in ring plains. Furthermore, geomorphic hazards can also be severe on inactive volcanoes.

Recent case studies, such as the protracted post-eruption crisis around Mount Pinatubo, indicate that despite some success in forecasting events and in mitigating hazardous volcanic processes at individual volcanoes, the losses and damage still remain too high. Even in developed countries, damage increases with respect to the national GNP because of the increasing value of the resources at stake. Over the past 50 years, the risk to society from volcanic eruptions has increased sharply due to population growth, more developed and diversified economies, and a more technologically advanced infrastructure. Correct land-use planning is fundamental in minimizing both loss of life and damage to property. Remote sensing technology, warning systems and GIS have emerged as the most promising tools to support the decision-making process. More people and decision-makers are facing problems of mitigating volcanic hazards, especially in large cities located near potentially destructive volcanoes, such as Naples and Mount Vesuvius, Arequipa and El Misti, Tacoma-Seattle and Mount Rainier (Chester *et al.*, 2001). With the continuing rise of global urbanization, the management of volcanic hazards in urban and densely populated areas will present major challenges for scientists and disaster workers in this century. Policy decisions will require improvements in awareness of potential threats as well as a timely and appropriate response. In addition to forecasting and monitoring a crisis, a coherent set of actions seeking a timely and appropriate response should: (1) use methods aimed at enhancing risk perception amongst the threatened population; (2) reach a common evaluation of what is acceptable risk for a given group or society; and (3) help to select strategies aimed to protect and prepare people and to rehabilitate housing and infrastructure.

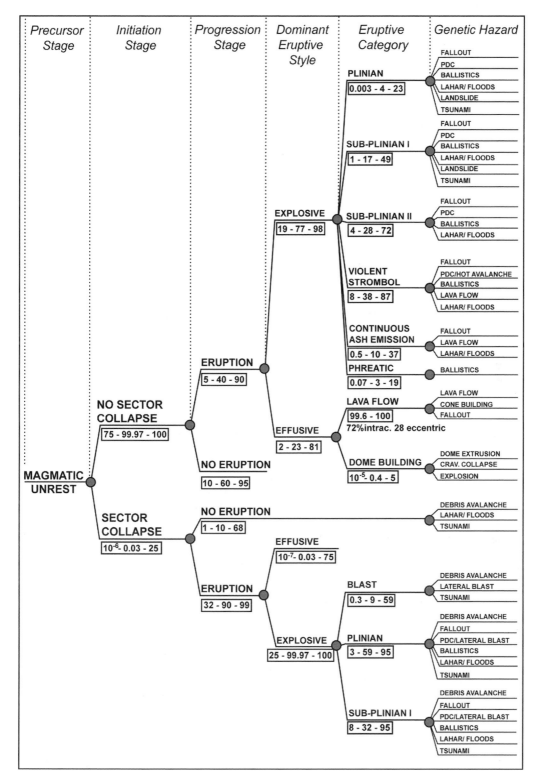

FIGURE 3.1. Event tree framework summarizing potential eruption scenarios for next volcanic crisis at Vesuvius and possible associated hazards that may develop. The whole tree is shown, but the paper focuses on the branches for the main five explosive eruptions. The hypothetical branch probabilities (and 5% and 95% credible intervals) were those obtained from the expert elicitation with EXPLORIS researchers. (After Neri et al., 2008. Developing an Event Tree for probabilistic hazard and risk assessment at Vesuvius. *Journal of Volcanology and Geothermal Research*, **178**(3), Figure 1.)

In recent years, the philosophy underpinning volcanic hazard and risk assessment has undergone a notable paradigm shift – a transformation from reliance on simple concepts, which provide a basis for selective deterministic evaluations, to the conviction that full-fledged probabilistic modelling is the most appropriate way to characterize the intrinsic uncertainties associated with volcanic hazards and risks. As part of this shift, consideration has to be given in applied volcanology to important radical probabilistic concepts, such as quantitative hazard assessment, 'treatment of uncertainties', short- and long-term hazard forecasting, and so on.

However, as good as it may be, science will be unable to resolve all difficulties in forecasting risks. Equally important in predicting hazard effects is a broad approach to the complexity of the vulnerability system. The process of risk assessment and the reduction of the effects of the eruptions will be ameliorated if the roots of vulnerability and the values at stake are measured and incorporated in a scenario. Vulnerability studies represent a field where questions and objectives have been profoundly revised to unravel roots of vulnerability outside the natural hazards system (Chester et al., 2001; Blaikie et al., 2004). A qualitative approach aims at determining a complex set of social, economic, demographic and cultural factors; physical, functional and technical factors; and institutional, political and administrative factors. Unravelling this set of factors will help us to understand how a society can cope with the effects of eruptions and post-eruption crises.

Acknowledgments

I have drawn heavily on my Chapter 11 of the textbook *Mountain Geomorphology* (2004) as well as on comprehensive reviews on volcano hazards by R. I. Tilling (2005) and H. U. Schmincke (2004). I thank them because I have freely excerpted, paraphrased or distilled their text. I thank Professor Goudie for polishing the English of this manuscript.

References

Aspinal, W. P. (2006). Structured elicitation of expert judgement for probabilistic hazard and risk assessment in volcanic eruptions. In H. M. Mader, S. G. Coles, C. B. Connor, and L. J. Connor (eds.), *Statistics in Volcanology*. Special Publications of IAVCEI, **1**, pp. 15–30. London: Geological Society.

Baxter, P. J., Aspinall, W. P., Neri, A. *et al.* (2008). Emergency planning and mitigation at Vesuvius: a new evidence-based approach. *Journal of Volcanology and Geothermal Research*, **178**(3), 454–473.

Beget, J. E. and Kienle, J. (1992). Cyclic formation of debris avalanches at Mount St Augustine volcano. *Nature*, **356**, 701–704.

Björnsson, H. (2004). Glacial lake outburst floods in mountain environments. In P. O. Owens and O. Slaymaker (eds.), *Mountain Geomorphology*. London: Arnold, pp. 165–184.

Blaikie, P., Cannon, T., Davis, I. and Wissner, B. (2004). *At Risk: Natural Hazards, People's Vulnerability and Disasters*. London: Routledge.

Blong, R. (1984). *Volcanic Hazards: A Sourcebook on the Effects of Eruptions*. Sydney: Academic Press.

Blong, R. (2000). Volcanic hazards and risk management. In H. Sigurdsson, B. Houghton, S. R. McNutt, H. Rymer, and J. Stix (eds.). *Encyclopedia of Volcanoes*. San Diego: Academic Press, pp. 1215–1227.

Carracedo, J. C. (1999). Growth, structure, instability and collapse of Canarian volcanoes and comparison with Hawaiian volcanoes. *Journal of Volcanology and Geothermal Research*, **94**(1–4), 1–19.

Casadevall, T. J. (ed.) (1994). *Volcanic Ash and Aviation Safety*. Proceedings of the 1st International Symposium on Volcanic Ash and Safety. US Geological Survey Bulletin 2047.

Chester, D. K., Degg, M., Duncan, A. M. and Guest, J. E. (2001). The increasing exposure of cities to the effects of volcanic eruptions: a global survey. *Environmental Hazards*, **2**, 89–103.

Chinen, T. and Kadomura, H. (1986). Post-eruption sediment budget of a small catchment on Mt. Usu, Hokkaido. *Zeitschrift für Geomorphologie*, Supplement-Band **60**, 217–232.

Collins, B. D. and Dunne, T. (1986). Erosion of tephra from the 1980 eruption of Mount St. Helens. *Geological Society of America Bulletin*, **97**, 896–905.

Connor, C. B., Stamatakos, J. A., Ferrill, D. A. *et al.* (2000). Geologic factors controlling patterns of small-volume basaltic volcanism: application to a volcanic hazards assessment at Yucca Mountain, Nevada. *Journal of Geophysical Research*, **105**, 417.

Cronin, S. J., Neall, V. E., Lecointre, J. A. and Palmer, A. S. (1996). Unusual "snow-slurry" lahars from Ruapehu volcano, New Zealand, September 1995. *Geology*, **24**(12), 1107–1110.

Evans, S. G., Hungr, O. and Clague, J. J. (2001). Dynamics of the 1984 rock avalanche and associated distal debris flow on Mount Cayley, British Columbia, Canada; implications for landslide hazard assessment on dissected volcanoes. *Engineering Geology*, **61**, 29–51.

Felpeto, A., Marti, J. and Ortiz, R. (2007). Automatic GIS-based system for volcanic hazard assessment. *Journal of Volcanology and Geothermal Research*, **166**, 106–116.

Fournier d'Albe, E. M. (1979). Objectives of volcanic monitoring and prediction. *Journal of the Geological Society of London*, **136**, 321–326.

Hinze, W. J., Marsh, D., Weiner, R. F. and Coleman, N. M. (2008). Evaluating igneous activity at Yucca Mountain. *EOS, Transactions, American Geophysical Union*, **89**(4), 29–30.

Hooper, D. M. and Mattioli, G. S. (2001). Kinematic modeling of pyroclastic flows produced by gravitational dome collapse at Soufrière Hills volcano, Montserrat. *Natural Hazards*, **23**, 65–86.

Iverson, R. M., Schilling, S. P. and Vallance, J. W. (1998). Objective delineation of lahar-inundation hazard zones. *Geological Society of America Bulletin*, **110**(8), 972–984.

Janda, R. J., Scott, K. M., Nolan, K. M. and Martinson, H. A. (1981). Lahar movements, effects, and deposits. In P. W. Lipman and D. R. Mullineaux (eds.), *The 1980 Eruptions of Mount St. Helens, Washington*. Geological Survey Professional Paper 1250, pp. 461–478.

Janda, R. J., Daag, A. S., Delos Reyes, P. J. *et al.* (1997). Assessment and response to lahar hazard around Mount Pinatubo, 1991 to 1993. In C. G. Newhall and R. S. Punongbayan (eds.), *Fire and Mud: Eruptions and Lahars of Mt Pinatubo, Philippines*. University of Washington Press, pp. 107–139.

Kataoka K. S., Urabe A., Manville V. and Kajiyama A. (2008). Breakout flood from an ignimbrite-dammed valley after the 5 ka Numazawako eruption, northeast Japan. *Geological Society of America Bulletin*, **120**(9/10), 1233–1247.

Keating, B. H. and McGuire, W. J. (2000). Island edifice failures and associated tsunami hazards. *Pure and Applied Geophysics*, **157**, 899–955.

Kelfoun, K., Druitt, T., van Wyk de Vries, B. and Guilbaud, M. N. (2008). Topographic reflection of the Socompa debris avalanche, Chile. *Bulletin of Volcanology*, **70**, 1169–1187.

Lipman, P. W. (2000). Calderas. In H. Sigurdsson, B. Houghton, S. R. McNutt, H. Rymer, and J. Stix (eds.), *Encyclopedia of Volcanoes*. San Diego: Academic Press, pp. 643–662.

Major, J. J., Janda, R. J. and Daag, A. S. (1997). Watershed disturbance and lahars on the east side of Mount Pinatubo during the mid-June 1991 eruptions. In C. G. Newhall and R. S. Punongbayan (eds.), *Fire and Mud: Eruptions and Lahars of Mt Pinatubo, Philippines*. University of Washington Press, pp. 895–918.

Major, J. J., Pierson, T. C., Dinehart, R. L. and Costa, J. E. (2000). Sediment yield following severe disturbance: A two decade perspective from Mount St. Helens. *Geology*, **28**(9), 819–822.

Manville, V., White, J. D. L., Houghton, B. F. and Wilson, C. J. N. (1999). Paleohydrology and sedimentology of a post-1.8 ka breakout flood from intracaldera Lake Taupo, North Island, New Zealand. *Geological Society of America Bulletin*, **111**(10), 1435–1447.

Marti, J., Aspinall, W. P., Sobradelo, R. *et al.* (2008). A long-term volcanic hazard event tree for Teide-Pico Viejo strato-volcanoes (Tenerife, Canary Islands). *Journal of Volcanology and Geothermal Research*, **178**(3), 543–552.

Marzocchi, W., Sandri, L. and Furlan, C. (2006). A quantitative model for volcanic hazard assessment. In H. M. Mader, S. G. Coles, C. B. Connor, and L. J. Connor (eds.), *Statistics in Volcanology*. Special Publications of IAVCEI, **1**, 31–37, London: Geological Society.

Moore, J. G., Normak, W. R. and Holcomb, R. T. (1994). Giant Hawaiian underwater landslides. *Science*, **264**, 46–47.

Nakada, S. (2000). Hazards from pyroclastic flows and surges. In H. Sigurdsson, B. Houghton, S. R. McNutt, H. Rymer, and J. Stix (eds.), *Encyclopedia of Volcanoes*, San Diego: Academic Press, pp. 945–955.

Neri, A., Aspinall, W. P., Cioni, R., *et al.* (2008). Developing an Event Tree for probabilistic hazard and risk assessment at Vesuvius. *Journal of Volcanology and Geothermal Research*, **178**(3), 397–415.

Newhall, C. G. and Hoblitt, R. P. (2002). Constructing event trees for volcanic crises. *Bulletin of Volcanology*, **64**, 3–20.

Ollier, C. (1988). *Volcanoes*. London: Basil Blackwell.

Parfitt, E. A., and Wilson, L. (2008). *Fundamentals of Physical Volcanology*. Oxford: Blackwell.

Pierson, T. C. (ed.) (1999). *Hydrologic Consequences of Hot-rock/ Snowpack Interactions at Mount St. Helens Volcano, Washington 1982–84*. U.S. Geological Survey Professional Paper 1586.

Pierson, T. C. and Costa, J. E. (1987). A rheologic classification of subaerial sediment-water flows. In J. E. Costa and G. E. Wieczorek (eds.), *Debris Flows/Avalanches: Process, Recognition, and Mitigation*. Geological Society of America, Review of Engineering Geology, 7, pp. 1–12.

Pierson, T. C. and Janda, R. J. (1994). Volcanic mixed avalanches: a disaster eruption-triggered mass-flow process at snow-clad volcanoes. *Geological Society of America Bulletin*, **106**, 1351–1358.

Pierson, T. C., Janda, R. J., Thouret, J. C. and Borrero, C. A. (1990). Perturbation and melting of snow and ice by the 13 November 1985 eruption of Nevado del Ruiz, Colombia, and consequent mobilization, flow, and deposition of lahars. *Journal of Volcanology and Geothermal Research*, **41**, 17–66.

Pierson, T. C., Janda, R. J., Umbal, J. V. and Daag, A. S. (1992). *Immediate and Long-term Hazards from Lahars and Excess Sedimentation in Rivers Draining Mount Pinatubo, Philippines*. U.S. Geological Survey Water-Resources Investigations Report 92–4039.

Rampino, M. R. and Self, S. (2000). Volcanism and biotic extinctions. In H. Sigurdsson *et al.* (eds.), *Encyclopedia of Volcanoes*. San Diego: Academic Press, pp. 1083–1091.

Rowland, S. K., Garbeil, H. and Harris, A. J. L. (2005). Lengths and hazards from channel-fed lava flows on Mauna Loa, Hawaii, determined from thermal and downslope modeling with FLOWGO. *Bulletin of Volcanology*, **67**, 634–647.

Schmincke, H. U. (2004). *Volcanism*. Heidelberg: Springer.

Scott, K. M., Vallance, J. W., Kerle, N., *et al.* (2005). Catastrophic precipitation-triggered lahar at Casita volcano, Nicaragua: occurrence, bulking and transformation. *Earth Surface Processes and Landforms*, **30**, 59–79.

Segerstrom, K. (1950). *Erosion Studies at Paricutin, State of Michoacan, Mexico*. U.S. Geological Survey Bulletin, 965-A.

Sheridan, M., Hubbard, B., Bursik, M. I. *et al.* (2001). Gauging short-term volcanic hazards at Popocateptl. *EOS, Transactions AGU*, **185**, 187–188.

Siebert, L. (1996). Hazards of large volcanic debris avalanches and associated eruptive phenomena. In R. Scarpa and R. I. Tilling (eds.), *Monitoring and Mitigation of Volcano Hazards*. Berlin: Springer Verlag, pp. 541–658.

Simkin, T. and Siebert, L. (1994). *Volcanoes of the World: A Regional Directory, Gazetteer, and Chronology of Volcanism During the Last 10,000 Years*. Tucson, AZ: Geoscience Press.

Slaymaker, O. (ed.) (1996). *Geomorphic Hazards*. New York: J. Wiley.

Spence, R., Komorowski, J. C., Saito, K. *et al.* (2008). Modelling the impact of a hypothetical sub-Plinian eruption at La Soufrière of Guadeloupe (Lesser Antilles). *Journal of Volcanology and Geothermal Research*, **178**(3), 516–528.

Tanguy, J.-C., Ribière, Ch., Scarth, A. and Tjetjep, W. S. (1998). Victims from volcanic eruptions: a revised database. *Bulletin of Volcanology*, **60**, 137–144.

Thierry, P., Stieltjes, L., Kouokam, E., Nguéya, P. and Salley, P. M. (2008). Multi-hazard risk mapping and assessment on an active volcano: the GRINP project at Mount Cameroon. *Natural Hazards*, **45**, 429–456.

Thouret, J. C. (1990). Effects of the 13 November 1985 eruption on the ice cap and snow pack of Nevado del Ruiz, Colombia. *Journal of Volcanology and Geothermal Research*, **41**(1–4), 177–201.

Thouret, J.-C. (1999). Volcanic geomorphology: an overview. *Earth-Science Reviews*, **47**, 95–131.

Thouret, J.-C. (2004). Geomorphic processes and hazards on volcanic mountains. In P. O. Owens and O. Slaymaker (eds.) *Mountain Geomorphology*. London: Arnold, pp. 242–273.

Thouret, J.-C., Vandemeulebrouck, J., Komorowski, J. C. and Valla, F. (1995). Volcano-glacier interactions: field survey, remote sensing and modelling – a case study (Nevado del Ruiz, Colombia). In O. Slaymaker (ed.), *Steepland Geomorphology*. New York: J. Wiley, pp. 63–88.

Thouret, J.-C., Ramirez, J., Naranjo, J. L. *et al.* (2007). The Nevado del Ruiz ice cap, Colombia, 21 years after: volcano-glacier interactions, meltwater generation, and lahar hazards. *Annals of Glaciology*, **45**, 115–127.

Tilling, R. I. (1989). Volcanic hazards and their mitigation: progress and problems. *Reviews of Geophysics*, **27**(2), 237–267.

Tilling, R. I. (2005). Volcanic hazards. In J. Marti and G. J. Ernst (eds.), *Volcanoes and Environment*. Cambridge: Cambridge University Press, pp. 55–89.

Ui, T., Takarada, S. and Yoshimoto, M. (2000). Debris avalanches. In H. Sigurdsson, B. Houghton, S. R. McNutt, H. Rymer and J. Stix (eds.), *Encyclopedia of Volcanoes*. San Diego: Academic Press, pp. 617–626.

Vallance, J. W. (2000). Lahars. In H. Sigurdsson, B. Houghton, S. R. McNutt, H. Rymer and J. Stix (eds.), *Encyclopedia of Volcanoes*. San Diego: Academic Press, pp. 601–616.

Van Wyk de Vries, B. and Francis, P. W. (1997). Catastrophic collapse at stratovolcanoes induced by gradual volcano spreading. *Nature*, **387**, 387–390.

Voight, B. (1990). The 1985 Nevado del Ruiz volcano catastrophe: anatomy and retrospection. *Journal of Volcanology and Geothermal Research*, **42**, 151–188.

Voight, B. and Soussa, J. (1994). Lessons from Ontake-san: a comparative analysis of debris avalanche dynamics. *Engineering Geology*, **38**, 261–297.

Wadge, G., Jackson, P., Bower, S. M., Woods, A. W. and Calder, E. (1998). Computer simulation of pyroclastic flows from dome collapse. *Geophysical Research Letters*, **25**(19), 3677–3680.

Waitt, R. B., Gardner, C. A., Pierson, T. C., Major, J. J. and Neal, C. A. (1994). Unusual ice diamicts emplaced during the December 15, 1989 eruption of Redoubt volcano, Alaska. *Journal of Volcanology and Geothermal Research*, **62**, 409–428.

Waldron, H. H. (1967). *Debris Flow and Erosion Control Problems Caused by the Ash Eruptions of Irazu Volcano, Costa Rica*. U.S. Geological Survey Bulletin, 1241-I.

Waythomas, C. F. (2001). Formation and failure of volcanic debris dams in the Chakachatna River valley associated with eruptions of the Spurr volcanic complex, Alaska. *Geomorphology*, **39**(3–4), 111–129.

White, J. D. L., Houghton, B. F., Hodgson, K. A. and Wilson, C. J. N. (1997). Delayed sedimentary response to the AD 1886 eruption of Tarawera, New Zealand. *Geology*, **25**(5), 459–462.

4 Mountain hazards

Olav Slaymaker

4.1 Introduction to mountain geomorphic hazards

4.1.1 Mountain geomorphic hazards defined

A geomorphic hazard results from any landform or landscape change that adversely affects the geomorphic stability of a site or drainage basin (Schumm, 1988) and that intersects the human use system with adverse socio-economic impacts (White, 1974). If there are no people affected, there is no hazard and if the landform or landscape is unchanged there is no geomorphic hazard. Barsch and Caine (1984) have described the distinctive relief typologies of major mountain systems. Mountain geosystems are not exceptionally fragile but they show a greater range of vulnerability to disturbance than many landscapes (Körner and Ohsawa, 2005) and their recovery rate after disturbance is often slow. During the past three decades, the world's population has doubled, the mountain regions' population has more than tripled and stresses on the physical and biological systems of mountain regions have intensified many fold. The combination of extreme geophysical events with exceptional population growth and land use modifications underlines the urgency of better understanding of these interactions and working out the implications for adaptation to and mitigation of the effects of drivers of change on landforms and landscapes. Geomorphic hazards intensify and risks multiply accordingly.

4.1.2 The major drivers of change and 'key' vulnerability

The three drivers of environmental change in mountains are relief, as a proxy for tectonics (Tucker and Slingerland, 1994), hydroclimate and runoff (Vandenberghe, 2002) and human activity (Coulthard and Macklin, 2001). Not only are they important in themselves but they are commonly so closely interrelated that it becomes difficult to rank their relative importance and, indeed, their status, whether dependent or independent. One of the greatest challenges facing mountain scientists is to separate environmental change caused by human activities from change that would have occurred without human interference (Marston, 2008).

Unfortunately, there has been inadequate representation of the interactive coupling between relief, land use and climate in the climate change discussions to date (Osmond et al., 2004). Based on a number of criteria in the literature such as magnitude, timing, persistence/reversibility, potential for adaptation, distributional aspects of the impacts, and likelihood and importance of the impacts, some of these vulnerabilities have been identified as 'key' (Füssel and Klein, 2006). It is the point of exceedance of thresholds, where non-linear processes cause a system to shift from one major state to another, that expresses this key vulnerability. From the perspective of geomorphologists, these key vulnerabilities are often expressed as increasing magnitude and/or frequency of geomorphic hazards.

4.1.3 The scale question

To enhance the complexity of the puzzle, the three drivers of change (relief, hydroclimate and human activity) operate variably at different spatial scales

For the purpose of the following discussion we have adopted a scale typology as follows:

a. Site scale hazards ($<10^{-1}$ km^2)
b. Drainage basin scale hazards (10^{-1} to 10^3 km^2) and
c. Global scale hazards ($>10^3$ km^2)

Geomorphological Hazards and Disaster Prevention, eds. Irasema Alcántara-Ayala and Andrew S. Goudie. Published by Cambridge University Press. © Cambridge University Press 2010.

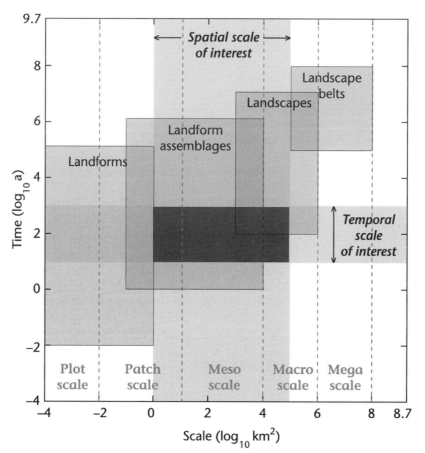

FIGURE 4.1. Spatial and temporal scales in geomorphology. On the x axis, the area of the surface of the Earth in km² is expressed as 8.7 logarithmic units; on the y axis, time since the origin of the Earth in years is expressed as 9.7 logarithmic units. (Slaymaker et al., 2009.)

We do not know where the breaks in the hierarchy should come. An example of the difficulty of establishing a scale typology for the study of landscapes under global environmental change is introduced as Figure 4.1. Motivation for defining preferred spatial and temporal scales of study can be found in Slaymaker et al. (2009), but in the present context it is sufficient to note that we have no theoretical justification for the separation of these scales of enquiry.

4.2 Site scale

4.2.1 Relief

The tradition of process geomorphology has focused most of its attention at smaller scales. In part, this is because site scale analysis is most amenable to quantitative treatment and, in part, because the geomorphic response to site scale disturbance is rapid. In mountains, relief is a major driving force, alongside of hydroclimate and human activity.

The driving forces associated with relief include: absolute elevation (which controls temperature and

precipitation); gradient (which controls the erosional force $g \sin \alpha$); concave slopes (which tend to concentrate water) and convex slopes (which tend to shed water). Aspect controls the amount of radiant energy received at the surface and leads to highly contrasted slope climates. The resisting forces associated with strength of surficial materials can be defined in numerous ways, such as cohesive and frictional strength, erodibility and modes of deformation (rheological properties defined by strain–time and stress–strain rate graphs).

4.2.2 Hydroclimate and runoff

Site scale hydroclimate is influenced by relief (see above) and also by the nature of the regolith. The regolith controls the preferred pathways of movement of water and sediment. The relative importance of weathering, sediment transport and depositional processes depends on the hydroclimate. Precipitation, snow storage, glacier storage, available soil moisture, groundwater storage, actual evapotranspiration and surface runoff are the components

of the hydrological cycle that influence and respond to environmental change. The magnitude, frequency and duration of storm events is vital information that often does not exist because the usual presentation of precipitation data is in the form of daily totals (Barry, 1992).

4.2.3 Human activity

Timber harvesting is a major land management practice whose precise influence on slope stability depends on the method of harvesting, density of residual trees and understory vegetation, rate and type of regeneration, site characteristics and patterns of water inflow after harvesting (Sidle *et al.*, 2002). Roads can be the focus of the highest rates of denudation in the landscape. Humans do accelerate slope failures through road building, especially when roads are situated in mid-slope locations instead of along ridge tops (Marston *et al.*, 1998). Roads in eastern Sikkim and western Garhwal have caused an average of two major landslides for every kilometer constructed. Road building in Nepal has produced up to 9,000 cubic meters of landslide per kilometer, and it has been estimated that, on average, each kilometer of road constructed will eventually trigger 1,000 tons of land lost from slope failures (Zurick and Karan, 1999).

4.2.4 Site scale hazards

Sites of initiation of rilling and gullying, the uppermost finger-tip tributaries of river networks, zero order basins (or colluvial bedrock depressions) and sites of mass movement initiation are all sites at which a threshold safety factor has been exceeded. The initiation of snow avalanches and debris avalanches are specific cases of such hazards.

Steep slopes and heavy snowfall at high elevations are the main factors affecting avalanche incidence. In countries such as Canada, Switzerland, Austria, France, Italy and Norway expenditures involve tens of millions of dollars annually. No clear trends have been identified in the frequency and number of avalanches in the Alps in the past century (Agrawala, 2007), but in many avalanche-prone areas the density of buildings and other investments has increased.

Permafrost degradation is likely to contribute to rockfall activity (Behm *et al.*, 2006) and may well trigger higher debris flow activity, but research into this link so far remains inconclusive. A recent study showed that frequency of debris flows originating from permafrost areas in Ritigraben (Swiss Alps) has been decreasing, although a lower frequency may also be associated with more intense events due to larger accumulation of materials between

events (Stoffel and Beniston, 2006). Fischer *et al.* (2006) have identified permafrost degradation as a cause of slope instabilities on Monte Rosa; and Gruber and Haeberli (2007) provide a comprehensive review of permafrost and slope instability.

Road systems are often critical in hazard generation; removal of vegetation from forest to agriculture to urban; removal of soil for urban and mining purposes; and gravel extraction all have potential for generating site scale hazards.

4.3 Drainage basin scale

4.3.1 The sediment cascade in mountains

The greater complexity of the drainage basin scale is expressed through emergent properties that do not exist at site scale. Some of these emergent properties are linear erosion, slope and channel coupling, and preferred surface and sub-surface pathways for movement of sediment and water. The literature has traditionally used such variables as basin area and drainage density to act as surrogates for these integrated effects.

The sediment cascade that results can be characterized in terms of four environments (after Caine, 1974) which are differentiated by dominant processes and forms as:

a. the mountain cryosphere system;
b. the coarse debris system;
c. the fine-grained sediment system; and
d. the geochemical system.

Note that the categories overlap and they are identified only in terms of their dominant characteristics.

Each of these components of the mountain geosystem is sensitive to environmental change, whether in response to relief, temperature, precipitation, runoff, sediment transport or land use changes. The cryosphere stores water and changes the timing and magnitude of runoff which erodes and transports sediment. Snow responds to environmental changes on a daily time scale; lake and river ice on an annual time scale; permafrost and glaciers on annual to century time scales; associated ecosystem responses are measured in decades to centuries; and sediment systems may take decades to millennia to respond (Figure 4.2)

4.3.2 Basin area

At the drainage basin scale, runoff intensity reaches a maximum within basins whose dimensions approximate those of the extreme event producing storm cells. Partly for this reason and also because, at smaller basin scales, slopes and

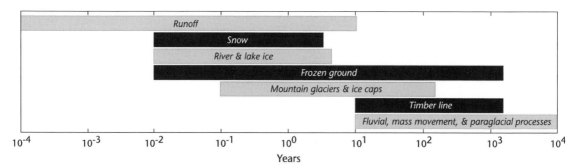

FIGURE 4.2. Components of the mountain cryosphere, ecosphere and sediment cascade and their response times following disturbance by hydroclimate or by human activity. (Slaymaker and Embleton-Hamann, 2009.)

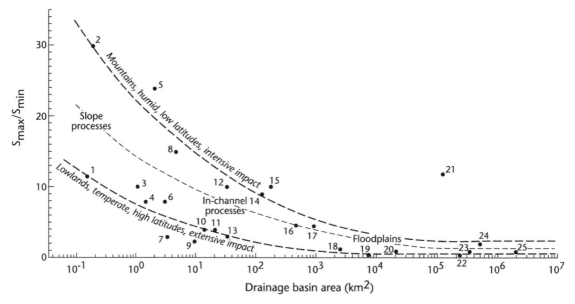

FIGURE 4.3. Relative magnitude of change between low or baseline sediment flux and maximum sediment flux (dimensionless ratio S_{max}/S_{min}) plotted against lake-catchment area (km²). (Dearing and Jones, 2003.)

channels are often coupled, debris torrent basins tend to average between 0.5 and 5 km². When basins reach 5 km² or greater, most slopes and channels are uncoupled and the sediment response time after extreme hydroclimatic events increases substantially. Systematic analysis of the relative magnitude of change in sediment flux in basins of different size shows that relative changes diminish with increasing basin size (Figure 4.3). This is because smaller basins are more sensitive and vulnerable to environmental change than the larger basins.

4.3.3 Drainage density

Drainage density (Dd) is influenced by mean annual precipitation, precipitation intensity, lithology, soil characteristics, relief, vegetation, human activity and stage of drainage network development. Although there are many ambiguities in the calculation of Dd (Schumm, 1997), data available suggest that values of Dd between 1 and 10 km^{-1} are representative for maturely dissected fluvial landscapes.

There is a wide range of values of drainage density reported in the literature. They range from the very high drainage densities recorded in the badlands of Perth Amboy, New Jersey (c. 600 km km^{-2}) (Schumm, 1956) to less than 1 on undissected plateau or plains surfaces. The influence of lithology and stage of network development has been clearly demonstrated in many studies but, as Chorley et al. (1984) point out, contemporary hydroclimate and surface properties seem to be dominant in controlling

TABLE 4.1. *Theoretical characteristics of basins with variable ruggedness number*

Dd (km * km^{-2})	Ht (m)	R^a	Coupling	Geomorphic process	Incidence of hazards
Mountains					
>10	>1000	>10	High	Debris flows	High
1–10	300–1000	*c.* 1	Intermediate	Fluvial	Intermediate
<<1	>>1000	<1	Low	Mass movement	High
Non-mountains					
>>10	<300	>3	High	Badlands	High
<1	<300	<0.3	Low	Desert	Low

^{a} R is ruggedness number, defined as $Dd \times Ht$, a dimensionless number.

differences of drainage density. There is a range of drainage density values from the American Southwest (Melton, 1957), Sri Lanka (Madduma Bandara 1971), the eastern Caribbean mountainous islands (Walsh 1985); the Seychelles (Walsh, 1996) and the continental USA (Hadley and Schumm, 1961; Chorley and Morgan, 1962). These variations reflect the direct and indirect influence of climatic variables. The direct influence is that of large rainstorm magnitude–frequency, as it is only the large runoff events associated with such rainstorms that are capable of eroding finger-tip sections of the drainage net. The indirect influence of climate is exerted via the influence of annual rainfall, rainfall seasonality and vegetation on soil characteristics.

Walsh (1996) suggests that there are three major questions outstanding: (1) what return period of rainstorm/runoff event is responsible for controlling the drainage network? (2) what are the processes responsible? and (3) how quickly do networks develop or adjust to changes in climate? Various drainage density models appear to be applicable. Different slope runoff processes and their spatial patterns, inheritance from collapsed pipe networks, channel development in landslide scars (Chorley *et al.*, 1984) or even the simultaneous operation of uncoupled pipe collapse and erosion by overland flow may be the dominant process. The essential point in this present discussion is that there would appear to be strong evidence for the relation between drainage density and hydrogeomorphological hazards, especially the incidence of large rainstorm and runoff events, whatever the precise mechanism involved.

4.3.4 The ruggedness number

The ruggedness number, R, defined as the product of drainage density (km^{-1}) and basin relief (m), ($Dd * Ht$) is one of the

three dimensionless numbers that describe the dynamics of drainage basin evolution (Church and Mark, 1980). It seems reasonable to suppose that the incidence and intensity of geomorphic hazards would also be related to the ruggedness number. Originally defined by Strahler (1952), the ruggedness number has been used in various guises by many students of morphometry, such as Schumm (1956) and Melton (1957) as an indicator of the relative dynamism of the basin, but rarely in the context of hazard studies. Kovanen and Slaymaker (2008) showed a strong relation between the Melton ruggedness number ($Ht * A^{-0.5}$) and debris fan slope; an inverse relation between debris fan slope and basin area, and a modest relation between basin area and debris fan area in the Nooksack basin, Cascade Ranges, USA. The fans, which are active at present, have a documented history of >7000 years of debris flow activity. The drainage basin ruggedness number is then a potential indicator of the hazardousness of a basin, especially of those hazards that are related to water movement and sediment mobilization, erosion on slopes and fluvial erosion.

In Table 4.1 we have postulated a direct relation between ruggedness number and extreme geomorphic events and an inverse relation between basin area and extreme geomorphic events.

4.3.5 Climate change

One of the most significant impacts of climate change (see also Chapters 5, 8 and 20) in glacierized basins may be the changing pattern of glacier melt runoff (Walsh *et al.*, 2005). Glaciers will provide extra runoff as the ice disappears. In most mountain regions, this will happen for a few decades and then cease. For those regions with very large glaciers the effect may last for a century or more. Kotlyakov *et al.* (1991) have provided estimates of change for Central Asia, which give a threefold increase of runoff from glaciers by

2050 and a reduction to two-thirds of present runoff by the year 2100. In the short term, a significant increase in the number of flood events in Norway is projected (Bogen, 2006). Bogen found that suspended sediment concentrations and volumes were dependent on the availability of sediments, the type and character of the erosion processes and the temporal development of the flood. In the Norwegian case, it appears that the glacier-controlled rivers are unlikely to respond dramatically in terms of sediment transport because of limited sediment availability. Nevertheless, in global sediment yield terms, Hallet *et al.* (1996) have conclusively demonstrated the importance of glacier melt waters. The global data and the regional Norwegian data demonstrate the difficulty of making generalizations about the probable effects of climate change on sediment transport in glacierized basins. Changing patterns of snowmelt resulting from climate change are also complex (Woo, 1996) and are sensitive both to elevation and to the seasonal and event distribution of temperature change and precipitation.

4.3.6 Human activity

There is growing evidence since the Third Assessment Report (IPCC, 2001b.) that adaptations that deal with non-climatic drivers are being implemented in both developed and developing countries. Examples of adaptations to land use change, such as construction, decommissioning and management of reservoirs, and adaptations to extreme sediment cascades and relief, as well as their interactions with climate include the following:

a. partial drainage of the Tsho Rolpa glacial lake in Nepal was designed to relieve the threat of GLOFs (glacial lake outburst floods) (Shrestha and Shrestha, 2004);
b. Sarez Lake in Tajikistan and the so-called 'quake lakes' in Szechwan Province, China are commanding national and international funding;
c. increased use of artificial snow making by the alpine ski industry in Europe, Australasia and North America.

Glacial retreat favors the formation of glacial lakes and ice avalanches, and disastrous events such as glacial lake outburst floods (GLOFs). GLOFs are the most destructive hazards originating from glaciers due to the large water volume and large areas covered. Luckily, glacial lakes, from which GLOFs originate, usually form slowly and can be monitored. McKillop and Clague (2007) have estimated the probability of the occurrence of GLOFs in southern British Columbia.

In mountain landscapes, extreme geophysical events interact with social systems in dramatic ways. The May

12, 2008 earthquake in Szechwan, China, which generated hundreds of large landslides and which blocked more than 30 large lakes, killing 75,000 people and causing millions of dollars in damage, is a case in point. Earthquakes and floods, which are not exclusively mountain hazards, account for more than 50% of the damage caused by natural hazards globally; and the highest damage in recent years in Switzerland, Austria and France, for example, was due to floods and windstorms (Agrawala, 2007).

The country of Tajikistan is part of the Pamir-Alai mountain system. Ninety-two percent of its land area of $c.140,000$ km^2 is mountainous; nearly 50% lies above 3,000 m and is classified as dry, cold desert. It is drained by the headwaters of the Amu Darya and Syr Darya rivers, the major feeders of the Aral Sea. This is a zone of intensive seismic activity and steeplands that are geomorphically highly active. Earthquakes with magnitudes exceeding 5 on the Richter scale have a recurrence interval of 75 days and the region is well known for the earthquake-triggered natural dams that have blocked large lakes, holding as much as 17 km^3 of water in the case of Sarez Lake, and presenting a permanent risk of catastrophic draining (Alford *et al.*, 2000). Rock falls and massive rock slides have accounted for the deaths of more than 100,000 victims during the twentieth century, and this in a country of barely 6 million people. Thirty-one percent of the country is said to be agricultural land and 13% is under forest, but overgrazing of the rangelands by sheep and rapid deforestation has encouraged widespread erosion on slopes and more frequent occurrence of mudflows and landslides.

4.3.7 Drainage basin scale hazards

High drainage densities have implications for slope profiles. A logical implication of fine dissection by the stream network is that slope angles will be steeper (and slope lengths shorter), even in areas of moderate altitudinal relief, than in other humid environments with lower drainage densities. This makes landsliding more likely and rainsplash erosion (heavily dependent on slope angle) more effective.

Japan is well known for its debris flow hazards (Figure 4.4) associated with the volcanic mountain landscapes and high drainage densities.

By contrast with the site scale hazards, drainage basin scale hazards involve coupled debris avalanches and debris flows and coupled slopes and channels in the headward parts of the basins, which is one degree of complexity. As drainage basins grow, slopes and channels are gradually decoupled and the rate at which this decoupling occurs is specific to the hydroclimate and nature of the regolith in

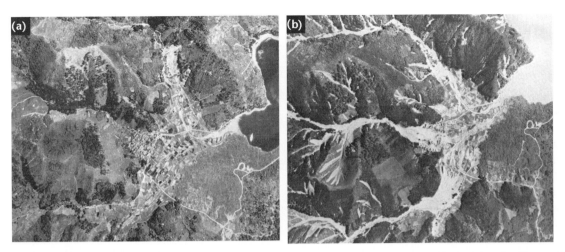

FIGURE 4.4. Debris flows as geomorphic hazard, illustrated from the effects of an intense rainstorm in 1966 at Ashiwada in Yamanashi Prefecture, Japan: (a) pre-1966; (b) post-Typhoon 26, 1966. (From Akagi, 1973.)

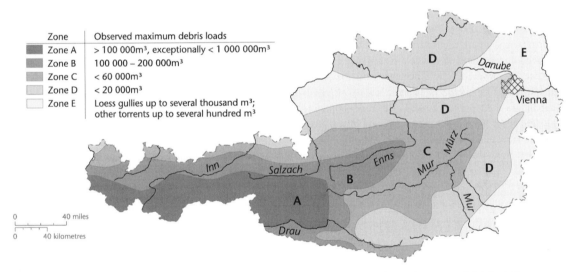

Zone	Observed maximum debris loads
Zone A	> 100 000m³, exceptionally < 1 000 000m³
Zone B	100 000 – 200 000m³
Zone C	< 60 000m³
Zone D	< 20 000m³
Zone E	Loess gullies up to several thousand m³; other torrents up to several hundred m³

FIGURE 4.5. Regional distribution of torrent and debris flow hazard in Austria, based on observed maximum debris load. (From Kronfellner-Kraus, 1989; Embleton-Hamann, 2007.)

each basin. The relative importance of hydroclimate, relief, regolith and land use is difficult to establish.

The debris flow hazard in Austria is instructive in this regard. Debris loads involved in about 2,000 torrent disasters from catchments up to 80 km² in size have been calculated and mapped (Figure 4.5).

The regional pattern of debris flow loads (declining from west to east) reflects rather closely the influence of relief and hydroclimate. However, individual torrent basins with intensive land use have a higher hazard and risk rating simply because of the scale and nature of socio-economic investment in the valley bottoms. Land use has become

more important in determining the magnitude of the hazard and the dimensions of the risk. It is also important to observe that land use in the form of afforestation can mitigate and even enhance the landscape (Figure 4.6)

If climate warming continues, destabilized mountain walls, increasing frequency of rock fall, incidence of rock avalanches, increase and enlargement of glacial lakes and destabilization of moraines damming these lakes are expected to accompany increased risk of outburst floods.

Geomorphic hazards associated with glacier retreat include rock avalanches, deep-seated slope sagging (sackung), debris flows, debris avalanches, debris slides, rock

FIGURE 4.6. Source area of the Gangbach. In the foreground is Aebnenegg, Canton Uri, near Altdorf, Switzerland. (a) A 33-hour rainfall event in June 13–14, 1910 caused extensive debris flow activity. (b) The same landscape in 1981 after the completion of afforestation between 1932 and 1960. (R. Kellerhals, personal communication, 2008.)

fall, moraine dam failures and glacier outburst floods (Geertsema *et al.*, 2006). The effects of glacier retreat on sediment transport are controversial. Schiefer *et al.* (2006) identified three periods of accelerated sedimentation in a montane lake in the Coast Mountains over the past 70 years: a period of intense rainstorms, a year of massive slope failure and a period of rapid glacier retreat between 1930 and 1946. The present period of rapid glacier retreat does not seem to be generating exceptional sedimentation events.

If climate warming is accompanied by increasing storminess (IPCC, 2007b) and intense rainfall, then peak stream discharges will increase and erosion, sediment transport and sediment deposition downstream will presumably also increase. Hazards will include flooding, possibly increasing both magnitude and frequency of floods and increased lateral instability of stream banks. A prominent indication of a change in extremes is the observed evidence of increases in heavy precipitation events over the mid-latitudes in the last 50 years, even in places where mean precipitation amounts are not increasing. (Kunkel, 2003). In Central and Southwest Asia the 1998–2003 drought provides an example of unanticipated effects of extreme events. Flash flooding occurred over hardened ground desiccated by prolonged drought in Tajikistan, central and southern Iran, and northern Afghanistan leading to accelerated erosion in early 2002 (IPCC, 2007a). In terms of impact on the landscape it is the extreme events (both high and low) and the freshet flows that mobilize most of the sediments and, to a lesser extent, the solutes. More frequent/intense occurrences of extreme weather events will exceed the capacity of many developing countries to cope.

The growth of cities in the mountain world places further stress on mountain stability. In Latin America, the Caribbean and countries in transition, nearly half of the mountain population lives in urban areas. In the case of tropical mountains by contrast, the towns are in the uplands and roads are often in ridge top locations. In the case of temperate mountain areas, the only space for urban agglomerations and transport routes is on the flat floor of broader valleys and mountain basins (Bätzing *et al.*, 1996). These used to be flood-prone wetland areas. Therefore all rivers needed to be turned into artificial channels in these areas. The consequence is that geomorphic hazards have been enhanced: concreting of the riverbanks and the waterproofing of urban areas represent aggravating factors for floods.

4.4 Global scale

4.4.1 Relief

At the larger mountain system and global scales, elevation and gradient are the most important relief elements insofar as they influence temperature, and precipitation. Elevation controls the incidence and intensity of freeze–thaw events as well as orographic precipitation, and many associated climatic effects. Gradient defines the gravitational driving force ($g \sin \alpha$) and influences radiation and precipitation receipt, wind regimes and snow. Erosion rates reported for the Nanga Parbat massif are among the highest measured (22 ± 11 m per 1,000 years) and reported rates of uplift for the Himalayas vary from 0.5 to 20 m per 1,000 years (Owen, 2004). Ahnert (1970) developed an equation relating denudation and local relief:

$$D = 0.1535h \qquad (4.1)$$

where D is denudation in mm/1,000 yrs and h is local relief in m/km.

Summerfield and Hulton (1994) analyzed 33 basins with an area >500,000 km^2 from every continent except Antarctica. Total denudation (suspended plus dissolved load) varied from 4 mm ka^{-1} (Kolyma in the Russian Far East) to 688 mm ka^{-1} (Brahmaputra). They found that more than 60% of the variance in total denudation was accounted for by basin relief ratio and runoff.

4.4.2 Disturbance regimes

There is an increasing sense that almost all mountain landscapes are transitional from one landscape-forming regime to another. This condition has been described by Hewitt (2006) as a disturbance regime landscape. By this, he meant that disturbances occur so frequently that the landscape gets no chance to equilibrate with contemporary processes. Mega-landslides in the Himalayas are the regional example that Hewitt has described. Note contrast with models of Brunsden (1993).

Mountain societies contain a higher incidence of poverty than elsewhere, and therefore have a lower adaptive capacity and a higher vulnerability to environmental change. Mountain peoples have been made more vulnerable to natural extreme events by vast numbers being uprooted and resettled in unfamiliar and more dangerous settings (Hewitt, 1997).

Many studies suggest that at least until 2050 land use change will be the dominant driver of change in human-dominated regions (UNEP, 2002). Not only are there geosystem disturbance regimes, such as those discussed by Hewitt (above), but land use change, fire and insect outbreaks can also be analyzed as disturbance regimes, using a shorter response time scale (Sala et al., 2005). Similarly, over-grazing, trampling and vegetation destabilization have been analyzed in this way in the Caucasus and Himalayas (IPCC, 2001b).

4.4.3 A conditionally unstable landscape

The landscape of British Columbia (B.C.) is in transition between the Last Glacial Maximum (LGM) and present and illustrates a specific kind of geosystem disturbance regime. The Cordilleran Ice Sheet covered almost the whole of the province before 14,500 BP and since that time the process of transition towards a fluvially dominated landscape has been on-going. Nevertheless, B.C.'s mountains still contain

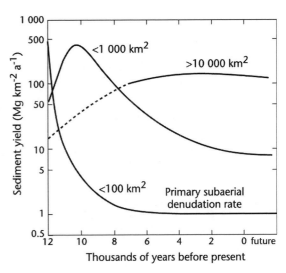

FIGURE 4.7. Temporal pattern of paraglacial sediment yield in formerly glaciated upland and valley sites in coastal British Columbia and Alaska. Values are for basins of order 100, 1000 and 10 000 km^2 based on the contemporary spatial pattern of sediment yield. (From Church, 1998.)

over 29,000 km^2 of glaciers and ice caps, in spite of warming since the Little Ice Age and marked glacier retreat. Schiefer et al. (2007) estimate total ice loss of 22.5 km^3 a^{-1} between 1990 and 2005.

The suite of landforms that evolved was characterized by 'non-glacial processes that are directly conditioned by glaciation' (Church and Ryder, 1972). A distinctive pattern of sediment yield (Figure 4.7) is interpreted as evidence of a transitional landscape with a relaxation time of the order of 10 ka (Slaymaker, 1987; Church and Slaymaker, 1989). This transitional landscape lasts until the glacially conditioned sediment stores are either removed or attain stability (Schumm and Rea, 1995). The landscape of B.C. is a disturbance regime landscape in so far as the post-glacial landscape has had insufficient time to recover from the effects of the last major disturbance, namely the LGM.

In B.C., and indeed in most of Canada, larger geomorphic systems have not yet achieved a form that is adjusted to the contemporary fluvial landscape (Church et al., 1999). The implication of this conclusion is that the B.C. mountain landscape is conditionally unstable and it can be anticipated that small changes in hydroclimate and/or land use may cause landscape change.

The regional climate projection of the Fourth Assessment Report (IPCC, 2007a) anticipates for B.C. a mean annual temperature increase and a mean annual precipitation increase, incorporating a summer decrease and a winter increase in precipitation. If this larger volume of moisture comes in the form of high magnitude, low frequency rainfall events, debris flow hazard may increase markedly. A decrease in snow season length and snow depth is also projected. The general tendency under climate warming will be an upslope shifting of hazard zones and widespread reduction in stability of formerly glaciated or perennially frozen slopes (Barsch, 1993; Ryder, 1998).

4.4.4 Human activity, population and land use

Human activity, in the form of population density and land use, is a direct driver of environmental change in mountains (Vorosmarty et al., 2003; Syvitski et al., 2005). It is not, however, the sheer numbers of people but aspects of population composition and distribution, especially the level of urbanization and household size, that exercise the greatest demands on the land (Lambin et al., 2001). High population densities in the developing world, for example, may lead to better management, such as in Kenya and Bolivia, described by Tiffen et al. (1994) as cases where the presence of more people has led to less erosion. The creation of infrastructure, especially roads, is a crucial step in triggering land use intensification. In developing countries, the largest mountain populations are found in the mid-elevation zones. In developed and transitional countries, by contrast, the lowest mountain zones are most heavily populated.

4.4.5 A typology of mountain systems sensitive to relief, hydroclimate and land use changes

If one is to get a realistic view of the incidence and source of geomorphic hazards, mountain systems should be differentiated not only in terms of relief and hydroclimate but in ways that reflect demography and land use. In this simplistic typology, we incorporate population density as a proxy for the intensity of the human signature on the landscape. Higher population densities lead to a higher pressure on land resources and intensified land use and therefore the human signature (though not necessarily a negative one) will be higher.

Polar mountains (population density <0.1/km²)

Svalbard, for example, has few permanent residents (around 2,300 as of 2000) and a few isolated mining activities, but it is estimated that c. 40,000 tourists visit each year (as of 2007). Sixty-five percent of the surface of Svalbard consists of protected areas. Climate and relief are the

dominant drivers of change and hazards associated with permafrost degradation and glacier retreat are most evident.

Low population density temperate mountains (population density 0.1–25/km²)

Mountains with a history of less dense settlement retain more of their traditional agriculture and forestry. Relief and hydroclimate are the most important drivers of environmental change. Formerly glaciated mountains are strongly controlled by the historical legacy of glaciation.

High population density temperate mountains (population density 25–75/km²)

In western Europe and Japan, mountain regions are experiencing increasing land use pressures because of competition between conservation use, mineral extraction and processing, recreation development and market oriented agriculture, forestry and livestock grazing. The human impact (both positive and negative) on these mountains far exceeds the documented effects of relief and hydroclimate.

Tropical mountains (population density 50–100/km²)

Many developing countries (defined as having a relatively low standard of living, an undeveloped industrial base and a moderate to low human development index) are located in tropical and semi-arid environments. In those regions, the mountain areas are usually cooler and/or wetter than the lowlands and more hospitable for living and commercial exploitation. They also have deeper soils and fewer diseases. Human encroachment has reduced vegetation cover, increasing erosion and siltation, thereby adversely affecting water quality and other resources. Direct anthropogenic influence on these mountain regions appears to greatly surpass climate effects.

4.4.6 Global scale hazards

Global scale geomorphic hazards necessarily involve a consideration of tectonic and historical legacy. Hovius et al. (1998) emphasizes geomorphic hazards controlled by tectonic activity in earthquake-dominated landscapes. Scaled for drainage basin area, sediment yields increase through almost five orders of magnitude from tectonic cratons (typical sediment yields of 100 t km^{-2} a^{-1}) to contractional mountains (up to 10,000 t km^{-2} a^{-1}). Tectonically active mountain belts not only provide relief to drive erosion processes but they also combine high regolith loss with the rapid uplift of new bedrock into the weathering zone to continually refresh these erosion processes and maintain high sediment yields. On the rapidly uplifting (5–7 mm a^{-1}) island of Taiwan, erosion appears driven by interaction of erodible substrates, rapid deformation in the form of frequent earthquakes and typhoon-driven runoff variability. Earthquakes produce sediment by rock mass shattering and landsliding, and landslides and debris flows are also triggered by typhoon-generated storm runoff which flushes sediments from the mountains (Dadson *et al.*, 2003). Thus changes in the frequency, magnitude and track positioning of typhoons in the Philippine Sea, western North Pacific Ocean, consequent upon climate change will have consequences for denudation rates; these processes are more important than simple relief and average precipitation controls (e.g. Andes: Aalto *et al.*, 2001; Himalayas: Finlayson *et al.*, 2002).

One of the implications of the fact that many mountain landscapes are disturbance regime landscapes is that they are exceptionally sensitive to environmental change and are in this sense geomorphically vulnerable. Geomorphic vulnerability is expressed not only by the frequent incidence of earthquake-triggered landslides, but, for example, by the fate of Himalayan glaciers, which cover 17% of the mountain area, and are predicted to shrink from the present 500,000 to 100,000 km^2 by 2035 (WWF, 2005). The glaciers on Mt. Kilimanjaro are likely to disappear by 2020 (Thompson *et al.*, 2002) and Bolivian glaciers are heading for the same fate (Thompson *et al.*, 1998).

Climate change can be expected to alter the magnitude and frequency of a wide variety of geomorphic processes (Holm *et al.*, 2004). Increased triggering of rock falls and landslides could result from increased groundwater seepage and pressure. Large landslides are propagated by increasing long-term rainfall whereas small landslides are triggered by high intensity rainfall. These tendencies will probably lead to enhanced sediment transport. Increased sediment input to glacier-fed rivers may lead to increased channel instability, erosion and flooding. The hazard zones related to most of these fluvial processes will extend a long way beyond the limits of the mountain area (Ashmore and Church, 2002). Rainfall amounts and intensities are the most important factors in water erosion and they affect slope stability, channel change and sediment transport. Increased precipitation intensity and variability is projected to increase the risk of floods and droughts in many areas. Changes in permafrost will affect river morphology through destabilizing of banks and slopes, increased erosion and sediment supply (Vandenberghe, 2002).

Agriculture has been the greatest force of land transformation on this planet (Lambin and Geist, 2006). Nearly a third of the Earth's land surface is currently being used for growing crops or grazing cattle (FAO, 2007). Much of this agricultural land has been created at the expense of natural forests, grasslands and wetlands. In Africa, for example, most of the mountains are under pressure from commercial and subsistence farming activities. In unprotected areas, mountain forests are cleared for cultivation of high altitude adapted cash crops such as tea, pyrethrum and coffee. Grazing and forestry are the predominant uses of mountain land in all regions. Extensive grazing has little impact on slope processes, but overgrazing can have severe impacts.

4.5 Conclusion in light of accelerating environmental change

It is possible to learn from ecology, where recent models have placed environmental change and system collapse as central to an understanding of contemporary change and where there are similar complexity problems. Panarchy is a metaphor designed to describe systems of ecosystems at varying spatial and temporal scales. The terminology developed for panarchy (Holling, 2001) is entirely ecological and needs to be translated for the needs of geomorphology. Holling suggests that complex systems are driven through adaptive cycles which exist at a range of spatial scales. The term adaptive is self-evident in ecological systems; in geomorphic systems we often speak of self-regulating systems (Phillips, 2003). Adaptive cycles are defined as consisting of four phases, namely exploitation (the environmental disturbance regime), conservation (the response), collapse (threshold exceedance and unpredictable behavior of the system) and reorganization (recovery). A geomorphic analogue would be a disturbance regime landscape characterized by both orderly evolution and system collapse. The duration of these phases of adaptation depends on the intrinsic strength of the system, the connectivity of the system and the time required for the recovery of the system (resilience).

The panarchy metaphor has a fascinating flexibility in dealing with complex systems. Although there are evident differences between geophysical systems and ecosystems, there are many parallels in the behavior of self-organizing systems that can assist in improving attempts to understand and manage the environment sensitively. The concepts and terminology of the panarchy model are consistent with geomorphic concepts such as complex response (Schumm, 1973), threshold exceedance, landscape sensitivity and barriers to change.

Many mountain hazards can be viewed in terms of scale, ruggedness number, lithologic strength and connectivity (criticality of the adaptive system). The risks and losses associated with these hazards depend on how close to a

TABLE 4.2. *Mountain hazards classified according to scale, ruggedness number, lithologic strength, connectivity and vulnerability (H is high and L is low)*

Scale	Resilience		Connectivity	Vulnerability	Hazard types	Risks
	Ruggedness	Lithology	Drainage density			
Site	H	L	H	H	Slope failure	Human life
	L	H	L	L	Debris flows	Structures
Drainage Basin	H	L	H	H	Extreme events	Community
	L	H	L	L	Land degradation	Infrastructure
Global	H	L	H	H	Human activity	Ways of life
	L	H	L	L	Climate change	Economic systems
					Eutrophication	Ecological integrity

condition of collapse the adaptive system is allowed to proceed (Table 4.2).

4.6 Conclusions

There appears to be a hierarchy of mountain hazards and in Holling's (2001) terms this is a panarchy of adaptive systems. Each adaptive system, at its own spatial scale, evolves towards a critical condition leading either to collapse (hazard) or to self-reorganization (adaptation or mitigation in socio-economic context). Climate change is just one of the drivers that operates on the adaptive system.

If the IPCC definition of vulnerability is seriously engaged, then geomorphologists also have to investigate adaptive systems. Reliance on reactive, autonomous adaptation to the cumulative effects of environmental change is ecologically and socio-economically costly. Planned and anticipatory adaptation strategies can provide multiple benefits. But there are limits on their implementation and effectiveness. Enhancement of adaptive capacity reduces the vulnerability of landscapes to environmental change, but adaptive capacity varies considerably among regions, cultures, and socio-economic groups. Improved understanding of geomorphic hazards at many temporal and spatial scales is urgently needed.

References

Aalto, R., Dunne, T. and Guyot, J. L. (2006). Geomorphic controls on Andean denudation rates. *Journal of Geology*, **114**, 85–99.

Agrawala, S. (ed.) (2007). *Climate Change in the European Alps: Adapting Winter Tourism and Natural Hazard Management*. Paris: Organization for Economic and Cultural Development.

Ahnert, F. (1970). Functional relationships between denudation, relief and uplift in large mid-latitude drainage basins. *American Journal of Science*, **268**, 243–263.

Akagi, M. (1973). *Sabo Works in Japan*. Tokyo: The National River Conservation-Sabo Society.

Alford, D., Cunha, S. F. and Ives, J. D. (2000). Mountain hazards and development assistance: Lake Sarez, Pamir Mountains, Tajikistan. *Mountain Research and Development*, **20**, 20–23.

Ashmore, P. and Church, M. (2002). *The Impact of Climate Change on Rivers and River Processes in Canada*. Geological Survey of Canada Bulletin 555. Ottawa: Geological Survey of Canada.

Barry, R. G. (1992). *Mountain Weather and Climate*. London: Routledge.

Barsch, D. (1993). Periglacial geomorphology in the 21st century. *Geomorphology*, **7**, 141–163.

Barsch, D. and Caine, N. (1984). The nature of mountain geomorphology. *Mountain Research and Development*, **4**, 287–298.

Bätzing, W., Perlik, M. and Dekleve, M. (1996). Urbanization and depopulation in the Alps. *Mountain Research and Development*, **16**, 335–350.

Behm, M., Raffeiner, G. and Schöner, W. (2006). *Auswirkungen der Klima- und Gletscheränderung auf den Alpinismus*. Vienna: Umweltdachverband.

Bogen, J. (2006). Sediment transport rates of major floods in glacial and non-glacial rivers in Norway in the present and future climate. In J. S. Rowan, R. W. Duck and A. Werritty (eds.), *Sediment Dynamics and the Hydrogeomorphology of Fluvial Systems*. IAHS Publication **306**, pp. 148–158.

Brunsden, D. (1993). The persistence of landforms. *Zeitschrift für Geomorphologie Supp. Band*, **93**, 13–28.

Caine, N. (1974). The geomorphic processes of the alpine environment. In J. D. Ives and R. G. Barry (eds.), *Arctic and Alpine Environments*. London: Methuen, pp. 721–748.

Chorley, R. J. and Morgan, M. A. (1962). Comparison of morphometric features, Unaka Mountains, Tennessee and North Carolina, and Dartmoor, England. *Bulletin of the Geological Society of America*, **73**, 17–34.

Chorley, R. J., Schumm, S. A. and Sugden, D. (1984). *Geomorphology*. London: Methuen.

Church, M. (1998). The landscape of the Pacific Northwest. In D. L. Hogan, P. J. Tschaplinski and S. Chatwin (eds.), *Carnation Creek and Queen Charlotte Islands Fish/Forestry Workshop: Applying Twenty Years of Coast Research to Management Solutions*. B.C. Land Management Handbook, Victoria, B.C. Forestry Research Branch, pp. 13–22.

Church, M. and Mark, D. (1980). On size and scale in geomorphology. *Progress in Physical Geography*, **4**, 342–390.

Church, M. and Ryder, J. M. (1972). Paraglacial sedimentation: a consideration of fluvial processes conditioned by glaciation. *Bulletin of the Geological Society of America*, **83**, 3059–3072.

Church, M. and Slaymaker, O. (1989). Disequilibrium of Holocene sediment yield in glaciated British Columbia. *Nature*, **337**, 452–454.

Church, M. *et al.* (1999). Fluvial clastic sediment yield in Canada: scaled analysis. *Canadian Journal of Earth Sciences*, **36**, 1267–1280.

Coulthard, T. J. and Macklin, M. G. (2001). How sensitive are river systems to climate and land use changes? A model based evaluation. *Journal of Quaternary Science*, **16**, 347–351.

Dadson, S. J. *et al.* (2003). Links between erosion, runoff variability and seismicity in the Taiwan orogen. *Nature*, **426**, 648–651.

Dearing, J. A. and Jones, R. T. (2003). Coupling temporal and spatial dimensions of global sediment flux through lake and marine sediment records. *Global and Planetary Change*, **39**, 147–168.

Embleton-Hamann, C. (2007). Geomorphological hazards in Austria. In A. Kellerer-Pirklbauer *et al.* (eds.), *Geomorphology for the Future*. Innsbruck: Innsbruck University Press, pp. 33–56.

FAO (2007). *Crop Prospects and Food Situation*. www.fao.org.

Finlayson, D. P., Montgomery, D. R. and Hallet, B. (2002). Spatial coincidence of rapid inferred erosion with young metamorphic massifs in the Himalayas. *Geology*, **30**, 219–222.

Fischer, L. *et al.* (2006). Geology, glacier retreat and permafrost degradation as controlling factors of slope instabilities in a high mountain rock wall: the Monte Rosa east face. *Natural Hazards and Earth System Sciences*, **6**, 761–772.

Füssel, H.-M. and Klein, R. J. T. (2006). Climate change vulnerability assessments: an evolution of conceptual thinking. *Climate Change*, **75**, 301–329.

Geertsema, M. *et al.* (2006). An overview of recent large catastrophic landslides in northern British Columbia. *Engineering Geology*, **83**, 120–143.

Gruber, S. and Haeberli, W. (2007). Permafrost in steep bedrock slopes and its temperature-related destabilization following climate change. *Journal of Geophysical Research*, **112**, (F2), doi: 10.1029/2006JF000547.

Hadley, R. F. and Schumm, S. A. (1961). *Sediment Sources and Drainage Basin Characteristics in Upper Cheyenne River Basin*. US Geological Survey Water Supply Paper 1531-B, Washington, D.C.: US Geological Survey, pp. 137–196.

Hallet, B., Hunter, L. and Bogen, J. (1996). Rates of erosion and sediment evacuation by glaciers: a review of field data and their implications. *Global and Planetary Change*, **12**, 213–235.

Hewitt, K. (1997). *Regions of Risk: A Geographical Introduction to Disasters*. Harlow, UK: Addison-Wesley Longman.

Hewitt, K. (2006). Disturbance regime landscapes: mountain drainage systems interrupted by large rockslides. *Progress in Physical Geography*, **30**, 365–393.

Holling, C. S. (2001). Understanding the complexity of economic, ecological and social systems. *Ecosystems*, **4**, 390–405.

Holm, H., Bovis, M. and Jakob, M. (2004). The landslide response of alpine basins to post-Little Ice Age glacial thinning and retreat in southwestern British Columbia. *Geomorphology*, **57**, 201–216.

Hovius, N. *et al.* (1998). Landslide-driven drainage network evolution in a pre-steady-state mountain belt: Finisterre Mountains, Papua New Guinea. *Geology*, **26**, 1071–1074.

IPCC (2007a). *Climate Change 2007: The Physical Science Basis*. Cambridge and New York: Cambridge University Press.

IPCC (2007b). *Climate Change 2007: Impacts, Adaptation and Vulnerability*. Cambridge and New York: Cambridge University Press.

Korner, C. and Ohsawa (2005). Mountain systems. In *Millennium Ecosystem Assessment, Ecosystems and Human Well-being: A Framework for Assessment*. Washington, D.C.: Island Press.

Kotlyakov, V. M. *et al.* (1991). The reaction of glaciers to impending climate change. *Polar Geography and Ecology*, **15**, 203–217.

Kovanen, D. J. and Slaymaker, O. (2008). The morphometric and stratigraphic framework for estimates of debris flow incidence in the North Cascades foothills, Washington State. *Geomorphology*, **99**, 224–245.

Kronfellner-Kraus, G. (1989). *Die Änderung der Feststofffrachten von Wildbächen*. Informations bericht 4/89 des Bayer. Landesamtes für Wasserwirtschaft (München), pp. 101–115.

Kunkel, K. E. (2003). North American trends in extreme precipitation. *Natural Hazards*, **29**, 291–305.

Lambin, E. F. and Geist, H. (eds.) (2006). *Land-use and Land-cover Change: Local Processes and Global Impacts*. Global Change: The IGBP Series. Berlin, Heidelberg: Springer-Verlag.

Lambin, E. F., Turner, B. L. II, Geist, H. J. *et al.* (2001). The causes of landuse and land-cover change: moving beyond the myths. *Global Environmental Change*, **11**, 261–269.

Madduma Bandara, C. M. (1971). The morphometry of dissection in the Central Highlands of Ceylon. Unpublished Ph.D. dissertation, University of Cambridge.

Marston, R. A. (2008). Land, life and environmental change in mountains. *Annals of the Association of American Geographers*, **98**, 507–520.

Marston, R. A., Miller, M. M. and Devkota, L. (1998). Geoecology and mass movement in the Manaslu-Ganesh and Langtang-Jugal Himalaya, Nepal. *Geomorphology*, **26**, 139–150.

McKillop, R. J. and Clague, J. (2007). Statistical, remote sensing-based approach for estimating the probability of catastrophic drainage from moraine-dammed lakes in south-western British Columbia. *Global and Planetary Change*, **56**, 153–171.

Melton, M. A. (1957). *An Analysis of the Relation Among Elements of Climate, Surface Properties and Geomorphology.* Office of Naval Research Project NR389–042, Technical Report 11, Department of Geology, Columbia University, New York.

Osmond, B. *et al.* (2004). Changing the way we think about global change research: scaling up in experimental ecosystem science. *Global Change Biology*, **10**, 393–407.

Owen, L. A. (2004). The Late Quaternary glaciation of Northern India. In J. Elhers and P. Gibbard (eds.), *Extent and Chronology of Glaciations* Volume 3. Amsterdam: Elsevier, pp. 201–210.

Phillips, J. D. (2003). Sources of non-linearity and complexity in geomorphic systems. *Progress in Physical Geography*, **27**, 1–23.

Ryder, J. M. (1998). *Geomorphological Processes in the Alpine Areas of Canada: The Effects of Climate Change and Their Impacts on Human Activities.* Geological Survey of Canada Bulletin 524, Ottawa: Geological Survey of Canada.

Sala, O. E. *et al.* (2005). Global biodiversity scenarios for the year 2100. *Science*, **287**, 1770–1774.

Schiefer, E., Menounos, B. and Slaymaker, O. (2006). Extreme sediment delivery events recorded in the contemporary sediment record of a montane lake, southern Coast Mountains, British Columbia. *Canadian Journal of Earth Sciences*, **43**, 1777–1790.

Schiefer, E., Menounos, B. and Wheate, R. (2007). Recent volume loss of British Columbia glaciers. *Geophysical Research Letters*, **34** (L16503), 1–6.

Schumm, S. A. (1956). The evolution of drainage systems and slopes in badlands at Perth Amboy, New Jersey. *Bulletin of the Geological Society of America*, **67**, 597–646.

Schumm, S. A. (1973). Geomorphic thresholds and complex response of drainage systems. In M. Morisawa, (ed.), *Fluvial Geomorphology.* Binghamton: Publications in Geomorphology 3, pp. 299–310.

Schumm, S. A. (1988). Geomorphic hazards: problems of prediction. *Zeitschrift für Geomorphologie Supplementband*, **67**, 17–24.

Schumm, S. A. (1997). Drainage density: problems of prediction and application. In D. R. Stoddart (ed.), *Process and Form in Geomorphology.* London: Routledge, pp. 15–45.

Schumm, S. A. and Rea, D. K. (1995). Sediment yield from disturbed earth systems. *Geology*, **23**, 391–394.

Shrestha, M. L. and Shrestha, A. B. (2004). Recent trends and potential climate change impacts on glacier retreat/glacier lakes in Nepal and potential adaptation measures. *OECD Global Forum on Sustainable Development: Development and Climate Change*, ENV/EPOC/GF/SD/RD(2004)6/FINAL, OECD, Paris.

Sidle, R. C. (ed.) (2002). *Environmental Changes and Geomorphic Hazards in Forests.* Wallingford and New York: Commonwealth Agricultural Bureaux International.

Slaymaker, O. (1987). Sediment and solute yields in British Columbia and Yukon: their geomorphic significance re-examined. In V. Gardiner (ed.), *International Geomorphology 1986, Vol.1.* Chichester: J. Wiley, pp. 925–945.

Slaymaker, O. and Embleton-Hamann, C. (2009). Mountain landscapes. Chapter 2 in O. Slaymaker, T. Spencer and C. Embleton-Hamann (eds.), *Geomorphology and Global Environmental Change.* Cambridge and New York: Cambridge University Press.

Slaymaker, O., Spencer, T. and Dadson, S. J. (2009). Introduction: the unfilled niche. Chapter 1 in O. Slaymaker, T. Spencer and C. Embleton-Hamann (eds.), *Geomorphology and Global Environmental Change.* Cambridge and New York: Cambridge University Press.

Stoffel, M. and Beniston, M. (2006). On the incidence of debris flows from the early Little Ice Age to a future greenhouse climate: a case study from the Swiss Alps. *Geophysical Research Letters*, **33**, L16404.

Strahler, A. N. (1952). Hypsometric (area-altitude) analysis of erosional topography. *Geological Society of America Bulletin*, **63**, 1117–1142.

Summerfield, M. A. and Hulton, N. J. (1994). Natural controls of fluvial denudation rates in world drainage basins. *Journal of Geophysical Research*, **99** (B7), 13,871–13,883.

Syvitski, J. P. M. *et al.* (2005). Impact of humans on the flux of terrestrial sediment to the global coastal ocean. *Science* **308**: 376–380.

Thompson, L. G. *et al.* (1998). A 25,000 year tropical climate history from Bolivian ice cores. *Science*, **282**, 1858–1864.

Thompson, L. G. *et al.* (2002). Kilimanjaro ice core records: evidence of Holocene change in tropical Africa. *Science*, **298**, 589–593.

Tiffen, M., Mortimore, M. and Gichuki, F. (1994). *More People Less Erosion: Environmental Recovery in Kenya.* Chichester: J. Wiley.

Tucker, G. E. and Slingerland, R. L. (1994). Erosional dynamics, flexural isostasy, and long-lived escarpments: a numerical modeling study. *Journal of Geophysical Research*, **99**, 12,229–12,243.

UNEP (2002). *Global Environment Outlook 3. Past, Present and Future Perspectives*. London: Earthscan.

Vandenberghe, J. (2002). The relation between climate and river processes, landforms and deposits during the Quaternary. *Quaternary International*, **91**, 17–23.

Vorosmarty, C. J. *et al.* (2003). Anthropogenic sediment retention: major global impact from registered river impoundments. *Global and Planetary Change*, **39**, 169–190.

Walsh, J. *et al.* (2005). Cryosphere and hydrology. In *Arctic Climate Impact Assessment*, Cambridge and New York: Cambridge University Press, pp. 183–242.

Walsh, R. P. D. (1985). The influence of climate, lithology and time on drainage density and relief development in the tropical volcanic terrain of the Windward Islands. In I. Douglas and T. Spencer (eds.), *Environmental Change and Tropical Geomorphology*, London: George Allen and Unwin, pp. 93–122.

Walsh R. P. D. (1996). Drainage density and network evolution in the humid tropics: evidence from the Seychelles and the Windward Islands. *Zeitschrift für Geomorphologie N. F. Supplement-Band*, **103**, 1–23.

White, G. F. (ed.) (1974). *Natural Hazards: Local, National, Global*. New York: Oxford University Press.

Woo, M. K. (1996). Hydrology of northern North America under global warming. In J. A. A. Jones *et al.* (eds.), *Regional Hydrological Responses to Climate Change*. Dordrecht: J. Kluwer Academic Publishers, pp. 73–86.

WWF (2005). *An Overview of Glaciers, Glacier Retreat, and Subsequent Impacts in Nepal, India and China*. Nepal Program.

Zurick, D. and Karan, P. P. (1999). *Himalaya: Life on the Edge of the World*. Baltimore and London: Johns Hopkins University Press.

5 Review and future challenges in snow avalanche risk analysis

Michael Bründl, Perry Bartelt, Jürg Schweizer, Margreth Keiler and Thomas Glade

5.1 Background

Snow avalanches pose a major threat to alpine communities because they affect safety in villages and on traffic routes. Therefore, dealing with avalanche danger has a long tradition in Alpine countries. In most countries, avalanches contribute only to a small degree to the overall risk of a country. For Switzerland, for example, avalanche risk represents only 2% of all risks (BABS, 2003).

5.1.1 Snow avalanche formation, geomorphology and land use planning

Snow avalanches are a type of fast-moving mass movement. They can also contain rocks, soil, vegetation or ice. Avalanche size is classified according to its destructive power (McClung and Schaerer, 2006). A medium-sized slab avalanche may involve 10,000 m³ of snow, equivalent to a mass of about 2,000 tons (snow density 200 kg/m³). Avalanche speeds vary between 50 and 200 km/h for large dry snow avalanches, whereas wet slides are denser and slower (20–100 km/h). If the avalanche path is steep, dry snow avalanches generate a powder cloud.

There are different types of snow avalanches (Table 5.1), and in particular two types of release: loose snow avalanches and slab avalanches. Loose snow avalanches start from a point, in a relatively cohesionless surface layer of either dry or wet snow. Initial failure is analogous to the rotational slip of cohesionless sands or soil, but occurs within a small volume (<1 m³) in comparison to much larger initiation volumes in soil slides. Snow slab avalanches involve the release of a cohesive slab over an extended plane of weakness, analogous to the planar failure of rock slopes rather than to the rotational failure of soil slopes. Depending on the type of avalanche (Table 5.1) the damage and required control process may vary significantly. In general, slab avalanches are most disastrous.

Most snow slab avalanches start naturally during or soon after snow storms. Failure is due to overloading and existing weakness in the snowpack. The existence of a weak layer below a cohesive slab layer is a prerequisite for a dry snow slab avalanche. Weak layers typically contain crystals originating from kinetic grain growth such as surface hoar or faceted crystals. Slab thickness is usually less than 1 m, typically about 0.5 m, but can reach several meters in the case of large disastrous avalanches. The observed ratio between width and thickness of the slab varies between 10 and 10^3, and is typically about 10^2. Snow avalanches start from terrain that favors snow accumulation and is steeper than about 30–45°. On terrain less than about 15° snow avalanches start to decelerate and finally stop.

Forest stands may hinder avalanche formation because redistributed snow from the crown prevents weak layer formation (Bründl et al., 1999; Bartelt and Stöckli, 2001) and stems in dense forests may stabilize the snowpack. However, if a snow avalanche starts above the timberline, the forest has only a marginal influence on the avalanche flow process.

Besides natural triggering by overloading or internal weakening of the snowpack, snow slab avalanches can also be triggered artificially – unlike most other rapid mass movements – by localized, rapid, near-surface loading by, for example, people (usually unintentionally) or intentionally by explosives used as part of avalanche control programs. Occasionally, snow avalanches have been triggered by large earthquakes (Stethem et al., 2003). In general, naturally released avalanches mainly threaten residents and infrastructure, whereas human-triggered avalanches are the main threat to recreationists.

Geomorphological Hazards and Disaster Prevention, eds. Irasema Alcántara-Ayala and Andrew S. Goudie. Published by Cambridge University Press. © Cambridge University Press 2010.

TABLE 5.1. *International snow avalanche classification*

Zone	Criterion	Characteristic and denomination	
Origin	Manner of starting	from a point	from a line
		Loose snow avalanche	*Slab avalanche*
	Position of failure layer	within the snowpack	on the ground
		Surface-layer avalanche	*Full-depth avalanche*
	Liquid water in snow	absent	present
		Dry-snow avalanche	*Wet-snow avalanche*
Transition	Form of path	open slope	gully or channel
		Unconfined avalanche	*Channelled avalanche*
	Form of movement	snowdust cloud	flowing along ground
		Powder snow avalanche	*Flowing snow avalanche*
Deposition	Surface roughness of deposit	coarse	fine
		Coarse deposit	*Fine deposit*
	Liquid water in deposit	absent	present
		Dry deposit	*Wet deposit*
	Contamination of deposit	no apparent contamination	rock debris, soil, branches, trees
		Clean deposit	*Contaminated deposit*

UNESCO, 1981

FIGURE 5.1. Grüenbödeli avalanche (Canton Grisons, Switzerland.) Note the results of the destructive forces and also the soil and rock mass transported by these snow avalanches. (Source: SLF.)

Avalanche formation is usually, e.g. in avalanche control programs, assessed heuristically by weighing the so-called contributing factors: terrain, precipitation, wind, temperature and snow stratification, i.e. the complex interaction between terrain, snowpack and meteorological conditions are explored (Schweizer *et al.*, 2003).

Objects in the deposition area are influenced by two major processes. First, the air pressure plume in front of a dry snow avalanche has a huge destructive power. Second, the snow in motion exerts high impact pressures on objects located in the run out path (Sovilla *et al.*, 2008). The destructive forces of avalanches are also enforced by transported debris (Figure 5.1). While in motion, entrained rock particles of up to boulder size or large woody debris can cause major destruction (e.g. SLF, 2000).

As a consequence of the above, large snow avalanches are high energy processes that contribute to the shaping of environments with steep topography. Material picked up on run-over areas may be transported over tens of meters as reported e.g. from Iceland (Decaulne and Saemundsson, 2006). For example, large rocks in valley floors often do not correspond to rock fall processes but have rather been transported by snow avalanches (Kristjansdottir, 1997).

From a geomorphic perspective, a specific characteristic of every snow avalanche is, however, that the deposited snow melts every spring. Although the geomorphic forming effectiveness of a single snow avalanche might not be large, large full-depth wet snow avalanches (after

UNESCO, 1981) occurring every spring can considerably contribute to material transport (Ward, 1985; Becht, 1995) and can influence the vegetation cover in tracks by generally reducing tree growth, size and density (Kulakowski et al., 2006; Rixen et al., 2007). However, there is still a lack of knowledge in determining the role of snow avalanches in the coupled geomorphic process chains.

Changing land use is of major importance for avalanche risk. Examples are deforestation, changes in agricultural practices (e.g. grass cuts), and especially the development of infrastructure and settlements (Fuchs and Bründl, 2005). Land use planning is an important part of avalanche risk management and different types of maps indicating hazard or even risk do exist. Based on local, regional and national legislative conditions, these maps have either indicative character with no executive power and related requirements (typical scale 1:25,000) or they are compulsory for any future land use development plan (e.g. for Canada refer to Jamieson and Stethem, 2002). Hazard zone maps on the scale 1:10,000 to 1:5,000, which are mandatory for land use planning in many European countries, such as Austria, France and Switzerland, depict highly endangered areas and less endangered areas indicating where building is allowed and where not. For particular regions, risk maps have been developed (e.g. Arnalds et al., 2004) showing the risk level for specific areas. Based on such risk maps, acceptable risk levels have been developed in close cooperation with affected social actors (Bell et al. 2006). According to the requirements of sustainable development, it would be highly advisable to consider snow avalanche hazard – and even risk – maps as a compulsory part of land use planning for every mountain region in the world.

5.1.2 General methodological framework of risk management

Risk management denotes a general framework, which aims at assessing, reducing and controlling the risks from different sources. An integrated part of risk management is the risk concept, which addresses the following basic questions (Kaplan and Garrick, 1981; Haimes, 2004):

- "What might happen?" (risk analysis);
- "What is allowed to happen?" (risk evaluation);
- "What needs to be done?" (planning and evaluation of mitigation measures).

The terms "risk analysis" and "risk evaluation" are often summarized within the term "risk assessment" (Crozier and Glade, 2005). The increasingly widespread use of this concept in different disciplines such as finance and health management, engineering and technology, biodiversity,

nuclear technology or terrorism prevention has been proven to support decision-making in complex systems (Hatfield and Hipel, 2002) and has therefore become a routine procedure (Klinke and Renn, 2002) in the planning and realization of projects involving risks.

Risk management stresses the integration of risks from different fields into a systematic integrated approach. Based on existing guidelines (e.g. AS/NZS, 2004) this demand has recently been outlined in the new ISO Standard 31000 (ISO, 2008). This standard follows the argument that risk management can only be effective if it is a part of all management functions and decisions (Figure 5.2).

A risk analysis consists of four parts (Bründl et al., 2009). The goal of a *hazard analysis* is to determine the scenarios that have to be taken into account in a risk analysis. Basic data are indications of the process in the terrain, topographic maps, aerial photographs and satellite images, which allow one to create a geomorphologic map of the phenomena in the area under analysis. The analysis of event register data or historical chronicles allows assuming possible scenarios. The intensity, defined as the physical impact of a process, is determined by modelling the process. The results of a hazard analysis are visualized by intensity maps.

In the *exposure analysis*, exposed persons and assets are determined regarding their location number, type, value and probability of exposure.

In the *consequence analysis*, the results of hazard and exposure analysis are combined and the damage or loss in case of events in the regarded scenarios is calculated. Finally, the *risk is calculated* as a product of the frequency of the given scenarios with the calculated damage of these scenarios. The risk is expressed as individual risk to single persons and/or societal or collective risk to assets and persons in the area under investigation. The individual risk is defined as the probability per year that a single person affected by the given scenarios will die. The societal risk is expressed as expected damage per year in monetary values per year or fatalities per year.

The hazard analysis is the fundamental part of a risk analysis. Investigations on the sensitivity of factors in the risk equation suggest that assumptions on the frequency and the physical impact of the considered process have the most significant influence on the calculated risk (Schaub, 2008). Therefore, in the following section an overview on recent trends in hazard analysis is given.

5.2 Review and recent trends in hazard analysis

Large snow avalanches are rare events and the prediction of disastrous avalanche events involves great uncertainty

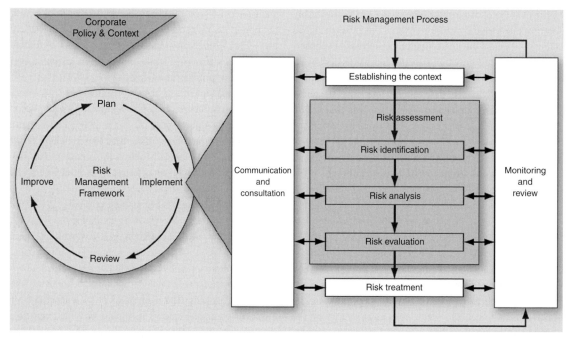

FIGURE 5.2. The systematic approach to risk management of ISO 31000. (Jaecklin, 2007.)

(Schweizer, 2008). Therefore, snow avalanche hazard analysis relies on defining hazard scenarios rather than well-described design events. Central to any snow avalanche hazard analysis is the definition of *hazard scenarios*. Scenario planning requires obtaining historical information (inventories of past events or case studies) and hands-on terrain information to develop an idea of what has and therefore what could happen at a particular site. In the end, the definition of an appropriate hazard scenario determines the quality of the entire hazard analysis since it tries to include how different factors (climatic conditions, release locations, wind blown snow, terrain, snow cover erosion, secondary releases, vegetation, etc.) can combine and lead to complex, historically unforeseen and therefore surprising and critical situations. Simple experience coupled with historical information are the primary resources that experts have at their disposal to define a reasonable hazard scenario.

Computer models can sometimes "support experience" if they can correctly simulate documented events. This provides the hazard expert with some confidence that the consequences of an undocumented situation can be predicted, or at least provide results that are thought provoking, causing the expert to modify his scenario. For example, the weight of different factors (e.g. location of starting zones or the inclusion of secondary release zones) may shift because of simulation results. Thus, the interplay between the computer model and the experience of the

hazard expert is rather complex. On one hand, an expert will only employ a computer model if it provides results that are "beyond" his experience; while, on the other hand, results that do not agree with experience are simply not trustworthy. The difference between these two cases is extremely small and is always resolved in the end by the judgement of the expert (not the computer model). Computer models that continually supply the expected result are confidence building but in the end superfluous. The decision whether a model "expands" the expert's experience, and is therefore useful because it has some predictive quality, or is simply used to satisfy the demands of the government authorities, who would prefer some "objective" calculation (and therefore non-expert, subjective analysis), is central to the future of risk analysis.

The application of computer models by hazard experts therefore has several fundamental consequences for avalanche and natural hazards engineering. First, computer models must contain enough detail (factors) to allow experts to test scenarios and, conversely, experts must have increasing knowledge of the numerical solution procedures at the core of the simulation programs. Model input must be flexible (and extremely user-friendly); solution algorithms robust, quick and stable. How the simulation results are affected by the mathematical approximation of terrain (slope averaging, grid spacing, smoothing of macro-sized roughness, etc.) must be known since these parameters influence whether or not the expert trusts the model results. The depiction of simulation results in the form of maps and three-dimensional visualizations is essential since the mass of numerical results provided by computer programs must be easily and quickly displayed and interpreted in a form experts can use.

However, at the very core of the expert–computer simulation problem is the "right" model physics. The model physics is *phenomenologically* right if the model provides the right answers – say run out distances or flow velocities – using flow parameters that have no direct experimental basis. The flow parameters have been determined over many years from model case studies, carried out under given assumptions (e.g. terrain averaging, no snow entrainment). The expert accepts these parameters because they do not transgress his experience and are, furthermore, usually recommended by the government authorities. A well-defined set of phenomenological model parameters is valuable since this set allows an expert to go "beyond" his experience. The phenomenological parameters are, however, always somewhat mysterious and intangible and therefore, in the end, uncertain. An expert will gladly use them, but is at the same time always troubled by them.

The model physics is *physically* right if the model provides the right answers using avalanche flow parameters that are experimentally based; that is, obtained from measurements, first with the material snow and second at the real scale. Such measurements in natural hazard science are unique (Sovilla *et al.*, 2006, 2007; Kern *et al.*, 2009). Clearly, physically correct models are needed for the future of avalanche science, but it should be stressed that physically correct models cause many problems:

(1) They disrupt the continuity of calculation procedures for hazard analyses. Authorities simply do not want to introduce new sets of parameters every few years since this will cause confusion in practice (perhaps with expensive re-reviews or even costly litigation). This means that new models must be carefully introduced – if at all – only after extensive testing and calibration. The lifetime of the model must be guaranteed to ensure some continuity – with the past as well as many years into the future. It is incorrect to suggest that older, phenomenological models are wrong because they have no experimental foundation. They are wrong only when they produce the wrong results – and this is not the case with well-known avalanche dynamics models such as the Voellmy–Salm model (Salm, 1993; Bartelt *et al.*, 1999).

(2) Physically "right" models usually require too exact a definition of release and entrainment conditions as well as terrain geometry. Whereas the older phenomenological models are generous and robust (since they provide the right answers under simplifying assumptions), physical models are rigid and less generous. For example, the Voellmy–Salm model could be used well without snow entrainment. Physical models will require the specification of snow cover and entrainment rates. The hazard expert might be grateful that an additional factor in his hazard scenario can be included in the hazard analysis, or the hazard expert might feel simply overwhelmed by the required detail of the problem. In the end the analysis could be less certain and more confusing – even though a physical model has been employed.

These considerations regarding the judgement of experts and the role of computer models have led to the development of the computer model RAMMS (Christen *et al.*, 2007). The program RAMMS is especially designed to aid experts in making judgements in the hazard analysis process and represents an important development in avalanche risk assessment:

(1) The program uses three-dimensional digital terrain models (in Switzerland for certain examples up to a 2 m resolution) to model complex mountain terrain. Many important questions in snow avalanche hazard analysis are related to avalanche motion on real terrain, such as multi-channel flows, channel widening and slope deviations in the run out zone.

(2) Both phenomenological models (Voellmy–Salm) and experimentally based physical models (Bartelt–Buser) are included in the program package. (For more information concerning the Bartelt–Buser model, see Bartelt *et al.*, 2006; Buser and Bartelt, 2009). This allows experts to exploit the advantages of both models and ensures model continuity. The comparison between well-calibrated phenomenological model results and more physically based models can be helpful in many situations, especially where the expert is uncertain.

Snow cover erosion and entrainment process models are included that allow a realistic representation of the avalanche mass balance (Sovilla *et al.*, 2007).

(3) The program system contains sophisticated output features that allow the results to be interpreted. Two-dimensional maps (Figure 5.3) as well as three-dimensional representations can be chosen by the user. Other useful features include the import of measured, historical avalanche data, or aerial photos that can be overlaid on the simulation results. This too helps the interpretation of the simulation results. The simulation results can likewise be easily exported to GIS systems.

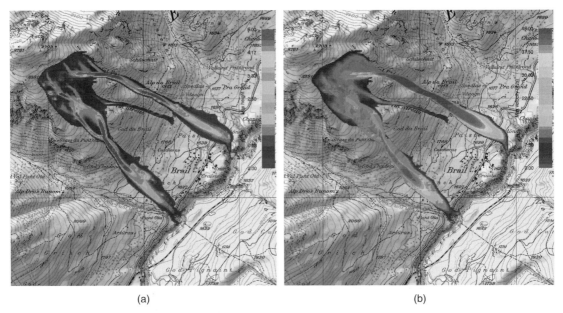

(a)   (b)

FIGURE 5.3. Calculations of an avalanche, which occurred in 1968 near Brail, canton of Grisons, Switzerland. (a) Calculated maximum flow heights. (b) Calculated maximum flow velocities. The simulations were validated with the original observations. The avalanche crossed the road before and after the village of Brail.

5.3 Methods of risk analysis

5.3.1 Definition of risk

The term risk is defined as "a measure of the probability and severity of loss to the elements at risk, usually expressed for a unit area, object, or activity, over a specified period of time" (e.g. Crozier and Glade, 2005). Generally expressed in a mathematical equation, this definition can be written as (e.g. Fuchs *et al.*, 2007; Bründl *et al.*, 2009):

$$R_{ij} = p_j \cdot p_{ij} \cdot A_i \cdot V_{ij} \qquad (5.1)$$

with R_{ij} as the risk to an object i in scenario j, p_j as the frequency of a scenario j, p_{ij} as the probability that an object i is present while scenario j is occurring, A_i as the value of an object or the number of exposed persons, and V_{ij} as the vulnerability of an object i in a scenario j. The frequency of a scenario can be approximated as the difference of the exceedance probability of two adjacent scenarios, e.g. $p_{10} = 0.1 - 0.033 = 0.067$ for a 10-year scenario (Bründl *et al.*, 2009; for an approximation from hazard maps see Rheinberger *et al.*, 2009). The total risk R for an area under investigation, e.g. for part of a village, is the sum of all risks to individual objects and the risks in all considered scenarios according to:

$$R = \sum_j \sum_i R_{ij} \qquad (5.2)$$

Equation (5.1) shows that there are several factors that influence the risk. On the one hand there are probabilities that an event happens and that an object is affected, on the other hand there are values at risk and their characteristics. In the following section these factors are briefly described.

5.3.2 Factors for calculation of risk

For estimating potential damage, the type, the number, the value and the vulnerability of the *elements at risk* that depended on the process intensity have to be assessed. The type of elements considered and the level of detail depends on the scope of the risk analysis (Bell and Glade, 2004) and the target scale. A characterization of each specific element at risk is often connected to an extensive data collection. Therefore, it must be decided whether each single object should be recorded or whether objects could be grouped in object classes. Elements at risk are either at fixed locations (e.g. buildings, power stations, etc.) or they are mobile (e.g. trains or cars). The value of these objects can be expressed in monetary units representing the recovery values. Special attention has to be given to persons at risk and their presence in or outside structures.

It is common to assess elements at risk only in their present form. It depends on the goal of the assessment whether an expected increase in the damage potential should be regarded or not. However, incorporating future variability bears uncertainty and should be carefully indicated.

An important factor for the estimation of risk is vulnerability. Although this term is always related to the consequences of a natural disaster or an event, it is defined in different ways, as shown in literature (e.g. Cutter, 1996; Weichselgartner, 2001; Birkmann, 2006, Fuchs et al., 2007). The definitions range from the social science perspective to those from engineering science. Thus, vulnerability is measured on a metric scale (as monetary units), or on an ordinal scale based on social values or perceptions and evaluations. From a natural science perspective, Uzielli et al. (2008) define vulnerability as a product of intensity during scenario j and the susceptibility of object i. Generally, vulnerability is considered as a function of a given intensity of a process, expressing the expected degree of loss for an element at risk (Varnes, 1984; Fell, 1994). The value ranges generally from 0 (no damage) to 1 (complete destruction).

The assessment of these "technical" vulnerability values involves the evaluation of several different parameters and factors such as building materials and techniques, state of maintenance, and the presence of protection structures (Fell, 1994; Fell and Hartford, 1997) in relation to the potential physical impact of a process (Fuchs et al., 2007). For risk analyses at the regional scale it might be appropriate to work with average values. For detailed local investigations it may be necessary to determine the vulnerability of an object individually by considering relevant characteristics (e.g. windows at the upslope side of buildings, local structural protection, etc., Holub and Fuchs, 2008).

The determination of vulnerability values for snow avalanche risk assessment is based on empirical data, but in some cases also on expert knowledge. Wilhelm (1997), Jónasson et al. (1999) and Barbolini et al. (2004a, b) suggested vulnerability functions based on empirical data and Borter (1999), Borter and Bart (1999), Bell and Glade (2004) and Kraus et al. (2006) propose average vulnerability values for buildings and exposed persons. Vulnerability curves can be distinguished with respect to the elements at risk, and also with respect to the avalanche type. Powder snow avalanches are characterized by higher flow depths than dense flow avalanches; consequently, whole buildings are affected and damage is distributed over the entire building. Vice versa, damage is located only at the lower floor levels if dense flow avalanches occur. An overview of the available vulnerability relations for snow avalanches is provided in Table 5.2.

Related to people, vulnerability is denoted as mortality rate λ, defined as the conditional probability that a person is killed given that the object in which the person is present is affected and damaged. The value ranges between 0 (no fatality) and 1 (all exposed and affected persons are killed).

TABLE 5.2. *Overview of available vulnerability relations for snow avalanches*

Vulnerability	Buildings		DA	Wilhelm (1997)
			PA	Barbolini et al. (2004a)
	People	Inside buildings	DA	Jónasson et al. (1999); Keylock and Barbolini (2001); Barbolini et al. (2004b)
			PA	Barbolini et al. (2004b)
		Outside buildings	DA	Barbolini et al. (2004b)
			PA	–

DA, dense flow avalanche; PA, powder avalanche
Cappabianca, 2008

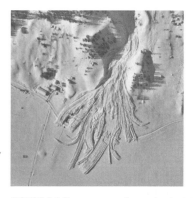

FIGURE 5.4. Run out area of an avalanche in Ulrichen, Switzerland (note the houses on the left side of the image as scale). The run out area is in most cases only partially affected by the deposition. This effect is considered in risk analysis by integrating a spatial occurrence probability. (Source: Federal Office for Topography, coordination unit for aerial photography.)

Observations of avalanche events show that single events in most cases do not cover the whole potential run out area (Figure 5.4). Therefore, the probability whether an object is hit or not is the conditional probability of the occurrence affected by the morphology of the terrain in the run out area, denoted as *spatial occurrence probability*. This conditional probability can be estimated by deriving an empirical probability based on past events using, for example, event trees for illustration. The other approach is to assume mean values for the spatial occurrence probability by analyzing past events for the run out location under investigation. Values in use in practice in Switzerland range from $p(e) = 0.10$ for a 30-year scenario, $p(e) = 0.50$ for a 100-year scenario, to $p(e) = 0.80$ for an extreme scenario with a return period of 300 years (BAFU, 2008).

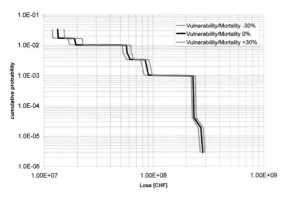

FIGURE 5.5. Example of a FN-diagram with an upper and a lower boundary reflecting the uncertainty of the factors vulnerability of buildings and mortality rate of persons in buildings (Bründl, 2008.)

5.3.3 Calculation and presentation of risk

The risk due to avalanches can be presented as societal or collective risk or individual risk. The *societal risk* is calculated by linking the frequency of scenarios with the consequence of the scenarios. The result is expressed as monetary units per year or number of fatalities per year. However, according to Kaplan and Garrick (1981) the societal risk cannot be expressed only as a single number. Therefore, the risk is usually presented with an FN-diagram (Figure 5.5). The area below the curve represents the societal risk given the assumed loss. The graph shows at what probability a certain damage is exceeded.

The *individual risk* is an indicator of how much risk a single, identifiable person has to bear as a consequence of the assumed scenarios. The total societal risk allows no statement of whether the risk for an individual person is above or below an accepted risk level. Fewer people with a high individual risk can result in the same societal risk as many people with a low individual risk. It is standard in risk assessment to calculate both the societal and the individual risk.

The calculation of risk in practice is based on the actual conditions, i.e. number and value of existing objects and persons, land use etc. However, land use is not constant over time. If factors in the risk equation are changing, it will have direct consequences on the risk.

5.4 Change in avalanche risk, influence of different risk factors

"The world is not static, and we discover new threats daily." This statement was made by Susan Cutter (2003, p. 4) to describe a world view, that seems sometimes to have been forgotten in risk research concerning natural hazards.

Beside this insensitivity to a dynamic perspective, natural hazards can no longer be seen as single, isolated events. Risk research requires an integrative approach to explain the interactions among natural and social systems. According to Cutter (2003) it requires a new way of viewing the world, one that integrates perspectives of science and social science and furthermore incorporates the dynamic (temporal and spatial) dimension.

A fundamental characteristic of risk resulting from natural hazards is the connectivity between the natural system (or geosystem, governing the physical part of the process) and the social system (including values at risk and vulnerability). Both systems are subject to continuous changes in space and time. Due to changes in the social system (such as varying risk awareness and acceptance, changing economic conditions or mobility behavior) new demands on the geosystems are entailed, resulting in, for example, different use of natural resources or engineering structures. This may produce the intended response of the geosystem due to changing process conditions, and additional unintended feedbacks, both of which in turn induce a reaction in the social system. If the run out area could be reduced to a specific magnitude, new development areas could be assigned in areas supposed to be safe.

The same development can be seen vice versa. Higher snow avalanche activity may give rise to a response of the social system, which may change the environmental settings for snow avalanches. Due to the dynamics of the geosystem and the social system, new interaction emerges and therefore enhanced connectivity can develop. Increasing connectivity is likely to induce higher complexity (Hufschmidt *et al.*, 2005). Complex systems imply two fundamental conditions: (1) the system consists of multiple interactive components and (2) these interactions give rise to emergent forms and properties, which are not reducible to the sum of the individual components of observed system. The geosystem, in our case snow avalanches, as well as the social system can be considered from this perspective as well as the interaction between these systems. Hence, rising losses related to natural hazard processes can not be solely connected either to the changes of the natural processes or to the development of the damage potential and the vulnerability. These losses are the result of increasing complexity. Thus, increasing knowledge about one part, such as the understanding of the hazard processes, elements at risk or vulnerability, without analyzing the interaction between these components and their dynamics does not help in finding useful management strategies to reduce risk.

The above-described concept differs from approaches used in science and practice. Risk analyses applied to natural hazards are in general static approaches (Jónasson *et al.*, 1999; Keylock *et al.*, 1999; Gächter and Bart, 2002; Bell and Glade, 2004) neglecting past risk levels and the history of evolution to the current situation under consideration as well as possible future risk levels. However, risk related to natural hazards is subject to temporal changes since the risk-influencing factors are variable over time, each factor with an individual evolution (Fuchs and Keiler, 2006). Therefore, identifying temporal changes of natural risk, as well as the underlying processes, contributes to an improved understanding of today's risk levels.

In the following, an overview of results of a study with the focus on temporal changes of snow avalanche risk between 1950 and 2000 (Keiler *et al.*, 2006) is presented. First, a brief description of the general change in avalanche activity, elements at risk and vulnerability since the 1950s in the Eastern Alps is given. In the second part, the evolution of risk is illustrated focusing on a few avalanche paths in the municipality of Galtür (Austria).

In the twentieth century, natural avalanche activity seems to have been neither significantly increasing nor decreasing, although the variability of events makes an exact statement difficult (Bader and Kunz, 1998; Schneebeli *et al.*, 1998; Laternser and Schneebeli, 2002, 2003). Thus, it can be assumed that changes in the natural processes are due to the construction of permanent mitigation measures in the release areas or run out areas of avalanche tracks. In Switzerland, about 1 billion euros has been invested for this purpose since 1950 (SLF, 2000). In the Galtur study, snow avalanches were simulated by the 3D model SAMOS (Sailer *et al.*, 2002; Sampl and Zwinger, 2004) considering the Austrian design event with a 150-year recurrence interval. Furthermore, the construction of supporting structures in the release area over time and their influence on processes area and occurring impacts were modelled (Keiler *et al.*, 2006).

In contrast to the snow avalanche activity, societies in the Alps have undergone considerable socio-economic changes since the mid twentieth century. This development reflects a shift from farming-based activities towards a tourism and leisure-time orientated economy (Bätzing, 1993). Contemporaneously, settlements and the population increased significantly in the Eastern Alps. A similar trend is outlined for the damage potential in Keiler (2004), Fuchs and Bründl (2005) and Keiler *et al.* (2005). Therefore, areas suitable for land development are relatively scarce in the Inner Alpine valleys, e.g. in Austria, only about 20% of the whole area is appropriate for development activities (BEV, 2004). The change in number and size of buildings is a good example to illustrate this development. For the long-term risk assessment, monetary values of buildings were calculated using the volume of the buildings and average

prices per cubic meter for new buildings, as used by insurance companies. Furthermore, details of the function and construction types were analyzed (Keiler, 2004; Keiler *et al.* 2006).

Regarding vulnerability approaches related to snow avalanches, it has to be stated that there is a lack of studies in general as well as on temporal changes of vulnerability in both natural science and social science. Changes in the construction method of buildings have a huge influence on vulnerability to the impact of snow avalanches. Therefore, the vulnerability functions for different construction methods related to snow avalanche pressure were used in this study, as outlined in Wilhelm (1997).

The following results (calculated using Formula (5.1)) show the changes of avalanche risk, expressed as the potential monetary loss of buildings resulting from the occurrence of the defined design event. This risk scenario illustrates the real-time change of the possible loss, taking into account changes of all three risk-influencing factors: (1) the shifts in the values at risk, (2) the varying vulnerability of buildings and (3) the construction of supporting structures.

In a comparison between 1950 and 2000, the development of risk of the three studied avalanche tracks differs considerably (see Figure 5.6). The risk related to the Großtal West and East avalanches doubled and nearly doubled, respectively. One of the avalanche tracks shows a steady increase, whereas the other one is characterized by a slight increase and decrease during this 50-year period. In contrast, the risk associated with the Gidisrinner avalanche in 2000 was just below the risk for the year 1950. Furthermore, the risk evolution of this avalanche path is shaped by the strongest increase and reduction of risk. Summing up the possible losses of all three avalanche tracks, the risk increased considerably until the 1980s, followed by a short period of

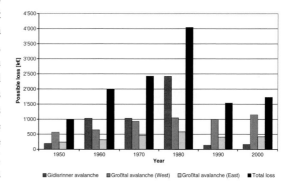

FIGURE 5.6. Development of the possible loss from 1950 to 2000 for the Gidisrinner and the Großtal avalanche (West and East). The black columns denote the total loss.

reduction due to mitigation measures, and then rose slightly in the last decade.

Considering that the development of the values at risk for all three avalanche tracks shows an increase of a factor of five between 1950 and 2000 (Keiler et al., 2006), no general trend could be determined for the calculated risk. Analyzing the different influencing factors and their interaction, this development is caused by different aspects. These aspects were the spatial distribution of the exposed objects, the values at risk, the occurring impact pressures and the related vulnerability of the objects, the effectiveness of the mitigation measures regarding both the extent of the run out zone and the reduction of the pressure as well as legal regulations. Small changes of one of these aspects can cause considerable differences in the resulting risk. These findings are consistent with recent studies in the Swiss Alps (Fuchs et al., 2004). The increase in risk can therefore not solely be attributed either to an overall increase in values at risk as suggested in different studies (White et al., 2001; Barbolini et al., 2002), or to a decrease in run out distance due to the construction of technical mitigation (Ammann, 2001); rather these losses are simultaneously a result of increasing complexity caused by changes of the geosystem, the social system, and the connectivity between these systems.

5.5 Conclusions: where to go from here; future challenges

Challenges for future work on snow avalanche risk assessment relate to different aspects. For avalanche formation and for defining the release conditions, the monitoring of snow accumulation and the modelling of both snowpack properties and stability will be major areas of research. Besides direct field measurements, which continue to be important as ground truth and for model verification, also applications of remote sensing technologies (e.g. Prokop, 2008; Schaffhauser et al., 2008) and modelling of the energy fluxes between atmosphere, snowpack and soil in complex alpine terrain (Lehning et al., 2008) are needed to improve hazard scenarios.

In hazard analysis the continuous improvement of snow avalanche models is essential since it is an important basis for risk assessment. Models that are able to accurately reconstruct the behavior of avalanches in three-dimensional terrain will be used in the near future. Real-scale experimental investigations of avalanches have placed the prediction of impact pressures (Sovilla et al., 2007) and extreme avalanche friction parameters (Buser and Bartelt, 2009) on a better foundation. Thus, it will be possible to formulate more realistic hazard scenarios and to better simulate mitigation methods such as dams.

However, the definition of initial conditions, solution parameters and grid resolution will also become more difficult. More than ever, practical experience will be required to interpret model results. Therefore, the development of other model approaches should not be forgotten (e.g. statistical run out modelling by Delparte et al. (2008)). In numerous regions of the world, very simple model applications are the only way to assess snow avalanche hazard.

Improved risk analyses in the future are also related to more realistic assumptions on the vulnerability of people and objects or the spatial distribution of avalanche deposits in the run out zone. Monitoring of damage-causing events and analyzing the dynamic interactions between process and objects is an issue that holds not only for snow avalanches but also for the assessment of other gravitational processes such as debris flow, rock fall or landslide.

Risk management for natural hazards includes an integrated use of different types of measures: technical (e.g. snow supporting structures), biological (e.g. protection forest), organizational (e.g. closure of roads or evacuation of buildings) and land use planning. Developing innovative methods and tools for optimization of the scarce financial resources for planning of measures is a field for future work (e.g. Cappabianca et al., 2008; Bründl et al., 2009).

In countries where the safety level is lower, the establishment of an observation and measurement network, the transmission of data, the building up of warning services adapted to local or regional conditions and the education of local or regional avalanche experts are key issues. Additionally, the coping capacity and the resilience of the potentially affected society have to be strengthened.

Snow avalanche hazard and risk has also to be framed within principle geomorphic hazard and risk assessments. Although the demand for a multi-hazard risk assessment has been conceptually addressed (e.g. Glade and von Elverfeldt, 2005) only minor work has been carried out in trying to link different geomorphic hazard assessments (e.g. Fuchs et al., 2001).

A major issue for the future is how changes in society (e.g. tourist resort development) and environment (e.g. climate change) cause temporal and spatial changes in avalanche risk. This also includes the changing consequences of snow avalanches due to suburban development on alluvial fans in valley floors or by expansion of lifeline networks. Even without changing snow avalanche activity, snow avalanche risk might change purely because of changes in land use. Dealing with these challenges will remain a central task for many countries in order to allow sustainable development of mountain regions in the future.

References

Ammann, W. (2001). Integrales Risikomanagement: der gemeinsame Weg in die Zukunft. *Bündnerwald*, **5**, 14–17.

Arnalds, Th., Jónasson, K. and Sigurðsson, S. Þ. (2004). Avalanche hazard zoning in Iceland based on individual risk. *Annals of Glaciology*, **38**, 285–290.

AS/NZS 4360, (2004). *Australian/New Zealand Standard. Risk Management.* Standards Australia.

BABS (2003). *KATARISK. Katastrophen und Notlagen in der Schweiz: Eine Risikobeurteilung aus der Sicht des Bevölkerungsschutzes.* Bern: Bundesamt für Bevölkerungsschutz.

Bader, S. and Kunz, P. (eds.) (1998). *Klimarisiken: Herausforderungen für die Schweiz.* Zürich: vdf Hochschulverlag.

BAFU (2008). *EconoMe: Wirtschaftlichkeit von Schutzmassnahmen gegen Naturgefahren.* www.econome. admin.ch, (accessed 16 December 2008).

Barbolini, M., Natale, L. and Savi, F. (2002). Effects on release conditions uncertainty on avalanche hazard mapping. *Natural Hazards*, **25**, 225–244.

Barbolini, M., Cappabianca, F. and Sailer, R. (2004a). Empirical estimate of vulnerability relations for use in snow avalanche risk assessment. In C. Brebbia, (ed.), *Risk Analysis IV.* Southampton: WIT, pp. 533–542.

Barbolini, M., Cappabianca, F. and Savi, F. (2004b). Risk assessment in avalanche-prone areas. *Annals of Glaciology*, **38**, 115–122.

Bartelt, P. and Stöckli, V. (2001). The influence of tree and branch fracture, overturning and debris entrainment on snow avalanche flow. *Annals of Glaciology*, **32**, 209–216.

Bartelt, P., Salm, B. and Gruber, U. (1999). Calculating dense-snow avalanche runout using a Voellmy-fluid model with active/passive longitudinal straining. *Journal of Glaciology*, **45**(150), 242–254.

Bartelt, P., Buser, O. and Platzer, K. (2006). Fluctuation-dissipation relations for granular snow avalanches. *Journal of Glaciology*, **52**(179), 631–643.

Bätzing, W. (1993). *Der sozio-ökonomische Strukturwandel des Alpenraums im 20. Jahrhundert.* Bern: Geographica Bernensia, P26.

Becht, M. (1995). Slope erosion processes in the Alps. In Slaymaker, O. (ed.), *Steepland Geomorphology.* New York: Wiley, pp. 45–61.

Bell, R. and Glade, T. (2004). Quantitative risk analysis for landslides: examples from Bíldudalur, NW Iceland. *Natural Hazards and Earth System Sciences*, **4**, 117–131.

Bell R., Glade T. and Danscheid M. (2006). Challenges in defining acceptable risk levels.- In W. Ammann, S. Dannenmann and L. Vulliet (eds.), *Coping with Risks Due to Natural Hazards in the 21st Century: "RISK 21".* 28 November – 3 December 2004, Monte Vérita (CH), Balkema: pp. 1–10.

BEV (2004). *Regionalinformation der Grundstücksdatenbank des Bundesamtes für Eich- und Vermessungswesen.* www. bev.at (accessed 15 January 2006).

Birkmann, J. (ed.) (2006). *Measuring Vulnerability to Natural Hazards: Towards Disaster Resilient Societies.* New York: United Nations University Press.

Borter, P. (1999). *Risikoanalysen bei gravitativen Naturgefahren: Methode.* Umwelt-Materialien 107/I, Bundesamt für Umwelt, Wald und Landschaft, BUWAL, Bern.

Borter, P. and Bart, R. (1999). *Risikoanalysen bei gravitativen Naturgefahren: Fallbeispiele und Daten.* Umwelt-Materialien 107/II, Bundesamt für Umwelt, Wald und Landschaft, BUWAL, Bern.

Bründl, M., Schneebeli, M. and Flühler, H. (1999). Routing of canopy drip in the snowpack below a spruce crown. *Journal of Hydrological Processes*, **13**, 49–58.

Bründl M., Krummenacher, B. and Merz, H. M. (2009). Decision making tools for natural hazard risk management: Examples from Switzerland. In S. Martorell, C. G. Soares and J. Barnett (eds.), *Safety, Reliability and Risk Analysis: Theory, Methods and Applications.* Leiden: CRC Press/Balkema, pp. 2773–2779.

Buser, O. and Bartelt, P. (2009). The production and decay of random energy in granular snow avalanches. *Journal of Glaciology*, **55**, 3–12.

Cappabianca, F. (2008). Empirical vulnerability function for use in snow avalanche risk assessment. In M. Naaim (ed.), *Vulnerability to Rapid Mass Movements*, IRASMOS Report D4, Grenoble, France. http://irasmos.slf.ch/pdf/WP4_D40_20080710.pdf (accessed 22 October 2009).

Cappabianca, F., Barbolini, M. and Natale, L. (2008). Snow avalanche risk assessment and mapping: a new method based on a combination of statistical analysis, avalanche dynamics simulation and empirically-based vulnerability relations integrated in a GIS platform. *Cold Regions Science and Technology*, **54**, 193–205.

Christen, M., Bartelt, P. and Gruber, U. (2007). Modelling avalanches. *GEOconnexion International*, **6**(4), 38–39.

Crozier, M. and Glade, T. (2005). Landslide hazard and risk: issues, concepts and approach. In T. Glade, T. Anderson and M. Crozier (eds.), *Landslide Hazard and Risk.* Chichester: John Wiley & Sons, pp. 1–40.

Cutter, S. (1996). Vulnerability to environmental hazards. *Progress in Human Geography* **20**, 529–539.

Cutter, S. (2003). The vulnerability of science and the science of vulnerability. *Annals of the Association of American Geographers*, **93**, 1–12.

Decaulne, A. and Saemundsson, Th. (2006). Geomorphic evidence for present-day snow-avalanche and debris-flow impact in the Icelandic Westfjords. *Geomorphology*, **80**, 80–93.

Delparte, D., Jamieson, B. and Waters, N. (2008). Statistical runout modeling of snow avalanches using GIS in Glacier National Park, Canada. *Cold Regions Science and Technology*, **54**(3), 183–192.

Fell, R. (1994). Landslide risk assessment and acceptable risk. *Canadian Geotechnical Journal*, **31**, 261–272.

Fell, R. and Hartford, D. (1997). Landslide risk management. In D. Cruden and R. Fell (eds.), *Landslide Risk Assessment.*

Proceedings of the International Workshop on Landslide Risk Assessment Honolulu, Hawaii, USA, 19–21 February 1997. Balkema: Rotterdam, pp. 51–109.

Fuchs, S. and Bründl, M. (2005). Damage potential and losses resulting from snow avalanches in settlements in the Canton of Grisons, Switzerland. *Natural Hazards*, **34**, 53–69.

Fuchs, S. and Keiler, M. (2006). Natural hazard risk depending on the variability of damage potential. In V. Popov and C. Brebbia (eds.), *Risk Analysis V: Simulation and Hazard Mitigation*, Wessex: WIT Press, pp. 13–22.

Fuchs, S., Keiler, M. and Zischg, A. (2001). *Risikoanalyse Suldental*. Innsbrucker Geographische Studien, Innsbruck.

Fuchs, S., Bründl, M. and Stötter, J. (2004). Development of avalanche risk between 1950 and 2000 in the municipality of Davos, Switzerland. *Natural Hazards and Earth System Sciences*, **4**(2), 263–275.

Fuchs, S., Heiss, K. and Hübl, J. (2007). Towards an empirical vulnerability function for use in debris flow risk assessment. *Natural Hazards and Earth System Sciences*, **7**, 495–506.

Gächter, M. and Bart, R. (2002). Risikoanalyse und Kostenwirksamkeit bei der Massnahmenplanung: Beispiel Diesbach. *Schweizerische Zeitschrift für Forstwesen*, **153**, 268–273.

Glade, T. and von Elverfeldt, K. (2005). MultiRISK: an innovative concept to model natural risks. In H. Oldrich, R. Fell, R. Coulture and E. Eberhardt (eds.), *International Conference on Landslide Risk Management*, Vancouver (CND), 31 May –3 June 2005, Rotterdam: Balkema, pp. 551–556.

Haimes, Y. Y. (2004). *Risk Modeling, Assessment, and Management*, 2nd edition. Hoboken, New Jersey: Wiley.

Hatfield, A. and Hipel, K. (2002). Risk and systems theory. *Risk Analysis*, **22**(6), 1043–1057.

Holub, M. and Fuchs, S. (2008). Benefits of local structural protection to mitigate torrent-related hazards. *WIT Transactions on Information and Communication Technologies*, **39**, 401–411.

Hufschmidt, G., Crozier, M. and Glade, T. (2005). Evolution of natural risk: research framework and perspectives. *Natural Hazards and Earth System Sciences*, **5**, 375–387.

International Organization for Standardization, ISO (2008). Draft International Standard ISO 31000. *Risk Management: Principles and Guidelines on Implementation*. http://www.broadleaf.com.au/pdfs/iso_31000/iso_iec_rm_princips.pdf (accessed 15 December 2008).

Jaecklin, A. (2007). Voll integriertes Risikomanagement. *MQ Management und Qualität* (11), 21–23. http://www.saq.ch/fileadmin/user_upload/mq/downloads/mq_2007_11_jaecklin.pdf (accessed 15 December 2008).

Jamieson, J. B. and Stethem, C. (2002). Snow avalanche hazards and management in Canada: challenges and progress. *Natural Hazards*, **26**(1), 35–53.

Jónasson, K., Sigurðsson, S. and Arnalds, Þ. (1999). *Estimation of Avalanche Risk*. Icelandic Meteorological Office, Reykjavík, Iceland, VÍ-R99001-ÚR01.

Kaplan, S. and Garrick, B. (1981). On the quantitative definition of risk. *Risk Analysis*, **1**(1), 11–27.

Keiler, M. (2004). Development of the damage potential resulting from avalanche risk in the period 1950–2000, case study Galtür. *Natural Hazards and Earth System Sciences*, **4**, 249–256.

Keiler, M., Zischg, A., Fuchs, S., Hama, M. and Stötter, J. (2005). Avalanche related damage potential: changes of persons and mobile values since the mid-twentieth century, case study Galtür. *Natural Hazards and Earth System Sciences*, **5**, 49–58.

Keiler, M., Sailer, R., Jörg, P. *et al.* (2006). Avalanche risk assessment: a multi-temporal approach, results from Galtür, Austria. *Natural Hazards and Earth System Sciences*, **6**, S. 637–651.

Kern, M., Bartelt, P. Sovilla, B. and Buser, O. (2009). Measured shear rates in large dry and wet snow avalanches. *Journal of Glaciology*, **55**, 327–338.

Keylock, C. and Barbolini, M. (2001). Snow avalanche impact pressure: vulnerability relations for use in risk assessment. *Canadian Geotechnical Journal*, **38**, 227–238.

Keylock, C., McClung, D. and Magnússon, M. (1999). Avalanche risk mapping by simulation. *Journal of Glaciology*, **45**, 303–314.

Klinke, A. and Renn, O. (2002). A new approach to risk evaluation and management: risk-based, precaution-based, and discourse-based strategies. *Risk Analysis*, **22**(6), 1071–1094.

Kraus, D., Hübl, J. and Rickenmann, D. (2006). Building vulnerability related to floods and debris flows: case studies. In W. Ammann, S. Dannenmann and L. Vulliet (eds.), *Coping with Risks Due to Natural Hazards in the 21st Century*. London: Taylor & Francis, pp. 181–190.

Kristjansdottir, G. B. (1997). *Jardfraedileg ummerki eftir snjoflod i botni Dyrafjardar*, BS Thesis, Department of Geology and Geography, University of Iceland (in Icelandic).

Kulakowski, D., Rixen, C. and Bebi, P. (2006). Changes in forest structure and in the relative importance of climatic stress as a result of suppression of avalanche disturbances. *Forest Ecology and Management*, **223**(1–3), 66–74, doi: 10.1016/j.foreco.2005.10.058.

Laternser, M. and Schneebeli, M. (2002). Temporal trend and spatial distribution of avalanche activity during the last 50 years in Switzerland. *Natural Hazards*, **27**(3), 201–230, doi: 10.1023/A:1020327312719.

Laternser, M. and Schneebeli, M. (2003). Long-term snow climate trends of the Swiss Alps (1931–99). *International Journal of Climatology*, **23**(7), 733–750, doi: 10.1002/joc.912.

Lehning, M., Löwe, H., Ryser, M. and Raderschall, N. (2008). Inhomogeneous precipitation distribution and snow transport in steep terrain. *Water Resources Research*, **44**, doi:10.1029/2007WR006545.

McClung, D. and Schaerer, P. (2006). *The Avalanche Handbook*, 3rd edition, Seattle, WA: The Mountaineers Books.

Prokop, A. (2008). Assessing the applicability of terrestrial laser scanning for spatial snow depth measurements, *Cold Regions Science and Technology*, **54**(3), 155–163.

Rheinberger, Ch., Bründl, M. and Rhyner, J. (2009). Dealing with the White Death: avalanche risk management for traffic routes. *Risk Analysis*, **29**(1), 76–94.

Rixen, C., Haag, S., Kulakowski, D. and Bebi, P. (2007). Natural avalanche disturbance shapes plant diversity and species composition in subalpine forest belt. *Journal of Vegetation Science*, **18**, 735–742.

Sailer, R., Rammer, L. and Sampl, P. (2002). Recalculation of an artificially released avalanche with SAMOS and validation with measurements from a pulsed Doppler radar. *Natural Hazards and Earth System Sciences*, **2**, 211–216.

Salm, B. (1993). Flow, flow transition and runout distances of flowing avalanches. *Annals of Glaciology*, **18**, 221–226.

Sampl, P. and Zwinger, T. (2004). Avalanche simulation with SAMOS. *Annals of Glaciology*, **38**, 393–396.

Schaffhauser, A., Adams, M., Fromm, R., *et al.* (2008). Remote sensing based retrieval of snow cover properties. *Cold Regions Science and Technology*, **54**(3), 164–175.

Schaub, Y. (2008). *Risikomanagement von Naturgefahren: Sensitivität der Risikoberechnung in Bezug auf die Eingabefaktoren und deren Bedeutung für die Massnahmenbewertung.* M.Sc. Thesis, Department of Geography, University of Zurich, Zurich.

Schneebeli, M., Laternser, M., Föhn, P. and Ammann, W. (1998). *Wechselwirkungen zwischen Klima, Lawinen und technischen Massnahmen.* Zürich: vdf Hochschulverlag.

Schweizer, J. (2008). On the predictability of snow avalanches. In C. Campbell, S. Conger and P. Haegeli (eds.), *Proceedings ISSW 2008, International Snow Science Workshop*, Whistler, Canada, 21–27 September 2008, pp. 688–692.

Schweizer, J., Jamieson, J. B. and Schneebeli, M. (2003). Snow avalanche formation. *Review of Geophysics*, **41**(4), 1016.

SLF (ed.) (2000). *Der Lawinenwinter 1999.* Davos: Eidgenössisches Institut für Schnee- und Lawinenforschung.

Sovilla, B., Burlando, P. and Bartelt, P. (2006). Field experiments and numerical modelling of mass entrainment in snow avalanches. *Journal of Geophysical Research*, **111**(F3), F03007, doi: 10.1029/2005JF000391.

Sovilla, B., Schaer, M., Kern, M. and Bartelt, P. (2007). Impact pressures and flow regimes in dense snow avalanches observed at the Vallée de la Sionne test site. *Journal of Geophysical Research*, doi:10.1029/2006JF000688.

Sovilla, B., Schaer, M. and Rammer, L. (2008). Measurements and analysis of full-scale avalanche impact pressure at the Vallée de la Sionne test site. *Cold Regions Science and Technology*, **51**(2–3), 122–137.

Stethem, C., Jamieson, B., Schaerer, P. *et al.* (2003). Snow avalanche hazard in Canada: a review. *Natural Hazards*, **28**, 487–515.

UNESCO (1981). *Avalanche Atlas: Illustrated International Avalanche Classification.* Paris, France: International Association of Hydrological Sciences, International Commission on Snow and Ice: Natural Hazards Series, Vol. 2.

Uzielli, M., Farrokh, N., Lacasse, S. and Kaynia, A. M. (2008). A conceptual framework for quantitative estimation of physical vulnerability to landslides. *Engineering Geology*, **102**(3–4), 251–256.

Varnes, D. (1984). *Landslide Hazard Zonation: A Review of Principles and Practice.* Paris: UNESCO.

Ward, R. G. V. (1985). Geomorphological evidence of avalanche activity in Scotland. *Geografiska Annaler*, **67A**, 247–256.

Weichselgartner, J. (2001). Disaster mitigation: the concept of vulnerability revisited. *Disaster Prevention and Management*, **10**(2): 85–94.

White, G., Burton, R. and Kates, I. (2001). Knowing better and loosing even more: the use of knowledge in hazards management. *Environmental Hazards*, **3**, 81–92.

Wilhelm, C. (1997). *Wirtschaftlichkeit im Lawinenschutz.* Davos: Eidgenössisches Institut für Schnee- und Lawinenforschung, Mitteilung 54.

6 Landslide hazards

David Petley

6.1 Introduction

Landslides are naturally occurring phenomena in every environment on Earth, including the tropics, the temperate regions and the high latitudes, and in the oceans. Unfortunately, this ubiquitous natural process represents a substantial hazard to humans because people and structures have a surprisingly low capacity to withstand the forces generated by mobile soil and/or rock. In consequence, there is a long recorded history of landslide disasters – for example, Nihon Shoki (the ancient chronicle of Japan), which was completed in the year AD 720, describes numerous landslides and failures associated with the Hakuho earthquake on 29 November AD 684, whilst the city of Helike in Greece is believed to have been submerged and destroyed as a result of a submarine landslide in 373 BC. Today, landslides continue to inflict a substantial economic and social toll, especially in mountainous, less developed countries, and there is a widely held but admittedly poorly quantified expert perception that the impacts associated with mass movements are increasing rapidly with time.

The term landslide is unfortunately something of a misnomer as many landslides do not in reality involve sliding. The word landslide is used to describe a range of processes that result in downward and outward movement of slope-forming material composed of rock, soil and artificial materials. In this context the term 'mass movement' might be preferable, but here the term landslide will be retained as it is in such common use in this context. In this chapter mass movements that are mostly formed from snow or ice are specifically excluded – these are discussed in Chapter 5.

Damaging landslides occur through a surprisingly wide range of magnitudes and invoke a large number of mechanisms. For example, the fall of a single piece of rock the size of a computer mouse can be enough to kill a person if it strikes them on the head at terminal velocity. On the other hand, the Seymareh landslide in the Zagros Mountains of Iran has a deposit with a volume of about $20\,km^3$, whilst some submarine landslides are now known to have a volume that may be as much as two orders of magnitude as large again. For this reason, landslide volumes are usually considered on a logarithmic scale. The wide range of landslide types is usually considered by classifying them according to the predominant material involved and the movement type (Table 6.1). A further refinement, which is important in the context of hazard causation, is to classify the landslides by movement rate as well. In general, rapid movement types are more likely to cause loss of life than are slow movements, whilst even slow rates of displacement (and low levels of total movement) can cause extensive damage to buildings and infrastructure.

Figure 6.1 presents the distribution of fatal landslides between January 2006 and December 2007 inclusive, based upon the Durham fatal landslide database (see Petley *et al.*, 2005a). Care is needed in the interpretation of such a map as the inclusion specifically of fatal landslides (rather than all landslides or all landslides that impact upon people) biases the data in particular ways. Most importantly, the data are skewed towards less developed countries where the level of mitigation against landslides might be lower and where the density of vulnerable economic assets might also be small. Nonetheless, it is clear that the recorded landslides form very distinct clusters, most notably in the following locations:

1. Along the southern edge of the Himalayan mountain chain;
2. In Central China;
3. In SW India;
4. Along the western boundary of the Philippine Sea plate through Japan, Taiwan and the Philippines;
5. In central Indonesia, particularly on the island of Java;
6. In the Caribbean and Central Mexico;

Geomorphological Hazards and Disaster Prevention, eds. Irasema Alcántara-Ayala and Andrew S. Goudie. Published by Cambridge University Press. © Cambridge University Press 2010.

TABLE 6.1. *A simplifed classification scheme for the main types of landslide movements*

Type of movement		Rock	Engineering soils	
			Coarse grained	Fine grained
Falls		Rock fall	Debris fall	Earth fall
Topples		Rock topple	Debris topple	Earth topple
Slides	Rotational	Rock slump	Debris slump	Earth slump
	Translational	Rockslide	Debris slide	Earth slide
Lateral spreads		Rock spread	Debris spread	Earth spread
Flows		Rock flow	Debris flow	Earth flow

Complex slope movements (i.e. combinations of two or more types)

After Varnes (1978)

7. On the western edge of the northern part of South America, especially in Colombia.

There is also a scattering of fatal landslides elsewhere, for example through Europe, the tropical parts of Africa and in North America. This distribution of intense fatal landslide activity reflects the juxtaposition of three key factors:

1. The occurrence of tectonic processes that in particular drive high rates of uplift and occasional seismic events;
2. The occurrence of high levels of precipitation, usually including both high annual precipitation totals and high short-term intensities;
3. The presence of a reasonably high population density.

Where one of these three factors is missing the occurrence of fatal landslides reduces markedly – thus for example Iran has a lower than expected landslide occurrence because of the absence of precipitation; Alaska has a lower than expected occurrence because of the absence of people; and Central Africa has a lower than expected occurrence because of the low levels of tectonic activity. A longer time-frame would change this picture slightly as landsliding in arid, tectonically active areas such as Iran is probably driven primarily by seismic activity with an additional input from low-frequency–high-magnitude rainfall events. Thus, if the data were collected over a sufficiently long period to capture

a number of large seismic events in Iran then the maps would take on a slightly different appearance.

6.2 Landslide causes and triggers

Based upon the geomorphic distribution, it is unsurprising that the vast majority of landslides are triggered by one or more of three key factors: precipitation, seismicity or the action of humans. However, for these processes to be able to induce the landslide there must first have been a series of other processes that have acted to prepare the slope for failure. These processes, which are usually termed 'causes' as opposed to triggers, include the following.

6.2.1 Geological causes

Geological causes are factors that make the materials that form a slope susceptible to failure. Key causes include, for example, materials that are weak or that are weathered; materials with strong joint sets, especially where they are orientated in such a way that they allow sliding to occur; and material combinations that cause water to be retained.

6.2.2 Morphological causes

The most obvious morphological cause is the slope angle. The key parameter is the angle of the slope in comparison with the strength of the material. Thus, it is not a straightforward relationship in which steeper slopes are less stable – in Norway, for example, slopes formed from unweathered gneiss are able to form cliffs that can stand vertically to elevations of many hundreds of metres along the margins of fjords. On the other hand, quick clays close to the water's edge can fail at slope angles as low as 10° when disturbed. A further key morphological factor can be the concavity or convexity of the slope, which can serve to concentrate water in key locations. Finally, in many high mountain areas the loss of glacial ice leaves slopes unsupported and thus prone to failure, whilst in coastal environments the under-cutting of cliffs can lead to reductions in stability.

6.2.3 Physical causes

A third group of causes are related to physical processes. For example, a slope might be more likely to fail if the groundwater level has been elevated by previous prolonged rainfall or by snowmelt. Similarly, the loss of tree cover may make a slope far more susceptible to shallow landslides. In California, for example, shallow landslides that sometimes transition into destructive debris flows are a particular problem in the wet season following large forest fires.

FIGURE 6.1. Map showing the locations of fatal landslides in 2006 and 2007, as recorded in the Durham University Landslide Database. Each black dot represents a single fatal landslide.

6.2.4 Human causes

The final group of factors is centred around human activities that can destabilise a slope. Examples include the removal of forestry through logging or firewood collection; over-steepening of slopes through cutting for road construction or through quarrying; and the leakage of pipes or swimming pools. For example, in Nepal the occurrence of fatal landslides has increased markedly in recent years as a result of the construction of rural roads with inadequate levels of slope stabilisation and water management. Slopes that have been destabilised by the road building then fail during intense precipitation events associated with the summer monsoon, blocking the road and causing substantial and increasing levels of loss of life (Petley *et al.*, 2007).

6.2.5 Causation vs. triggering

In most cases the final failure of a slope occurs as a result of a clear trigger. On a day-to-day basis the most common trigger is precipitation, sometimes supplemented by the effects of snowmelt. Precipitation serves to increase the pore pressures within a slope, reducing the resistance to movement. In temperate and cold environments this occurs primarily through increases in groundwater level. Thus, unless there has been a marked change in a causal factor as outlined above, landslides usually occur during precipitation that is low frequency–high magnitude, in terms of

intensity and/or duration. However, if there has been a major change to the causal factors – for example, if there has been extensive recent deforestation – then extensive landsliding can occur in non-exceptional rainfall events.

Thus, from a human perspective it is clear that landslide hazard is associated with a complex range of interrelated factors, which can be conceptualised as a 'chain of events' (Figure 6.2). For the landslide disaster to occur there must be the juxtaposition of a number of causes plus the final trigger and the element at risk. From a hazard management perspective this can be helpful as the occurrence of the damaging landslide can be prevented through any one of a number of interventions, each of which addresses either a key cause, the trigger or the elements at risk (Figure 6.2). Of course this is a somewhat simplistic way to view landslide hazard reduction, but in the real world the level of hazard can often be reduced through a number of approaches. Thus, in Nepal, for example, in areas in which good quality engineering input is utilised the slope hazard can be reduced by using one or a combination of the following approaches:

1. Selecting an alignment for the road that avoids areas of known existing instability, or areas in which materials are known to be problematic (black schist often causes problems in humid tropical and subtropical environments for example), or areas with a slope angle greater than a pre-determined value;

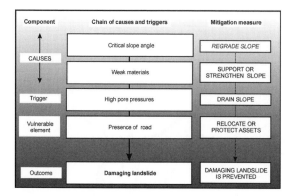

FIGURE 6.2. Conceptual diagram showing how the causes, a trigger and the existence of a vulnerable element conspire to create a slope accident. On the right, possible mitigating approaches are shown. Any one of these, applied properly, can prevent the accident from occurring.

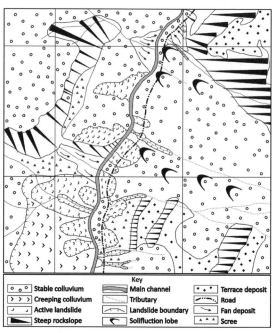

FIGURE 6.3. A terrain map used to identify the locations of existing landslides along the Arniko Highway in Nepal. The grid squares are 1 km.

2. Managing the water on the slope through the construction of effective drainage;
3. The construction of walls to support slopes that have been cut;
4. The planting of local vegetation species that can help to increase the strength of the soil or to draw down the water table. Vetiver grass (*Chrysopogon zizanioides*) is widely used in this capacity for example as it rapidly grows roots that extend to a depth of up to four metres.

Thus, key aspects of landslide hazard reduction are to identify and understand the causes; to identify and understand the triggering processes and the key thresholds at which they occur; and to try to ensure that human elements are not put at risk.

6.3 The role of geomorphology in landslide hazard management

Landslide management is a well-developed science, and where sufficient resources are available most small- to medium-sized landslides can be managed or mitigated if they have been identified and characterised properly and if sufficient resources are available. The management of landslides is a multi-disciplinary task, with key inputs from geotechnical engineers, engineering geologists, biologists, meteorologists, planners and others in addition to geomorphologists. Indeed, landslide management is rarely trusted to geomorphologists alone, instead the key role is to act as part of a multi-disciplinary team. The nature of the management of a landslide hazard depends to a large degree on the nature and timing of the movement, and whether the failure in question is a first time event or a reactivation of an

existing landslide. However, geomorphology is a key aspect of all stages of the landslide management process.

6.4 Terrain mapping

A key aspect of landslide hazard management from a geomorphic perspective is that of terrain mapping. Terrain mapping, which is increasingly a key part of the early stages of infrastructure projects, especially in mountainous terrain and in areas affected by neotectonic processes, is most commonly used to identify areas of existing or past slope instability. In many cases, this provides a key input into route selection for roads, railway lines and in particular for pipelines. For example, route selection for the Dharan–Dhankuta highway in Nepal was undertaken primarily on the basis of geomorphological mapping (Brunsden *et al.*, 1975). The value of this approach has been demonstrated by the remarkable stability of this road in comparison with similar projects for which this approach was not adopted (Hearn, 2002). Sometimes it is impossible to route the corridor away from areas with identified instability, in which case the terrain mapping is used as an input into more detailed geotechnical investigation. Terrain mapping is also often used as a first order hazard assessment technique, allowing the location of existing failures that might threaten an asset. For example, this approach was used to assess the likelihood of landslides along the access roads to Kathmandu in Nepal in the event of a large earthquake (Figure 6.3).

In recent years, the availability of high resolution digital aerial photography, satellite imagery, LiDAR data, and digital terrain models has allowed increasing levels of sophistication in terrain mapping. The development of increasingly sophisticated automated feature extraction will continue to enhance these capabilities. Despite this, terrain mapping is an under-used tool in infrastructure projects. Expensive failures, such as the reactivation of an ancient landslide that was clearly visible on aerial photographs at Ok Tedi in Papua New Guinea in 1984 (Griffiths *et al.*, 2004), show that there are substantial advantages in the undertaking of good terrain mapping. The resultant legal case, which reached the Supreme Court of Papua New Guinea in 1989, was for approximately £575 million (Griffiths *et al.*, 2004). The case was settled out of court.

The greatest weakness of terrain analysis continues to be the subjective nature of the process. Fookes and Dale (1992) examined six independent interpretations of the pre-failure condition of the Ok Tedi site described above, showing that even with highly skilled geomorphologists substantially different interpretations resulted. Similarly, when different analysts looked at identical aerial images of the site, surprisingly different interpretations resulted (Fookes *et al.*, 1991). Nonetheless, terrain mapping remains a core tool in the development of infrastructure projects.

Brunsden (2002) advocated that terrain mapping should be the basis of a much more integrated geomorphological analysis of potential instability. He encouraged a move from an essentially two-dimensional analysis of landslides – i.e. the production of landslide maps – into a four-dimensional analysis that used geomorphological tools to understand the three-dimensional structure of an unstable slope, plus its development through time. Thus, for example, modern dating techniques can be used to understand the evolution of the slope, which can then be benchmarked against the increasingly high quality climatic and in some cases seismic histories that are now available. In consequence the relationship between movement and causal/triggering factors can be elucidated, which gives an improved ability to understand future behaviour. This is becoming increasingly important in the context of climate change, which means that existing magnitude/frequency catalogues for triggering events may not be relevant.

6.5 Susceptibility analysis

The inherent subjectivity of terrain mapping has led to a multitude of attempts to develop more reliable, quantitative/semi-quantitative landslide hazard mapping

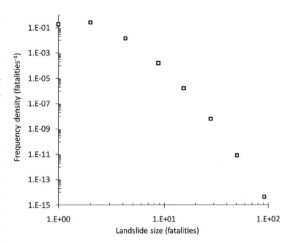

FIGURE 6.4. Landslide frequency (here represented by a frequency density function) plotted against magnitude (here represented by number of fatalities per event) for Nepal. (After Petley *et al.* 2007.)

techniques. The development of GIS has undoubtedly facilitated this approach as it hypothetically at least allows a consistent, high resolution approach to be applied to large areas. In all cases a simple algorithm is used to derive a score that indicates the susceptibility of a slope to failure. The simplest algorithms are based upon a small number of parameters (for example slope angle and material), and the parameters themselves are arranged in classes (e.g. slope angles of 0° to 5°, 5° to 15°, etc.). Summation is commonly used to derive an overall score that indicates the susceptibility to failure. In the most basic applications the score is determined on a slope by slope basis using a proforma that is completed by an operator. However, GIS-based approaches utilise algorithms that combine data from thematic layers. More complex systems utilise multiple input parameters and algorithms that are, for example, based upon empirical models of slope behaviour, such as the infinite slope equation and soil slope hydrology equations. In recent years, as processing power has increased, the use of probabilistic approaches, often based upon Monte Carlo simulations, has become popular.

Geomorphology plays a key role in all aspects of the development of susceptibility analyses. For example, geomorphologists have been at the cutting edge of the development of both the classification-based approach and the more quantitative analyses, and in the interpretation of their outcomes. Globally, the most commonly used approach is the CHASM model developed by Malcolm Anderson and colleagues at Bristol (e.g. Brooks *et al.*, 2004) and subsequently applied in many different parts of the world. Furthermore, geomorphologists often implement the model runs and analyse the outcomes. However, serious questions about the reliability of this approach given the simplifications that are inherent in the construction of any algorithm, and in the input data, remain. The greatest concern with such analyses lies in the ability to verify or refute the outcomes. Thus, there is still much to do to improve the approach.

6.6 Hazard and stability analyses

In some cases, susceptibility models are developed to full hazard and even risk models. A hazard model requires that the frequency (i.e. probability) of occurrence is quantified and that the magnitude is also determined. In the case of landslides, magnitude generally refers both to the size of the landslide (i.e. the volume and surface area) and the rate of movement, as both are critical to the actual impact of the event. The development of magnitude–frequency relationships for landslides remains a fascinating area of research, not least because of the surprising similarities in these relationships between different areas and as a result of different triggers (see Malamud *et al.*, 2004, for example). The cause of this consistent pattern of self-organisation, which is applicable to landslide losses as well as the landslides themselves (Figure 6.4) remains unclear, but represents an interesting potential input into hazard evaluation.

A key aspect of the evaluation of the frequency of landslides is the reconstruction of landslide chronologies using archival data. The principle is that our conventional scientific records are too short to provide a proper representation of the occurrence of landslides, and that the occurrence of landslides under environments that are different to the present can only be properly evaluated using long-term records. A range of approaches have been developed by the geomorphology community to allow this, and these continue to improve with time as, in particular, dating methods become better constrained. The use of long-term archive records of movement extracted from non-scientific reports is a key approach – for example, the relationship between movement and rainfall patterns for the Ventnor landslide on the Isle of Wight has been constrained using such a technique (Ibsen and Brunsden, 1996). Such datasets are limited by human records, which in a geomorphological sense are short. Extension of these records has been achieved using various dating techniques. For example, Borgatti and Soldati (2002) used carbon dating of wood entrained in landslide debris to establish the temporal occurrence of landslides in the Alps that could then be related to palaeoclimate reconstructions for the Holocene, establishing the link between climate and landslides. Whilst being very powerful, such techniques are expensive and time-consuming as many dates are required. Furthermore, uncertainties remain as landslides tend to be triggered by weather events whereas palaeoclimate data provide an indication of climate. Similar studies using other dating

FIGURE 6.5. The town of Taihape in North Island, New Zealand, which is built upon a slow moving earthflow. Monitoring is used to ensure that the town remains secure.

techniques have been used to establish landslide chronologies, including lichenometry and dendrochronology (see Lang, 1999, for a detailed review). In the last few years the availability of cosmogenic isotope dating, which allows the time of unroofing of rock surfaces to be established, has allowed the dating of large rock avalanches and rockfalls (Mitchell *et al.*, 2007). There is increasing interest in the use of this technique for the determination of the chronology of large earthquakes using rock avalanches as the indicator of the seismic event (Korup *et al.*, 2004).

Estimating the likely volume or surface area of a potential failure remains very troublesome, especially if runout distance is to be included. Rates of movement are generally analysed using numerical models. This is easier for a single landslide in a site that is well characterised, but is very difficult where hazard is being assessed for a larger area. Even in the case of a well-constrained specific site, reliable modelling of known failures still requires that the model is tuned with specific parameters that are not physically representative. Thus, predictive modelling has a poor level of reliability.

6.7 Monitoring, behaviour prediction and warning systems

In many parts of the world, people live on or are affected by landslides that cannot be easily mitigated. Examples include settlements that are located on slow moving or inactive landslides (the towns of Ventnor in the UK and Taihape in New Zealand (Figure 6.5) are examples where this is the case); transportation routes that cross landslides (for example the Ashcroft landslides in Canada), and landslides that threaten other economic assets, such as reservoir bank failures that imperil dams through displacement waves. Frequently the landslides are either too large or too numerous to be effectively mitigated and the assets cannot be relocated. In this case an increasingly common approach to the management of the slope hazards is the use of monitoring systems, some of which are used to provide warnings.

Warning systems over large areas tend to be based upon measurement and analysis of the thresholds at which potential triggers start to induce landslide movement. Considerable work has been undertaken in particular on the development of thresholds of rainfall for landslide activation. Generally the approach used is to examine rainfall events that are known to have caused landslides and to compare them with events in which no landslides have occurred. Usually, the best relationships are found by looking at a combination of medium-term precipitation (perhaps rainfall total over the previous month) and short-term rainfall (over the last few hours). Thus, the threshold for movement in terms of short-term rainfall is usually lower if the previous few days have been wet than if they have been dry. Such approaches are the basis of warning systems in a number of places – for example Japan and Hong Kong both operate large-scale systems for landslide warning. However, implementation of such systems is rather complex as rainfall is very spatially variable. In Hong Kong, which has a surface area of just $1,092 \, km^2$, the warning system is based upon 110 rain gauges plus the use of a Doppler rain radar system. Even then, the system is operated with caveats, in particular that unexpected intense rainfall, perhaps associated with a rapidly developing convective system, can induce landslides before a warning can be issued. Nonetheless, the development of appropriate rainfall thresholds has been and continues to be a key task for geomorphologists. In Malaysia for example, which has a surface area of about $330,000 \, km^2$, there are plans to ultimately create a nationwide slope warning system based upon rainfall thresholds, although this will take years to implement and will be expensive to maintain. Similar warning systems for earthquake-induced slope hazards are in their infancy, but the new Taiwan High Speed Rail Line has earthquake acceleration sensors mounted along the length of the track that automatically stop the trains if ground accelerations exceed 40 gals, partly to ensure that trains do not hit failures induced in the earthworks alongside the track. Given the low shaking threshold, it can be hoped that in the event of a large earthquake the sensors would initiate the stopping sequence before the main earthquake waves arrive at the track, and thus before any embankment failures, occurred. However, such systems have rarely been tested by large earthquakes, so we wait to see their effectiveness.

Warning systems on individual slopes generally take a quite different approach, in this case being based upon the detection of landslide movement. Three approaches are generally adopted:

1. Sensors, often based upon vibration or using echo sounders to detect changes in sediment volume, are placed in the path along which a landslide is expected to move, allowing a warning signal to be issued. Such warning systems generally provide a warning over only a very short period (seconds to minutes), allowing emergency evacuations along pre-determined routes or the closure of key transportation routes such as roads. Geomorphology plays a key role in all aspects of the development of such systems, including the identification of hazardous slopes that require warning systems; the determination of the optimum monitoring locations and sensor type; the selection of appropriate thresholds; and the determination of safe escape routes and zones. Such warning systems have proven to be successful in

many locations – for example, the town of Funes in the Dolomite mountains of northern Italy has been protected for over a decade against catastrophic debris flows using such a system (Petley *et al.*, 2005b).

2. An alternative approach is to measure the conditions in the slope that are considered likely to trigger failure. In most cases this is based upon a calculation of the stability of the slope using a static equilibrium (i.e. factor of safety) calculation. This allows the critical groundwater level (or pressure condition) at which failure is considered likely to be determined. This is often backed up with records of movement, analysed in conjunction with measurements or models of the groundwater conditions at the time. Monitoring of the groundwater level using piezometers then allows a warning to be made when the groundwater approaches this critical depth.

3. The final, increasingly popular, approach is to detect the early (precursory) stages of landslide movement, generally using movement sensors. The rationale is that catastrophic failure is usually presaged by a period of accelerating movements of the slope. Thus, once a predetermined movement rate is reached, or when a specific pattern of acceleration is observed, a warning is sounded. Early systems focussed upon the use of inclinometers located in boreholes, which measure the movement between the base of the landslide and the underlying bed, or extensometers that measure movement across the back scar at the landslide head. Increasingly, however, technologically based approaches are being adopted, typically using robotised theodolites in conjunction with reflective targets located on the landslide body (Petley *et al.*, 2005b) or high resolution differential GPS receivers located on the landslide mass (Tagliavini *et al.*, 2007). These approaches are proving capable of providing warnings, and also have the advantage of generating detailed datasets on the movement patterns of the landslide that allows a better understanding of its dynamics. A series of new experimental approaches are also being developed, most notably slope radar, which is increasingly being used to provide warnings on large open-cast pit slopes, and satellite-based radar interferometry (InSAR) (Catani *et al.*, 2005). The latter shows some potential but remains problematic, not least because displacement–time plots generally have an unrealistic linear trend, which suggests that crucial elements of movement are not being resolved.

A key challenge for the geomorphological community is the development of enhanced understanding of warning thresholds and the development of new techniques for the analysis of movement patterns. In particular, we still understand precursory movement patterns very poorly. An even greater challenge is that of behaviour prediction – i.e. can we determine the likely future movement of a landslide? Generally such problems are associated with large mass, slow moving movements that seem to have the potential to move rapidly. The slow movement implies a factor of safety very close to unity, so increased movement often seems very possible. An example is the Ventnor landslide on the Isle of Wight in southern England. This landslide has built upon it a town with a permanent population of 7,000 people and economic assets with a replacement cost of over £150 million. The landslide moves continually at a rate of a few millimetres to centimetres per annum, and the geomorphology evolves as a result. Opinion remains divided as to whether the slope is likely to evolve into a large-scale, rapid failure and if so, under what circumstances. However, Carey *et al.* (2007) used a combination of geomorphological analysis, interpretation of movement and piezometer data, and novel laboratory testing to examine the mechanisms of movement and thus to forecast likely behaviour, suggesting that a very rapid failure event is unlikely. Such approaches will increasingly represent a frontier of landslide research that builds upon the availability of good-quality, real-time movement data and, increasingly, models that reliably represent the full range of processes that occur within a slope.

6.8 Secondary hazards and sediment production

Landslides are frequently considered to be secondary hazards associated with a primary event such as a typhoon or an earthquake. However, it is increasingly clear that landslides generate their own set of comparatively poorly understood secondary hazards, most notably dam-break floods and tsunamis. The former was strongly demonstrated by the 2008 Wenchuan earthquake in Sichuan province in China, which induced a very large number of landslides that caused catastrophic damage – for example, the town of Beichuan was almost completely destroyed by two rock slope large failures. In the aftermath of the earthquake, however, there was great concern associated with the existence of 44 valley-blocking landslides, each of which had the capacity to cause a substantial and very destructive dam-break flood. The mitigation of such sites requires a high level of input from geomorphologists, most notably through:

1. The identification of sites in which there is high potential for the occurrence of valley-blocking landslides, most notably through terrain analysis (see above);
2. The analysis of the dynamics of a potential dam-break flood. Two key approaches are used: first, using

statistical relationships derived from previous events; and second, through flood modelling. In both cases information on the dimensional and material properties of the dam is critical;

3. The identification of safe locations for populations downstream that need to be relocated;
4. The design of safe mitigation measures, most notably the construction of a spillway to safely drain the lake.

In the case of Wenchuan, the authorities successfully drained all of the dangerous landslide lakes without the reported loss of a single life. However, the experience has caused many other earthquake-prone countries to reflect upon their own capabilities in this area. It is likely that there will need to be considerable investment in this field to allow a repeat of the Chinese achievements elsewhere.

The further substantial secondary hazard is that of the generation of a tsunami from a terrestrial or a submarine landslide. Here the threat is very real, as a number of well-documented cases demonstrate (Bardet *et al.*, 2003). Unfortunately, the discussion of these hazards is sometimes sidetracked by over-blown descriptions of extreme landslide events for which there is no real physical evidence, in which single landslides are postulated as having the potential to generate tsunamis that could devastate whole ocean basins. Although such scenarios have little or no scientific credibility, the potential for serious localised impacts of large landslides into water bodies is well established. For example, in 1997 a tsunami associated with a $Mw = 7.0$ earthquake in Papua New Guinea struck the Sissano Lagoon, killing over 1,000 people. It is now clear that the source of this tsunami was a submarine slump triggered by the earthquake (Lynett *et al.*, 2003). Currently, our understanding of the likelihood of such tsunamis is comparatively poor, not least because they are high magnitude but low frequency events. Considerable work is now being undertaken both to model the occurrence of such events and to map and date deposits left by them.

The 1999 Chi-Chi earthquake in Taiwan probably represents the most intensely studied landslide event on record (e.g. Chang *et al.*, 2007). One aspect of these landslides that has been particularly interesting is the emphasis placed on understanding the ways in which the mass failures contribute to the erosional mass balance of tectonically active mountain chains (Lin *et al.*, 2006). There have been two key components of this work. First, considerable effort has been expended in trying to understand how patterns of landsliding evolve in the aftermath of large earthquakes. In Taiwan it has been clear in a number of catchments that whilst the number of landslides associated with the initial earthquake is high, the number increases dramatically in the first extreme rainfall

event after the temblor. In many catchments the number of landslides more than doubled. This of course has profound implications for hazard management in the aftermath of the earthquake – in Taiwan the Central Cross Island Highway, which is the main arterial route across the Central Mountain chain, has been repeatedly destroyed by landslides in the aftermath of the earthquake, at considerable cost. Nine years after the earthquake the occurrence of landslides is still well above the background level. Similar effects have been noted elsewhere (Keefer, 1994), but considerable further work is required to understand this process properly.

A linked issue is that of sediment mobility. The landslides triggered by large events can release huge quantities of sediment into the fluvial system. At the same time however large valley-blocking landslides can cause sediment to be deposited and stored within the channel. Understanding the interaction between these two processes, and their relationship to periods that are not affected by recent large events, is a key aspect of work in geomorphology. In the case of Taiwan, sediment transport increased dramatically in the aftermath of the earthquake, especially during large flood events. The rivers responded by aggrading – in places the river bed level has increased by as much as 30 m. Dadson *et al.* (2003) demonstrated that across Taiwan as a whole there is a correlation between seismic moment (i.e. earthquake energy release) and sediment production, presumably as a result of the occurrence of landslides. These variations in sediment release and transportation associated with earthquakes have important implications for the understanding of the evolution of hazards in affected areas – in the Tachia River Valley in Taiwan the landslides associated with the Chi-Chi earthquake and the subsequent sediment disasters are estimated to have cost a total of about US$968 million (Table 6.2).

TABLE 6.2. *Estimated costs of landslide and sediment induced damage in the aftermath of the 1999 Chi-Chi earthquake*

Item	Cost (US dollars)
Repairs to hydroelectric power infrastructure	$300 million
Additional power generation costs	$320 million
Initial road repairs	$53 million
Additional transportation costs for agriculture	$95 million
Estimated costs to rebuild Central Cross Island Highway	$200 million
Total	$968 million

6.9 Conclusions

Landsliding is a natural geomorphological process that acts primarily to balance uplift. Human activities exacerbate this situation considerably, increasing the spatial density and temporal frequency of failures. As such, landslides represent a hazard in all inhabited areas with slopes. Geomorphologists play a key role in the management of these hazards. Indeed, in recent years many UK-based engineering consultants have formed geomorphology units specifically to make use of the expertise that geomorphologists can bring to infrastructure projects in potentially unstable areas. However, many challenges remain, not least to gain a better understanding of the mechanisms of landslides in mountainous, tropical environments and to find effective ways to manage landslides in less developed countries.

References

Bardet, J.-P., Synolakis, C. E., Davies, H. L., Imamura, F. and Okal, E. A. (2003). Landslide tsunamis: recent findings and research directions. *Pure and Applied Geophysics*, **160**, 1793–1809.

Borgatti L. and Soldati M. (2002). The influence of Holocene climatic changes on landslide occurrence in Europe. In J. Rybar, J. Stemberk and P. Wagner (eds.), *Landslides*. Rotterdam: Balkema, pp. 111–116.

Brooks, S. M., Crozier, M. J., Glade, T. W. and Anderson, M. G. (2004). Towards establishing climatic thresholds for slope instability: use of a physically-based combined soil hydrology-slope stability model. *Pure and Applied Geophysics*, **161**, 881–905.

Brunsden, D. (2002). Geomorphological roulette for engineers and planners: some insights into an old game. *Quarterly Journal of Engineering Geology and Hydrogeology*, **35**, 101–142.

Brunsden, D., Doornkamp, J. C., Fookes, P. G., Jones, D. K. C. and Kelly, J. M. H. (1975). Large scale geomorphological mapping and highway engineering design. *Quarterly Journal of Engineering Geology*, **8**, 227–225.

Carey, J. M., Moore, R., Petley, D. N. and Siddle, H. J. (2007). Pre-failure behaviour of slope materials and their significance in the progressive failure of landslides. In R. McInnes, J. Jakeways, H. Fairbank and E. Mathie (eds.), *Landslides and Climate Change: Challenges and Solutions*. London: Taylor and Francis, pp. 207–215.

Catani, F., Farina. P., Moretti, S., Nico, G. and Strozzi, T. (2005). On the application of SAR interferometry to geomorphological studies: estimation of landform attributes and mass movements. *Geomorphology*, **66**, 119–131.

Chang, K. T., Chiang, S. H. and Hsu, M. L. (2007). Modeling typhoon- and earthquake-induced landslides in a mountainous watershed using logistic regression. *Geomorphology*, **89**, 335–347.

Dadson, S., Hovius, N., Chen, H., *et al.* (2003). Links between erosion, runoff variability and seismicity in the Taiwan orogen. *Nature*, **426**, 648–651.

Fookes, P. G. and Dale, S. G. (1992). Comparison of interpretations of a major landslide at an earthfill dam site in Papua New Guinea. *Quarterly Journal of Engineering Geology and Hydrogeology*, **25**, 313–330.

Fookes, P. G., Dale, S. G and Land, J. M. (1991). Some observations on a comparative aerial-photography interpretation of a landslipped area. *Quarterly Journal of Engineering Geology and Hydrogeology*, **24**, 249–265.

Griffiths, J. S., Hutchinson, J. N., Brunsden, D., Petley, D. J. and Fookes, P. G. (2004). The reactivation of a landslide during the construction of the Ok Ma tailings dam, Papua New Guinea. *Quarterly Journal of Engineering Geology and Hydrogeology*, **37**, 173–186.

Hearn, G. J. (2002). Engineering geomorphology for road design in unstable mountainous areas: lessons learnt after 25 years in Nepal. *Quarterly Journal of Engineering Geology and Hydrogeology*, **35**, 143–154.

Ibsen, M-L. and Brunsden, D. (1996). The nature, use and problems of historical archives for the temporal occurrence of landslides, with specific reference to the south coast of Britain, Ventnor, Isle of Wight. *Geomorphology*, **15**, 241–258.

Keefer, D. K. (1994). The importance of earthquake-induced landslides to long-term slope erosion and slope-failure hazards in seismically active regions. *Geomorphology*, **10**, 265–284.

Korup, O., McSaveney, M. J. and Davies, T. R. H. (2004). Sediment generation and delivery from large historic landslides in the Southern Alps, New Zealand. *Geomorphology*, **61**, 189–207.

Lang, A. (1999). Classic and new dating methods for assessing the temporal occurrence of mass movements. *Geomorphology*, **30**, 33–52.

Lin, J-C, Petley, D. N., Jen, C-H. and Hsu, M-L. (2006). Slope movements in a dynamic environment: A case study of Tachia river, Central Taiwan. *Quaternary International*, **147**, 103–112.

Lynett, P. J., Borrero J. C., Liu, P. L.-F. and Synolakis, C. E. (2003). Field survey and numerical simulations: a review of the 1998 Papua New Guinea earthquake and tsunami. *Pure and Applied Geophysics*, **160**, 2119–2146.

Malamud, B. D., Turcotte, D. L., Guzzetti, F. and Reichenbach, P. (2004). Landslide inventories and their statistical properties. *Earth Surface Processes and Landforms*, **29**, 687–711.

Mitchell, W. A., McSaveney, M., Zondervan, A. *et al.* (2007). The Keylong Serai rock avalanche, NW Indian Himalaya: geomorphology and palaeoseismic implications. *Landslides*, **4**, 245–254.

Petley, D. N., Dunning, S. A. and Rosser, N. J. (2005a). The analysis of global landslide risk through the creation of a database of worldwide landslide fatalities. In O. Hungr, R. Fell, R. Couture and E. Eberhardt, (eds.), *Landslide Risk Management*, Amsterdam: A. T. Balkema, pp. 367–374.

Petley, D. N., Mantovani, F., Bulmer, M. H. K. and Zannoni, F. (2005b). The interpretation of landslide monitoring data for movement forecasting. *Geomorphology*, **66**, 133–147.

Petley, D. N., Hearn, G. J., Hart, A. *et al.* (2007). Trends in landslide occurrence in Nepal. *Natural Hazards*, **43**, 23–44.

Tagliavini, F., Mantovani, M., Marcato, G., Pasuto, A. and Silvano, S. (2007). Validation of landslide hazard assessment by means of GPS monitoring technique: a case study in the Dolomites (Eastern Alps, Italy). *Natural Hazards and Earth System Sciences*, **7**, 185–193.

Varnes D. J. (1978). Slope movement types and processes. In R. L. Schuster and R. J. Krizek (eds.), *Landslides, Analysis and Control*. Transportation Research Board Sp. Rep. No. 176, National Academy of Sciences, pp. 11–33.

7 Catastrophic landslides and sedimentary budgets

Monique Fort, Etienne Cossart and Gilles Arnaud-Fassetta

Landslides are a dominant geomorphic process affecting mountain slopes worldwide (see also Chapters 6 and 8). They represent a major sediment source that can supply a large amount of unstable debris to river channels and may affect the fluvial sediment yield. A distinctive part of the geomorphic evolution of active mountain belts, catastrophic landslides generally develop very rapidly so that they are among the most powerful natural hazards on Earth. By temporarily or persistently impounding river channels, they delay or block the delivery of sediments, affect the (dis)continuity of the cachment-scale sediment cascade, and exert a control over the fluvial valley systems (Hewitt, 2002; Korup *et al.*, 2004). As a consequence, landslides may generate indirect hazards along fluvial systems, so that they represent a major threat to settlements, infrastructures and catchment management that may be felt over long distances from the unstable area (Plafker and Eriksen, 1978; Li *et al.*, 1986; Costa, 1991; Korup, 2005; Hewitt *et al.*, 2008). Their magnitudes, together with their causative factors, suggest the difficulty, if not the impossibility, of preventing and/or to predicting them.

In this brief overview, we shall firstly define catastrophic landslides and their geomorphic impacts, with special attention given to landslide-induced dams; we shall move on to their influence on sediment budgets, at local and basin-wide scales, before eventually considering a few actions that may help to minimize the vulnerability and risks for the potentially affected populations.

7.1 Catastrophic landslides: definition, modes of emplacement and geomorphic significance

7.1.1 Definition

The term landslide covers a large array of features. Catastrophic landslides (also referred to as super-large or giant landslides in the literature) are considered here as low frequency, massive rock slope failures characterized by their magnitude, the rapidity of their emplacement and by their spatial and temporal impacts on geomorphic systems. More specifically, the magnitude of catastrophic landslides refers either to the depth of the sliding place, to the scar area or the total disturbed area, including the deposition zone, or to their volume (Table 7.1). The displaced material generally covers at least five orders of magnitude between 10^5 and 10^{10} m^3 (Evans *et al.*, 2006; Hewitt *et al.*, 2008).

Determination of landslide volumes can be assessed directly from field data. Alternatively, at a basin-wide scale, digital elevation models (DEMs) are useful tools but they require good positional accuracy (Korup, 2005). Aerial photo measurements (based mostly on the total affected area) allow the production of an exhaustive inventory, and time series analysis, yet their use necessitates photogrammetric modelling and ground truthing; also, the fast recovery by the vegetation in some (tropical) mountains may lead to an underestimate of the final area/volume (Brardinoni and Church, 2004; Koi *et al.*, 2008) and introduce biases in magnitude/frequency estimation. Despite this limitation, a number of analyses of landslide magnitude/frequency relations have shown they are scale invariant and they obey a power law of the general form $N(A) \approx A^{-b}$, where A is landslide area, $N(A)$ the number of events of greater than a given volume, and b is a constant (Hovius *et al.*, 2007). This equation is used to quantify the distribution of landslides in space and time, and hence their density and recurrence.

7.1.2 Occurrence and modes of emplacement of catastrophic landslides

Landslide occurrence depends on various controls: climate, slope steepness, relief amplitude, bedrock geology, failure plane orientation, vegetation and landuse cover, etc.

Geomorphological Hazards and Disaster Prevention, eds. Irasema Alcántara-Ayala and Andrew S. Goudie. Published by Cambridge University Press. © Cambridge University Press 2010.

TABLE 7.1 *Selected catastrophic landslides of the twentieth century and their impacts*

Year	Location	Landslide volume (m³)	Trigger	Impacts	Comments
1911	Usoi rockslide (Tadjikistan)	2.2×10^9	M 7.4 Pamir earthquake	54 killed; Lake Sarez, the largest natural landslide dam in the world (>60km long, >500m depth, volume 17×10^9 m³) across the Murgab River	Concerns about a possible failure of the dam, that would affect >5 million people along the Bartang–Pyanj–Amu Darya watershed
1933	Diexi, Sichuan Province (China)	150×10^6	M 7.5 earthquake	>2,400 killed; dam of the Min River for 45 days (lake volume: 400×10^6 m³)	Draining of the lake by overspilling; the flood propagated >250km downstream (average velocity 5.5–7m/s)
1959	Malpasset, Fréjus (France)	30×10^6	Heavy precipitation	423 killed; railway damaged along 2.5km; 50 farms destroyed	Inadequate artificial dam, built against weaken bedrock (foliated and tectonized schists + non-identified faults)
1963	Vaiont rockslide (Italy)	270×10^6	Precipitation + high water level followed by rapid lake drawdown	2,000 killed; rockslide failed in the 150×10^6 m³ Vaiont reservoir, resulting in catastrophic flood (city of Langarone badly damaged, together with 4 other villages)	A 100m high wave of water overtopped the doubly curved arch of hydroelectric power dam; possible reactivation of a relict landslide; brittle failure of clays in depth
1967	Tanggudong debris slide, Sichuan (China)	68×10^6		Dam of the Yalong River; lake volume 600×10^6 m³	Catastrophic failure of the dam
1970	Nevado Huascaran rock debris/avalanche (Peru)	30–50×10^6	M 7.7 earthquake	>18,000 killed; city of Yungay destroyed	Average velocity 280km/hr; same peak already affected by the same type of failure in 1962 (4,000–5,000 killed)
1980	Mount St Helens debris avalanche (USA)	2.8×10^9	Volcanic eruption	5–10 killed, most people evacuated; destruction of infrastructure; Spirit Lake (258×10^9 m³) dammed	Rotational rock slide, followed by 23km long debris avalanche; average velocity 125km/hr; dam artificially stabilized by an outlet tunnel
1983	Thistle debris slide (USA)	21×10^6	Snowmelt and heavy rain	Destruction of infrastructure; dam of the Spanish Fork River; lake volume 78×10^6 m³	Heavy economic losses ($US 600 million); lake permanently drained by human intervention
1983	Sale Mountain landslide, Gansu Province (China)	30×10^6	No obvious trigger	237 killed, 3 villages destroyed; toe of the displaced mass across the 800m wide valley of the Baxie river	Mudstone and loess material; composite landslide, with progressive failure and increasing velocity up to 20 m/s
1984	Mount Ontake (Japan)	36×10^6	M 6.8 earthquake	Travel distance of 13km down to adjacent valleys	Failure of the southeastern flank of Mt Ontake volcano, transformed into a lahar (estimated volume 56×10^6 m³); velocity 22–36m/s

Year	Event	Volume (m³)	Trigger	Description	Details
1985	Bairaman debris avalanche (Papua New Guinea)	180×10^6	M 7.1 earthquake	210 m high landslide dam; lake (volume 50×10^6 m³) artificially drained; villagers evacuated	Debris flow volume 120×10^6 m³; average velocity 20 km/hr; affected 39 km Bairaman River down to Salomon Sea
1985	Catastrophic lahar, Nevado del Ruiz (Columbia)	90×10^6	Snowmelt lahar triggered by volcanic eruption	>23,000 killed (buried city of Armero, 40 km distant); small eruption (<5×10^6 m³ magma ejected), but large amount of meltwater (38–44×10^6 m³) released in 20–90 min	Lahar velocity in the Lagunillas valley, 50 to 80 km/hr; peak flow velocities 5–15 m/s; lahar reached 104 km in 4 hours; economic losses ($US 1 billion); former destructive lahar in 1845, with same areas affected
1987	Val Pola landslide (Italy)	40×10^6	Cumulative and unusually severe rainfall	27 killed and destruction of many buildings ($US 400 million); damming of the Adda River valley	The rock avalanche produced a wave of muddy water 2.7 km upstream; velocity 248–310 km/h; spillway tunnel constructed
1991	Randa rockfall (Switzerland)	20×10^6	Melting of permafrost?	Dam of the Vispa River, flooding of the Randa village	Continuous rockfall lasting several hours, with exponential acceleration before final movement
1993	La Josephina landslide (Ecuador)	20–44×10^6	Rainfall	Dam of the Paute River; lake volume 177×10^6 m³; drained out catastrophically; 71 killed, most people evacuated before lake breakout	Flood propagated >60 km downstream; peak discharge: 9,000–14,000 m³/s; 13×10^9 m³ solid load; velocity 5–20 m/s; heavy economic losses ($US 147 million)
1999	Malpa rockfall and debris flow (India)	1×10^6	High intensity rainfall	221 killed; Malpa River dammed, followed by an outburst flood	Former landslides at the same site; proximity of the Main Himalayan Thrust Fault
1999	Mt Adams, South Westland (New Zealand)	10–15×10^6	No obvious trigger	Dam formed in the lower Poerua River gorges; massive fanhead aggradation at the mountain range front	Outburst flood within less than one week (peak discharge 1,000–3,000 m³/s); persistent postfailure aggradation
2000	Yigong debris slide and debris flow, Tibet Province (China)	300×10^6	Excessive rainfall and snow-melt waters	Landslide dam of the Yigong River; natural breach caused catastrophic flood (peak discharge 120,000 m³/s); loss of property and infrastructure downstream to India	Failure very similar to the 1900 Yigong, large (5.1×10^8) landslide and resulting dam and lake
2007	Guinsaugon rockslide–debris avalanche, Leyte Island, Philippines	15×10^9	No specific trigger but heavy rainfall in the preceding days	Disintegration of slidden mass; average thickness of debris 10 m; runout distance 3,800 m; >1,000 casualties when Guinsaugon village, located on flat valley floor, was buried under 4–7 m deep debris.	Failure of 450 m high, forested rock slope located within active Philippine fault zone (sheared and brecciated rocks); runout distance enhanced by friction reduction when debris ran over flooded paddy fields; estimated and simulated mean velocity 27–38 m/s and 35 m/s respectively.

(Cruden and Varnes, 1996; Dikau *et al.*, 1996). The world-wide distribution of catastrophic landslides shows a good correlation with mean local relief greater than 1000 m (which makes up over 5% of Earth's land surface; Korup *et al.*, 2007), with seismo-tectonically active mountains (e.g. Central and High Asia, New Zealand Alps, Coastal Ranges) or volcanic belts (Pacific Rim), and with recently deglaciated mountain slopes (Evans and Clague, 1994).

A large number of catastrophic rock failures consist of rock avalanches (or sturzstroms; Hsü 1975) and/or rock-slides, even if other processes such as translational rock-slides, rotational slumps or spreads cannot be excluded. The mode of emplacement of rock avalanches can be very complex: it generally involves an initial failure from steep mountain walls of cohesive rock mass that descends several hundreds or thousands of metres and is disintegrated and crushed, a process that gives great momentum to the resulting debris (Hewitt, 2002). The nature and extent of the resulting deposits depend on the geological setting and valley morphology (Costa and Schuster, 1988). In topographically confined settings, the deposits may pile up on and run up the opposing slope, or split into separate lobes both upstream and downstream of the impacted slope, all situations that favour the blockage of the valley and the formation of a lake. Alternatively, wherever there is no spatial confinement, the crushed rocks may travel long distances and evolve, in the presence of water (or snow, or ice), into giant debris flows (Plafker and Eriksen, 1978; Fort, 1987; Shang *et al.*, 2003). These failures may also incorporate a variety of earth materials entrained in their path, and the runout process may affect, or be affected by, the type of substrate (Hewitt, 2002; Schneider *et al.*, 2004).

Earthquakes and/or high intensity precipitation are the most efficient triggering factors for rapid collapse. The failure site may be controlled by major tectonic lines (Figure 7.1; Fort, 2000), or be considered as a direct adjustment to high rates of rock uplift and correlated river incision (Burbank *et al.*, 1996). The failure may be preceded by a phase during which slow, deep-seated deformation leads to the opening of tension cracks, weakening of the shear strength of the rock mass, and eventually to high pore-water pressures. Other unusual mechanisms such as acoustic fluidization during sliding, or internal, self-accelerating rock fracture (Kilburn and Petley, 2003) may also favour large-scale rock slope failures (Hewitt *et al.*, 2008). Additionally, anthropogenic actions may be the cause of large hillslope destabilization (e.g. Vaiont reservoir; Table 7.1).

These very large rock mass failures are generally considered as a major denudational process of active orogens, and as formative events influencing landscape development (Brunsden and Jones, 1984; Fort, 1988, 2000; Hewitt,

1988, 1998, 2002; Fort and Peulvast, 1995; Burbank *et al.*, 1996; Korup *et al.*, 2004, 2007). More specifically, they give rise to complex sedimentary assemblages and to specific constructional and erosional landforms that create 'interrupted valley landsystems' distinctive of these large-scale landslides (Hewitt, 2006).

7.2 Geomorphic impacts of catastrophic landslides

Catastrophic landslides have direct geomorphic impacts at the local, basin-wide and mountain-belt scales, as synthesized by Costa and Schuster (1988), Hewitt (2002, 2006) and Korup (2005). We want to stress here local and regional impacts, which are the most meaningful in terms of hazards and risks for the population. Geomorphic impacts and their consequences on water and sediment fluxes and budgets will be appreciated in considering different interactions between landslides and fluvial systems.

7.2.1 Interactions with river systems

Three different situations are illustrated: partial blockage of the valley by the landslide, complete damming and upstream water ponding, and catastrophic collapse of the landslide dam (Figure 7.2).

In the case of partial blockage (Figure 7.2, case 1), the landslide mass forces the river to divert its course to the opposite bank; this in turn leads to a change in transverse and longitudinal channel geometry, channel pattern and morphology. Wherever the opposite bank consists of soft material (slope or alluvial deposits), this diversion triggers bank erosion and further destabilizes the entire hillslope (Figure 7.2B). Upstream of the landslide mass, which acts as the local base level, braiding of the river and aggradation predominate. These may indirectly increase overflow and channel avulsion-shifting frequencies, and flood hazards for the adjacent settlements (Figure 7.2A). Across the landslide mass, the narrower cross section of the river favours erosion of the landslide debris and increases sediment fluxes downstream, whereas the larger blocks exceeding the competence of the stream power form a debris lag, which armours the channel bed and prevents further erosion during regular high flows.

When the volume of the landslide is sufficient to block the valley entirely (Figure 7.2, case 2 and Figure 7.2D), a lake will instantaneously start filling up, whereas the downstream part is starved of water. Lake depth depends on valley geometry and landslide height; the latter may reach hundreds of metres (e.g. Lake Sarez; Table 7.1). Landslide dams may be ephemeral (a few minutes to a few hours), or

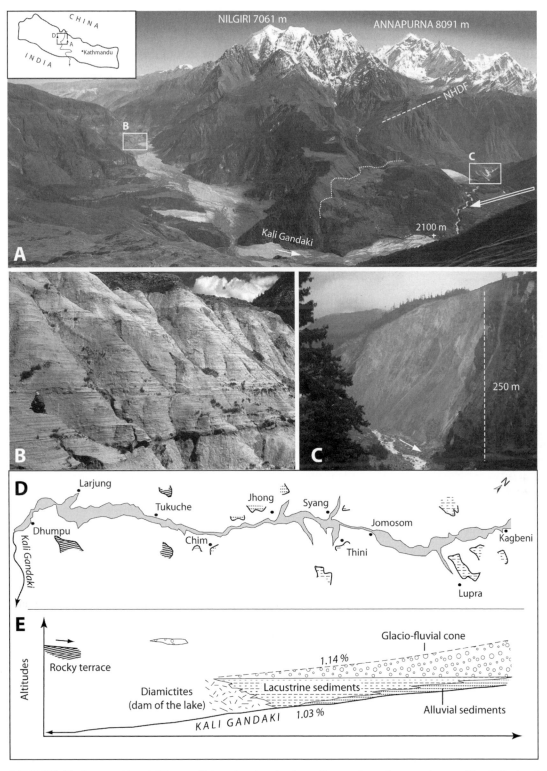

FIGURE 7.1. The large, prehistoric Dhumpu–Kalopani rock avalanche and its morpho-sedimentary impacts (Nepal Himalayas). The collapsed mass failed along the North Himalayan Detachment Fault (NHDF) and blocked durably the Kali Gandaki River, which had carved the deepest gorges in the world across the Annapurna (8091 m) and Dhaulagiri (8172 m) ranges. The 23 km long, >200 m deep Marpha lake developed and filled in with sediments brought from upstream and from glaciated tributary valleys. The braided pattern of the river course reflects the still persisting role of the rock-avalanche barrier in the denudational evolution of the mountain.

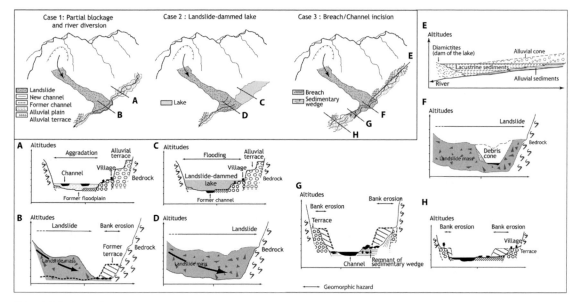

FIGURE 7.2. Types of geomorphic impacts of landslide dams on hydrosystems.

short term (a few weeks or months), or persistent (several thousand years), a duration that directly influences their potentially catastrophic nature and their control on sediment fluxes and budgets (Figure 7.1 and Figure 7.2E). Dam longevity is a function of its stability, which depends on the size and shape of the dam, the characteristics of the geological material composing the dam (material properties, grain size distribution), the volume and rate of water and sediment inflow to the newly formed lake, and the rate of seepage through the dam (Costa and Schuster, 1988). Water ponding occurs up to the dam height, resulting in upstream backwater flooding that can be damaging for settlements and infrastructures (Figure 7.2C). The rapidity of the water level rise depends on the river inflow versus the size and shape of the inundated valley; it generally leaves enough time for evacuation of the threatened populations to take place.

In contrast, catastrophic downstream flooding (Figure 7.2, case 3) will occur following a rapid failure of the landslide dam, caused either by overtopping and immediate retrogressive incision across the dam (Figure 7.2F), or by increasing shear stress because of seepage at the base of the dam (that can be reinforced by the weight of the impounded water mass upstream), or by displacement waves triggered by landslide failures directly into the lake. The catastrophic nature of the (flash)flood, involving the release of a large amount of sediments and water, is recorded by extreme peak discharges (usually back-calculated from field surveys and hydraulic equations) and high velocities that permit the transport of sediments

derived from the dam and from the upstream lacustrine reservoir. The geometry and extent of the coarse debris wedge formed immediately downstream of the dam (Figure 7.2G) suggest both the competence of the debris laden waters (often in the form of debris flows) and their rapid evolution in both time and space into hyper-saturated flows and then into 'normal' high flows, in relation to progressive deposition of the coarser material in the flood plain (Figure 7.2H).

The catastrophic draining of a landslide-dammed lake and associated flooding may create secondary impacts. Firstly, new landslides may develop, both upstream of the lake, following the rapid drawdown of the water, and across and downstream of the dam, where the propagation of the flood contributes over very large distances to a sudden rise of water, bank undercutting and hillslope instabilities (e.g. Yigong debris flow; Table 7.1), which may in turn give rise to new dams and flooded areas. Secondly, specific aggradational landforms may develop, such as 'barrier-defended' terraces upstream of the dam, with the sedimentary facies reflecting the architecture and sequences of debris inputs from the trunk river and adjacent tributaries and slopes (Hewitt, 2002). Thirdly, a distinctive morphology of 'interrupted valleys' appears as a legacy of past catastrophic events, consisting of partial obstructions by landslide barriers (e.g. boulder accumulations resting in the middle of river channels) or in persistent impoundments that 'create a chronically fragmented drainage system' (Hewitt, 2006). This is the case for the Dhumpu–Chooya–Kalopani rock avalanche that dammed the upper Kali

Gandaki valley (Nepal Himalaya) some 70,000 years ago and is still influencing the bed profile and planform pattern (braiding) of the river channel (Fort, 2000) (Figure 7.1). This influence is related not only to the stability of the dam, but also depends on the stream power of the river being sufficient to breach the dam. This is a function of the longitudinal profile, and of the liquid and solid discharge of the river (parameters controlled by climate, autocyclic processes, slope and tectonics).

7.2.2 Impacts on sedimentary fluxes and budgets

In the frame of sedimentary budgets, landslide masses form a particular type of sediment store, characterized by their volume and calibre of materials, and by their distance from the channel (Slaymaker, 2006). As already mentioned, large landslides impact adjacent channels and valleys, either by accelerating, or slowing down, or even interrupting, the conveyance of sediment to the downstream reservoirs. These different impacts may occur successively in time, depending on the magnitude of the landslides and their capacity to act or not as an efficient barrier; they may randomly affect the functioning of the sediment fluxes, and contribute to the formation of a cascading sequence of intramontane storage units (Korup, 2005).

Local scale

An illustration is provided by the sedimentary budget of the upper Cerveyrette catchment (Southern French Alps). This was directly influenced by a series of earth-flows that developed after glacial retreat, and was caused by the combined effects of post-glacial debuttressing and lique-faction of the serpentinite bedrock, in a context of seismic activity (Cossart and Fort, 2008) (Figure 7.3A). The larg-est slide-earth flow of the Chenaillet ($c. 3 \times 10^7 \, \text{m}^3$; 3.5 km long) efficiently dammed the valley, so that the Bourget Plain developed in response to the impoundment. Longitudinally, the trap was infilled with alluvial gravels and silts fed by alpine periglacial and/or still-glaciated hillslopes. This debris accumulated as a prograding fan delta encroaching upon the lake that developed in the distal part of the valley, closest to the dam. Lacustrine deposits are interfingered with colluvial debris derived from the steep slopes of Mount Lasseron as scree and avalanche cones.

The sediment budget was estimated by combined field surveys, DEM and GIS approaches (Figure 7.3B; Cossart and Fort, 2008). The sediment stored in the Bourget Plain represents approximately the same volume ($22.3 \times 10^6 \, \text{m}^3$) as the Chenaillet landslide dam ($21 \times 10^6 \, \text{m}^3$), whereas

debris still blankets the hillslopes ($18.4 \times 10^6 \, \text{m}^3$) without reaching the Bourget Plain. This store developed between the Late Pleistocene and $c.$ 5,000 BP; it first interrupted, then considerably reduced the sediment fluxes downstream. As soon as the blockage of the valley occurred, the down-stream part of the Cerveyrette torrential river started adjust-ing its longitudinal profile by retrogressive erosion. The present situation corresponds to a total sediment removal and export of only $10^6 \, \text{m}^3$ ($c.$ 1/60 of the debris). However, sporadic dissection of the landslide dam triggered by extreme meteorological events is directly threatening the Cervières village, as during the 1957 >100-years recurrence flood that destabilized the entire hydrosystem. This exam-ple shows how, in alpine headwater contexts, landslide dams are persistent features controlling the sediment fluxes long after their occurrence.

Regional scale

Large landslides may contribute to a significant increase in sediment yield in a very short period following failure, not only by direct massive input of landslide debris into the fluvial system, but also by secondary hillslope pro-cesses that rework the landslide material before vegetation regrowth. Such landslide-derived sediment pulses result in aggradation in the downstream reaches from the land-slide site, and are often accompanied by metamorphosis and/or change in the river course that may induce off-site hazards and damaging impacts to downstream settlements and infrastructures. For instance, Korup et al. (2004) calculated that the 1999 Mount Adams rock avalanche (New Zealand) produced a specific sediment yield in excess of approximately $75,700 \pm 4,600 \, \text{t km}^{-2} \, \text{a}^{-1}$, as expressed by the massive fanhead aggradation and the opening of a major avulsion channel at the mountain range front. This sediment yield represents a sediment discharge of $2.5 \times 10^6 \, \text{m}^3 \, \text{a}^{-1}$ calculated for the three years following failure, an amount that rapidly declined after the event. In the Nepal Himalayas, Fort (1987) described a giant ($>4 \times 10^9 \, \text{m}^3$) collapse of the south face of the Annapurna IV peak that occurred about 500 years ago, and filled the 35 km distant Pokhara valley under a 60 to 100 m thick gravel aggradation, which blocked and caused the flooding of adjacent tributary valleys. The calculated annual contribution of Annapurna rockslide-derived sediment is in the order of $4 \times 10^6 \, \text{m}^3 \, \text{a}^{-1}$ and represents, for the upper catchment, a sediment yield of $22,860 \, \text{m}^3 \, \text{a}^{-1} \, \text{km}^{-2}$ (Fort, 1987; Fort and Peulvast, 1995). This is a figure averaged over a 500-year period, which in fact is exceptionally high if one considers the fact that the aggradation took place in a very short time, i.e. 'instantaneously' after the failure.

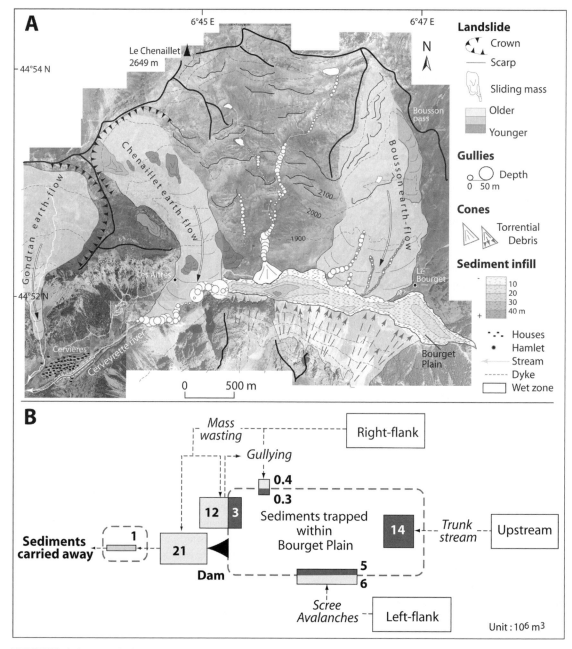

FIGURE 7.3. Sedimentary budget assessment in response to the Chenaillet earth-flow dam (Southern French Alps). (Modified and completed after Cossart and Fort, 2008.)

7.3 Forecasting and preventing

Catastrophic rock failures are generally preceded by periods of accelerated creep, which may result in observable slope deformation and 'sackung' related features: development and widening of tension cracks, buckling, increased rockfall activity and toppling, or break-out across bedding. Measured stress drops in crustal rock, and related fracturing could provide a physical basis for quantitatively forecasting catastrophic slope failure (Kilburn and Petley, 2003). However, monitoring of all potential threatening slopes is not realistic, especially in developing countries where other priorities (food, shelter, employment) come well ahead of unpredictable natural hazards.

Other forecasting methods are based on statistical assessment of landslide susceptibility; they rely on a series of variables (landslide inventory with the help of GIS, terrain predisposing factors, neo-predictive variables with a geomorphological meaning) that are tested (simulations) and eventually evaluated by expert judgement (Thiery *et al.*, 2007). Another similar approach is based on geotechnical modelling and 3-D model calibration set up from a known and well-studied event (quantitative data), as was done after the 1987 Val Pola landslide, one of the most destructive and costly natural disasters that has occurred in Italy during recent decades (Costa, 1991; Crosta *et al.*, 2004). In their study, Crosta *et al.* (2004) showed how modelling techniques help one to understand the rheology of such a failure, and to predict its timing in highlighting the transformation of potential energy to kinematic energy.

The power law curve of landslide magnitude/frequency can also be used for landslide hazard assessment (Evans *et al.*, 2006), and as input into a quantitative risk calculation if combined with vulnerability data, as was done to assess rockfall risk along transportation corridors (Hungr *et al.*, 1999).

To prevent the failure of landslide dams and resulting catastrophic floods, the most commonly used control measure consists of the construction of spillways either across the landslide crest or across the adjacent bedrock. Alternatively, drainage by siphon pipes, pump systems or diversion tunnels (Val Pola, Mount St Helens, Bairaman; Table 7.1) are short-term measures to control lake level. More radical methods consist in large-scale blasting, as was done across the landslide dams that were formed near Beichuan city after the 18 May 2008 Sichuan earthquake in China. However, the disaster may occur before adequate control measures can be completed (more particularly in remote and rugged areas that render transport of heavy equipment very difficult), or because high rapid inflow to the impoundment often exceeds general predictions and causes an early dam failure (Shang *et al.*, 2003). In the case of Lake Sarez and the related Usoi landslide dam (Tadjikistan), the highest dam on Earth, and despite extensive observations and technical studies that suggest a satisfactory stability of the dam against sliding, the possibility of a catastrophic outburst flood that would destroy the many villages and infrastructures of the Bartang–Pyanj–Amu Darya catchment, situated between the lake and the Aral Sea, cannot be entirely ruled out (>5 million people would be affected). Combined measures are thus being implemented, consisting of (i) the monitoring of the stability of the dam and slopes surrounding the lake, (ii) an early warning system to alert inhabitants of the upper Amu Darya valley, together with (iii) the modelling of a series of flood scenarios to determine the degree of risk and vulnerability of downstream villages and infrastructures (Alford and Schuster, 2000). Whatever these measures, a simple doubling of the present mean streamflow volume of $2000 \, \text{m}^3 \, \text{s}^{-1}$ of the Bartang River would readily destroy large portions of existing roads, low-lying villages and agricultural land for more than 100 km downstream from the dam, whereas a lake outburst flood would generate an instantaneous, catastrophic peak flow of about one million cubic metres per second (Alford and Schuster, 2000).

7.4 Conclusions

Despite their low frequency, large landslides are natural hazards that induce geomorphic impacts that can badly damage human settlements (Table 7.1). They are generally associated with episodes of extreme rainfall and/or with earthquakes, which are the natural hazards that cause the highest human losses on Earth. The rapid failure of a landslide dam will cause catastrophic downstream flooding whereas a long-lasting dam and its resulting filling by sediments will mostly affect mountain valley morphology and the sediment cascade.

The development of large-scale, catastrophic landslides should be put in the broader, long-term perspective of the evolution of a mountain range. In the context of the overall balance between erosion and rock uplift rates (Burbank *et al.*, 1996; Montgomery, 2001), superficial variables such as sediment fluxes and hillslope angle respond to both climate forcing and sporadic, catastrophic mountain slope collapse. This results in punctuated epicycles of aggradation and/or river incision that offset the equilibrium state at a shorter, 10^4–10^5 year time-scale (Fort, 1988; Pratt-Sitaula *et al.*, 2004). These high magnitude, low frequency failures are the very 'formative events' of the present morphology of active mountains and accomplish the main part of the denudation process of these orogens.

References

Alford, D. and Schuster, R. L. (2000). Introduction and summary. In *Usoi Landslide Dam and Lake Sarez: An Assessment of Hazard and Risk in the Pamir Mountains, Tajikistan*. United Nations, ISDR Prevention Series n°1, pp. 1–18.

Brardinoni, F. and Church, M. (2004). Representing the landslide magnitude-frequency relation: Capilano river basin, British Columbia. *Earth Surface Processes and Landforms*, **29**, 115–124.

Brunsden, D. and Jones, D. K. C. (1984). The geomorphology of high magnitude-low frequency events in the Karakoram mountains. In K. J. Miller (ed.), *The International Karakoram Project*, vol. 1, Cambridge: Cambridge University Press, pp. 383–388.

Burbank, D. W., Leland, J., Fielding, E. *et al.* (1996). Bedrock incision, rock uplift and threshold hill slopes in the northwestern Himalayas. *Nature*, **379**, 505–510.

Cossart, E. and Fort, M. (2008). Consequences of landslide dams on alpine river valleys: examples and typology from the French Southern Alps. *Norsk Geografisk Tidsskrift (Norwegian Journal of Geography)*, **62**, 75–88.

Costa, J. E. (1991). Nature, mechanics, and mitigation of the Val Pola landslide, Valtellina, Italy 1987–1988. *Zeitschrift für Geomorphologie*, **35**, 15–38.

Costa, J. E. and Schuster, R. L. (1988). The formation and failure of natural dams. *Geological Society of America Bulletin*, **100**, 1054–1068.

Crosta, G. B., Chen, H. and Lee, C. F. (2004). Replay of the 1987 Val Pola Landslide, Italian Alps. *Geomorphology*, **60**, 127–146.

Cruden, D. M. and Varnes, D. J. (1996). Landslide types and processes. In *Landslides, Investigation and Mitigation*. Washington, DC: National Academy Press, pp. 36–75.

Dikau, R., Brunsden, D., Schrott, L. and Ibsen, M. (eds.) (1996). *Landslide Recognition: Identification, Movement and Causes*. New York; Chichester: Wiley.

Evans, S. G. and Clague, J. J. (1994). Recent climatic change and catastrophic geomorphic processes in mountain environments. *Geomorphology*, **10**, 107–128.

Evans, S. G., Scarascia Mugnozza, G., Strom, A. L. *et al.* (2006). Landslides from massive rock slope failure and associated phenomena. In S. G. Evans *et al.* (eds.), *Landslides From Massive Rock Slope Failure*, Dordrecht: Springer, pp. 3–52.

Fort, M. (1987). Sporadic morphogenesis in a continental subduction setting: an example from the Annapurna Range, Nepal Himalaya. *Zeitschrift für Geomorphologie Supplement-Band*, **63**, 9–36.

Fort, M. (1988). Catastrophic sedimentation and morphogenesis along the High Himalayan front: implications for palaeoenvironmental reconstruction. In P. Whyte, J. S. Aigner, N. G. Jablonski *et al.* (eds.), *The Palaeoenvironment of East Asia from Mid-Tertiary*. Centre of Asian Studies, University of Hong Kong, pp.170–194.

Fort, M. (2000). Glaciers and mass wasting processes: their influence on the shaping of the Kali Gandaki valley (higher Himalaya of Nepal). *Quaternary International*, **65/66**, 101–119.

Fort, M. and Peulvast, J.-P. (1995). Catastrophic mass-movements and morphogenesis in the Peri-Tibetan Ranges: examples from West Kunlun, East Pamir and Ladakh. In O. Slaymaker (ed.), *Steepland Geomorphology*. Chichester: Wiley, pp. 171–198.

Hewitt, K. (1988). Catastrophic landslide deposits in the Karakoram Himalaya. *Science*, **242**, 64–67.

Hewitt, K. (1998). Catastrophic landslides and their effects on the Upper Indus streams, Karakoram Himalaya, northern Pakistan. *Geomorphology*, **26**, 47–80.

Hewitt, K. (2002). Postglacial landforms and sediment associations in a landslide-fragmented river system: the Transhimalayan Indus stream, Central Asia. In K. Hewitt, M. L. Byrne, M. English and G. Young (eds.), *Landscapes of Transition: Landform Assemblages and Transformations in Cold Regions*. Dordrecht: Kluwer, pp. 63–91.

Hewitt, K. (2006). Disturbance regime landscapes: mountain drainage systems interrupted by large rockslides. *Progress in Physical Geography*, **30**, 365–393.

Hewitt, K., Clague, J. J. and Orwin, J. F. (2008). Legacies of catastrophic rock slope failures in mountain landscapes. *Earth-Science Reviews*, **87**, 1–38.

Hovius, N., Stark, C. P. and Allen, P. A. (2007). Sediment fluxes from a mountain belt derived by landslide mapping. *Geology*, **25**, 231–234.

Hsü, K. J. (1975). Catastrophic debris streams (sturzstroms) generated by rockfalls. *Geological Society of America Bulletin*, **86**, 129–140.

Hungr, O., Evans, S. G. and Hazzard, J. (1999). Magnitude and frequency of rockfalls and rock slides along the main transportation corridors of Southern British Columbia. *Canadian Geotechnical Journal*, **36**, 224–238.

Kilburn, C. R. J. and Petley, D. N. (2003). Forecasting giant, catastrophic slope collapse: lessons from Vajont, Northern Italy. *Geomorphology*, **54**, 21–32.

Koi, T., Hotta, N., Ishigaki, I. *et al.* (2008). Prolonged impact of earthquake-induced landslides on sediment yield in a mountain watershed: the Tanzawa region, Japan. *Geomorphology*, **101**, 692–702.

Korup, O. (2005). Geomorphic imprint of landslides on alpine river systems, southwest New Zealand. *Earth Surface Processes and Landforms*, **30**, 783–800.

Korup, O., McSaveney, M. and Davies, T. R. H. (2004). Sediment generation and delivery from large historic landslides in the Southern Alps New Zealand. *Geomorphology*, **61**, 189–207.

Korup, O., Clague, J. J., Hermanns, R. L., *et al.* (2007). Giant landslides, topography, and erosion. *Earth and Planetary Science Letters*, **261**, 578–589.

Li, T., Schuster, R. L. and Wu, J. (1986). Landslide dams in south-central China. In R. L. Schuster (ed.), *Landslides Dams: Processes, Risk, and Mitigation*. American Society of Civil Engineers, Geological Special Publication, 3, pp. 21–41.

Montgomery, D. R. (2001). Slope distributions, threshold hillslopes, and steady-state topography. *American Journal of Science*, **301**, 432–452.

Plafker, G. and Ericksen, G. E. (1978). Nevados Huascaran avalanches, Peru. In B. Voight (ed.), *Rockslides and Avalanches: I, Natural Phenomena*. Amsterdam: Elsevier, pp. 277–314.

Pratt-Sitaula, B., Burbank, D. W., Heimsath, A. and Ohja, T. (2004). Landscape disequilibrium on 1000–10,000 year scales, Marsyandi River, Nepal, Central Himalaya. *Geomorphology*, **58**, 223–241.

Schneider, J. L., Pollet, N., Chapron, E., Wessels, M. and Wassmer, P. (2004). Signature of Rhine Valley sturzstrom dam failures in Holocene sediments of Lake Constance, Germany. *Sedimentary Geology*, **169**, 1–2, 75–91.

Shang, Y., Yang, Z., Li, L. *et al.* (2003). A super-large landslide in Tibet in 2000: background, occurrence, disaster and origin. *Geomorphology*, **54**, 225–243.

Slaymaker, O. (2006). Towards the identification of scaling relations in drainage basin sediment budgets. *Geomorphology*, **80**, 8–19.

Thiery, Y., Malet, J.-P., Sterlacchini, S., Puissant, A. and Maquaire, O. (2007). Landslide susceptibility assessment by bivariate methods at large scales: application to a complex mountainous environment. *Geomorphology*, **92**, 38–59.

8 Landslides and climatic change

Lisa Borgatti and Mauro Soldati

8.1 Introduction

Climate change refers to a statistically significant variation in either the mean state of the climate or in its variability, persisting for an extended period, typically decades or longer. Climate change on Earth may be due to natural internal processes or external forcing, or to persistent anthropogenic perturbation of the atmosphere composition or of land use. The Fourth Assessment Report of the Intergovernmental Panel on Climate Change (IPCC, 2007) states that the warming of the climate system is unequivocal. Moreover, there is a high level of confidence that this warming is a result of human activities releasing greenhouse gases to the atmosphere from the burning of fossil fuels, deforestation and agricultural activities. A range of future greenhouse gas emission scenarios are also presented, based on estimates of economic growth, technological development and international cooperation. In all scenarios temperatures continue to rise worldwide, with global mean temperatures averaging plus 2 to 4 °C by the end of the century, accompanied by changes in the amounts and patterns of precipitation. The predicted rate of warming seems to be faster than ever recorded and in particular over the last 2,000 years, and also since the Earth was exiting the Little Ice Age. There will also be an increase in the frequency and intensity of extreme temperature and precipitation events at any time of the year, regardless of the season.

If the evidence of climate change is unequivocal, is climate change triggering more landslides, or will it in the future? Intuitively, yes: climate changes have the potential to modify the stability of slopes, both natural and constructed. This issue is important and urgent, no matter what are the actual causes of climate changes. It has been proved that most landslides are caused by saturated soil moisture conditions and by loss in soil strength (Wieczorek,

1996), triggered by climatically controlled processes, such as intense and/or prolonged precipitation events, rapid snowmelt, glacier thinning, permafrost degradation or river migration, depending on geomorphological settings. If climate change leads to increased frequency and/or magnitude of these events and processes, the frequency and/or magnitude of landslides in a region will be similarly influenced (Crozier and Glade, 1999) (see also Chapters 6 and 7). Thus, we can expect more instability, as a consequence of the increasing number of short but intense events, as well as by increasing cumulative rainfall etc.

Although the frequency and/or magnitude of landslides may increase with the anticipated climate change, the regional distribution of landslides is not expected to change significantly, as many of the primary factors controlling landslide susceptibility, such as geology, physiography and slope, remain relatively constant. Potentially unstable areas, however, could for example include slopes presently underlain by degrading permafrost (Harris et al., 2001). A sort of domino effect may also be expected: increased landslide activity may lead to increased sediment loads and channel instability in rivers (Korup et al., 2004).

Currently, where the best practice in landslide risk mitigation is established, the design and management of infrastructure and urban assets affected by natural slope instability is carried out on the basis of specified standards and guidelines that assume static environmental conditions. However, the rate of dominant input parameters (i.e., precipitation and temperature) is now clearly changing. Hence, a review of this approach is demanded, as the assumption of a steady climate state can be misleading (see for example Winter et al., 2005).

Assessment of climate change's impact is currently speculative and is still difficult to unravel from 'pure' anthropogenic effects, in some cases driven by climate changes

Geomorphological Hazards and Disaster Prevention, eds. Irasema Alcántara-Ayala and Andrew S. Goudie. Published by Cambridge University Press. © Cambridge University Press 2010.

themselves. Accordingly, compound effects of climate change and other human actions (where humans are both co-triggers and element at risk) must be accounted for. In this sense, some studies seem to highlight the fact that climate change is important, but land use change is even more important in the case of slope instability (Glade, 2003).

It is also worth stressing the fact that predicted temperature changes could influence the response of slopes through increased evapotranspiration, leading to a change in the triggering precipitation thresholds. This could help counterbalance the impact of changes in precipitation rates and patterns. In this sense some regions could experience fewer landslides as a consequence of climate change. Moreover, in arid regions wet periods can lead to an increase of vegetation cover that could create more stable conditions.

In this framework, there is an urgent need for models that discriminate between and incorporate both human and climate effects. Analysis of sources of uncertainty in the models is also needed to establish the factors that contribute to the predicted changes in slope instability. Assessment of these factors can provide an indication of the potential impact of climate change in different landslide-prone areas where susceptible features are found, such as weak rocks, areas close to sea level, climatically sensitive areas at high latitudes and high altitudes and so forth.

8.2 Conceptual framework

Research on climate and related impacts has addressed the assessment of the effects of climatic variability (Viles and Goudie, 2003) and climate change (Goudie, 1992) on geomorphic processes and hazards, including landslides.

Future climates, predicted by general circulation models (GCMs), have been utilized for the assessment of slope instability processes, both at a regional (Dehn et al., 2000) and at a local scale (Dixon and Brook, 2007), downscaling climate change time series in hydrological models and slope stability models in the Italian and French Alps, in southeast Spain and in southern England (Brooks, 1997; Buma and Dehn, 1998, 2000; Dehn, 1999; Dehn and Buma, 1999; Collison et al., 2000; van Beek, 2002; Malet et al., 2005). In this sense, it has to be underlined that landslides are localized phenomena, often occurring in upland areas where rainfall patterns are complex. Global climate models work at larger spatial scales and the outputs are difficult to interpret at the slope scale.

Effects of past climate changes on formative events on slopes have been assessed by field evidence (Corominas and Moya, 1999), and by physically based models (Brooks

et al., 1999). Establishing links between climate and past landslide activity is indeed very difficult. This is primarily due to the relatively few records of landslide events (imprecise dates, incomplete databases) dating back to the last century, to the Little Ice Age and to the Holocene. There have been some attempts to estimate these links especially for the Alps (see for instance Matthews et al., 1997, and the synthesis of Soldati et al., 2004).

Landslide occurrence has been exploited as a climate proxy itself (Matthews et al., 1997; Borgatti et al., 2007). This approach is hindered by a number of biases, due to the complexity of climate–landslide coupling, which is much more complicated at a long temporal scale (Crozier, 1997).

Landslides are considered to be natural, multi-dimensional and non-linear dynamical systems with complex behaviour in space and time (Brunsden, 1999). The evolution of the slope system is coupled with other geomorphic processes, and is sensitive to both inherited and present controls. Mass movements are therefore four-dimensional phenomena and display complex temporal development. Each phase of movement or surge is the near-immediate response to an external trigger that increases the stress in the slope or reduces the strength of the slope material (Wieczorek, 1996). The main landslide triggers are intense rainfall, snow melting, earthquake shaking, volcanic eruption, stream and wave erosion. Nevertheless, mass movements may also occur without any evident near-immediate trigger.

Climate is related to landslides via the highly non-linear soil–water system and there are no unique relationships between climate conditions and landslides. It can be assumed that the relationship between climate, in particular positive (or negative) moisture balance, and landslide activity (or inactivity) exists at every time scale. Field observation of present-day activity and historical records shows that first-time failures of large landslides follow a complex hydrological and mechanical behaviour (Corominas, 2001). In fact, first-time failures are the result of long-term evolutionary processes on the slope, rather than the near-immediate response to a specific trigger (Noverraz et al., 1998). On the other hand, the influence of moisture balance is evident in the case of reactivations of dormant landslides, the acceleration of active movements and in the triggering of shallow slope failures. In particular, changes in the hydrological balance, resulting from the temporal distribution of temperature and rainfall, and the resulting evapotranspiration, directly influence the hydrological regime of slopes, which in turn governs the type, the rate and the temporal and spatial evolution of mass movements. Consequently, in the analysis of the relationships between landslides and climate, it is necessary to focus on temperature changes as well as on the timing, frequency and magnitude of rainfall.

At a long temporal scale, the relationship between landslide activity and triggering mechanisms can be established from the temporal clustering of dated landslides. In order to discriminate between the climatic and non-climatic factors, a multi-disciplinary approach directed to the appraisal of the paleoenvironmental conditions at the time of the landslides must be adopted (see Borgatti et al., 2007). Therefore, besides the development of a landslide events record, other proxies have to be considered in order to disentangle the possible interactions between the slope-system, climate and humans (Figure 8.1).

At a short time scale, it is widely accepted that high duration/intensity rainfall events are the most important triggering mechanism of landslides worldwide (Wieczorek, 1996; Corominas, 2001). Physically based models of rainfall-induced landslides have been used to model and understand this complex interaction and to derive thresholds under different boundary conditions (Van Asch and Buma, 1997). The derivation of a general rule, that takes into account the rainfall thresholds related to landslide activity, has been attempted since the pioneering work by Caine (1980) and many others (Fukuoka, 1980; Crozier, 1986; Jibson, 1989; Crozier, 1999). It is now accepted that there are no universal rainfall thresholds that can be associated with landslides (Dikau and Schrott, 1999), but the variability of these thresholds is well known (see the comprehensive review of worldwide thresholds by Guzzetti et al., 2008).

The frequency, magnitude and type of landslide may be related to different rainfall conditions, thus implying that similar climatic conditions can be easily associated with different patterns of landslide distribution (Van Asch et al., 1999).

At the same time, different types of slope movements are related to distinct hydrological conditions. Shallow translational slides are usually triggered by the water infiltration

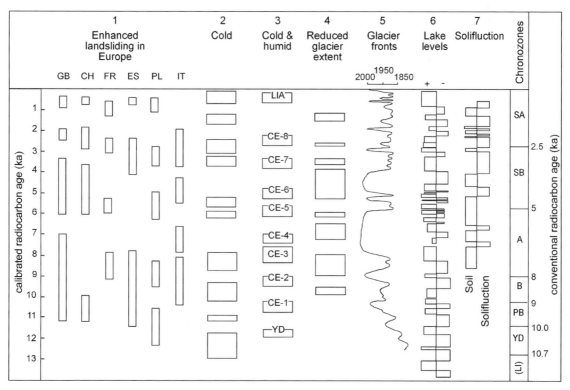

FIGURE 8.1. Temporal occurrence of landslides in Europe and comparison of different Late Glacial and Holocene paleoclimatic record at different spatial scales (modified after Corsini et al., 2000; Borgatti et al., 2007). 1. Enhanced slope instability events in Europe (González Diez, et al. 1996; Ibsen and Brunsden, 1997; Lateltin, 1997; Margielewski, et al. 2001). 2. Cold and humid periods in the Alps and on the Swiss Plateau (Haas et al., 1998; Tinner and Amman, 2001; Tinner and Kalterieder, 2005). 3. Phases of reduced glacier extent, recorded by the retreats of the Unteraar and other Swiss glaciers (Hormes et al., 2001). 4. Glacier fluctuations in the Swiss Alps, with reference to the glacier front stages of 1850, 1920 and 2000 (Maisch et al., 2000). 5. Mid-European lake levels (Magny, 1999) as palaeohydrological indicators. 7. Solifluction phases in the Alps (Gamper, 1993). Ll: Late-glacial Interstadial, not a chronozone and therefore shown in brackets; YD: Younger Dryas; PB: Preboreal; B: Boreal; A: Atlantic; SB: Subboreal; SA: Subatlantic; LIA: Little Ice Age; CE-n: Central European cold-humid phases (Haas et al., 2008).

in unconsolidated slope deposits, which cover near-impermeable rock masses. Increasing soil saturation is responsible for the reduction of the shear strength of the soil, by the temporary rise in pore water pressure, and by the loss of the apparent cohesion of the soil (Wieczorek, 1996). Translational slides, rotational slides as well as complex and composite slope movements are triggered by long-duration rainfall and subsequent modification of the regional groundwater tables and subsequent shear strength reduction (Van Asch *et al.*, 1999). Such hydrological conditions can occur as a consequence of rainfall periods 40–90 days long. Slope movements due to bank erosion are mostly activated during flash flood episodes, related to very intense and concentrated rainfall inducing increased stream discharge and erosion along river banks.

Landslide activity can be recurrent on the same slope. In particular, in the case of slow-moving landslides, phases of activity can last for periods of different duration, even thousands of years, depending on local and global climate conditions (Soldati *et al.*, 2004).

8.3 Landslides and climate: state of the art

Climate oscillations have been recognized at different time scales, from interannual to millennial (and more) and at different spatial scales (see Viles and Goudie, 2003, for a review from a geomorphological work standpoint). Moreover, these modes of variability were also operating in the past, with interactions and teleconnections among them.

One can argue to what temporal and spatial extent climate variability and climate change are having an impact on slope instability. As climatologists need to look at past climates to validate models, landslide scientists have to look at archives that can be both historical and 'natural'. The temporal and spatial scale of changes and the related slope instability phenomena have to be taken into account with different methodologies that can tackle time resolution and landform persistence in the landscape etc.

8.3.1 Landslides and long-term climate changes, at the millennial scale

Based on the pronounced imprint of millennial-scale climate change on surface processes and landscape evolution, previous investigations have clearly shown that, from the Late Glacial to the present, climate has influenced slope evolution, either directly or indirectly, and that slope processes may be considered geomorphological indicators of climate changes (Goudie, 1992).

Temporal clustering of ancient landslide events has in fact been reported from both southern and northern Europe (Starkel, 1991; Frenzel *et al.*, 1993; González Díez *et al.*, 1996; Panizza *et al.*, 1996; Ibsen and Brunsden, 1997; Lateltin *et al.*, 1997; Schoeneich *et al.*, 1997; Alexandrowicz and Alexandrowicz, 1999; Dikau and Schrott, 1999; Margielewski, 2001; Dapples *et al.*, 2002; Bertolini *et al.*, 2004; Schmidt and Dikau, 2004; Soldati *et al.*, 2004; Bigot-Cormier *et al.*, 2005; Soldati and Borgatti, 2009). Apart from large landslides, debris flow records are also considered to reflect the increased occurrence of heavy rainstorms during the Holocene (Sletten *et al.*, 2003).

As far as the late Holocene is concerned, landslide movements were intensified towards the close of the Little Ice Age and were related to episodes of increased rainfall. Matthews *et al.* (1997) demonstrated an increase in landslide activity all over Europe during the Little Ice Age, characterized by increased rainfall combined with a lowering of temperatures (Figure 8.1). Lichenometric studies and data from written documents make it possible to determine phases of intensified landslide movements in this period (Bajgier-Kowalska, 2008).

Recently, case studies from Africa (Thomas, 1999; Busche, 2001), northern and southern America (Bovis and Jones, 1992; Trauth *et al.*, 2000, 2003; Smith, 2001) and Asia (Sidle *et al.*, 2004) have also been reported. Post-Little Ice Age glacial retreat is one of many factors influencing landslide activity in British Columbia (Holm *et al.*, 2004). Time-series analysis reveals periods of more humid and more variable climates at the time of clustering of landslide events in the Andes (Trauth *et al.*, 2003).

Some studies have focused on the understanding of these relationships, relating them to relatively known, also proxy-derived, past climate series exploited for projections of future unknown climates (Schmidt and Dikau, 2004). The modelling of groundwater-controlled landslides shows that the highest slope instability occurs at the transition from the more humid Little Ice Age to the drier recent climate. The intensity of this impact, however, varies with the sensitivity of the geomorphic system, i.e., local landforms and lithology, and cannot be related to a specific hillslope (Schmidt and Dikau, 2004).

8.3.2 Landslides and short-term climate variability, at the interannual to decadal scale

The spatial distribution of rainfall, as well as its seasonal and interannual variability, may be explained in terms of the global circulation and regional climate factors (e.g., latitude, orography, oceanic and continental influences). The relationship between low-frequency atmospheric and oceanic circulation oscillations (NAO and ENSO, among many others) controlling rainfall and landslide activity has been attempted for many areas of the world.

The North Atlantic Oscillation (NAO) corresponds to one of the most important large-scale modes of atmospheric circulation in the winter season over the entire Northern Hemisphere (Hurrell, 1995). Several studies have established links between the NAO phase and precipitation in Western Europe and the Mediterranean Basin. In particular, Trigo *et al.* (2002, 2005) and then Marques *et al.* (2008) showed the extensive influence of the NAO phases on the timing and frequency of major seasonal and monthly rainfall episodes and associated landslide activity in Portugal.

The El Niño Southern Oscillation (ENSO) climatic phenomenon consists of slow manifestations of ocean and atmosphere interactions starting in the Pacific Ocean (Walker, 1923) that are the source of interannual climatic variability at a global scale. The ENSO cycle is a system that comprises a warm phase, called El Niño, and the opposite phase, cold episodes named La Niña, occurring every few years (Philander, 1999). This climatic phenomenon has been widely referred to around the world because of its impact on economies and human activities, such as agriculture and fishing in developing countries. El Niño events such as 1982–3 and 1997–8 had a major impact in California and coastal Central America (Reynolds *et al.* 1997; Coe *et al.*, 1998; Godt, 1999), but the teleconnections were global, with effects in Europe, the Atlantic area and Asia. Furthermore, El Niño has been associated with greater landslide occurrence around the world (Ellen and Wieczorek, 1988; Wieczorek *et al.*, 1989; Cayan and Webb, 1992; Glantz, 1995; Godt *et al.*, 1997; Reynolds *et al.*, 1997; Ngecu and Mathu, 1999). In southern America, the temporal distribution and frequency of landslides triggered by precipitation seem to be strictly conditioned by the ENSO climatic cycle (Moreiras, 2005), and also at longer time scales (Trauth *et al.*, 2003).

Some authors associate intensified Asian summer-monsoon circulation phases with enhanced precipitation, discharge and sediment flux leading to an increase in pore-water pressure, lateral scouring of rivers, and over-steepening of hillslopes, eventually resulting in failure of slopes and exceptionally large mass movements (Bookhagen *et al.*, 2005).

An increasing number of cyclones, hurricanes and extreme storm events have also been recorded, triggering a large number of landslides, especially shallow landslides in coastal regions.

8.4 Conclusions: landslides in a changing environment; issues and perspectives

The increasing frequency of extreme weather events has highlighted our vulnerability to the impact of climate changes, and has resulted in enormous human and economic loss. Among natural disasters, landslides are ranked seventh as far as casualties are concerned, after windstorms, floods, droughts, earthquakes, volcanoes and extreme temperatures, claiming 800 to 1,000 lives on average in each of the last 20 years. In the past century, Asia suffered by far the highest number of landslide events any world region, but landslides in North, Central and South America have caused the most deaths and injuries (more than 25,000). Landslides in Europe are the most expensive, causing damage in the order of millions of euros.

If climate change predictions are accurate, more intense and extreme rainfall is expected that, coupled with population growth, can drastically increase landslide-associated casualties, especially in developing countries, where both population and agriculture pressure on land resources often lead to exploitation of unstable slopes. Climate change may trigger landslides in other ways. The landslide that claimed 107 lives in Egypt in 2008 was blamed on rock mass destabilization due to temperature regime changes.

Understanding the relationship between landslides and climate change is therefore crucially important in planning a proactive approach to hazard and risk management. Advances in geohazard modelling and prediction, as well as in real-time monitoring technology, enable us to be better prepared for the impacts of climate changes but, besides climate change impacts, there is still a need for effective risk management and informed planning policy to improve the safety and sustainability of communities at risk.

Besides the intrinsic complexity of slope instability phenomena, many factors may produce changes in both frequency and magnitude of landslide events at different spatial and temporal scales. Mass wasting processes are primarily controlled by steady geological and structural predisposing factors, but the effects of different environmental changes (temperature, rainfall, vegetation etc.) and human impact all have to be taken into account together.

At a long time scale (i.e., millennial), despite the inherent difficulties in correlating proxy records of climate changes and landslide records, which are mainly due to different spatial scales (local, regional and global), dissimilar time resolutions and several dating constraints, some remarkable indications are apparent. On the one hand, landslides have proved to provide records of climate variability at a range of temporal and spatial scales, being geomorphological proxies themselves. On the other hand, the periods of past

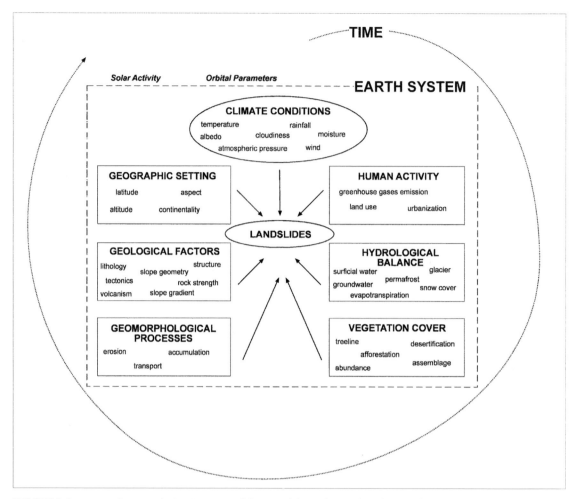

FIGURE 8.2. Preparatory factors and triggering causes of slope instability with special emphasis on climate conditions.

enhanced slope instability found worldwide display a positive correlation with indicators of cold and humid climates, suggesting that these phases could have been climatically driven, and that, in particular, a positive moisture balance could have played a major role in conditioning landslide activity at the hundred to thousand years time scale.

In addition, in formerly glaciated mountain belts, deglaciation and permafrost melting may result in long-term climatic effects, opposite to the actual climate trends. Also the impacts of cold spells should be stressed, as in the case of the Little Ice Age, which has left a clear signature in the landscape.

At a decadal time scale, the relationships between magnitude of the NAO- and ENSO- related precipitation and mass wasting is clear and allows for the development of models to be used for water resource and risk assessment. Improved GCMs give the possibility of modelling future landslide activity based on future precipitation scenarios and past landslide archives.

At present not only climate is changing, but also all the environmental variables that are climate-related at unprecedented or at least not witnessed rates of change, with strong positive and negative feedback mechanisms (Figure 8.2). At the same time, society is changing, with growing needs, sensibility, information and vulnerability. The phrase 'global change' is comprehensive and in this context strong and even opposite regional changes are becoming evident. Observable evidence from the far and near past suggests an increase in landslide activity, but in the same old sites, therefore adaptation and avoidance strategies are possible and desirable.

References

Alexandrowicz, S. W. and Alexandrowicz, Z. (1999). Recurrent Holocene landslides: a case study of the Krynica landslide in the Polish Carpatians. *The Holocene*, **9**, 91–99.

Bajgier-Kowalska, M. (2008). Lichenometric dating of landslide episodes in the Western part of the Polish Flysch Carpathians. *Catena*, **72**, 224–234.

Bertolini, G., Casagli, N., Ermini, L. and Malaguti, C. (2004). Radiocarbon data on Lateglacial and Holocene landslides in the Northern Apennines. *Natural Hazards*, **31**, 645–662.

Bigot-Cormier, F., Braucher, R., Bourles, D. *et al.* (2005). Chronological constraints on processes leading to large active landslides. *Earth and Planetary Science Letters*, **235** (1–2), 141–150.

Bookhagen, B., Thiede, R. C. and Strecker, M. R. (2005). Late Quaternary intensified monsoon phases control landscape evolution in the northwest Himalaya. *Geology*, **33**(2), 149–152.

Borgatti, L., Ravazzi, C., Donegana, M. *et al.* (2007). A lacustrine record of early Holocene watershed events and vegetation history, Corvara in Badia, Dolomites, Italy. *Journal of Quaternary Science*, **22**, 173–189.

Bovis, M. J. and Jones, P. (1992). Holocene history of earth flow mass movement in south-central British Columbia: the influence of hydroclimatic changes. *Canadian Journal of Earth Sciences*, **29**, 1746–1755.

Brooks, S. (1997). Modelling the role of climatic change in landslide initiation for different soils during the Holocene. In J. A. Matthews, D. Brunsden, B. Frenzel, B. Gläser and M. M. Weiß (eds.), *Rapid Mass Movement as a Source of Climatic Evidence for the Holocene*. Palaeoclimate Research, vol.19, Jena: Gustav Fischer, pp. 207–222.

Brooks, S. M., Anderson, M. G., Wilkinson, P. and Ennion, T. (1999). Exploring the potential for physically-based models and contemporary slope processes to examine the causes of Holocene mass movement. In S. Hergarten and H. J. Neugebauer (eds.), *Process Modelling and Landform Evolution*. Lecture Notes in Earth Sciences, 78, Heidelberg: Springer.

Brunsden, D. (1999). Some geomorphological considerations for the future development of landslide models. *Geomorphology*, **30**(1–2), 13–24.

Buma, J. and Dehn, M. (1998). A method for predicting the impact of climate change on slope stability. *Environmental Geology*, **35**(2–3), 190–196.

Buma, J. and Dehn, M. (2000). Impact of climate change on a landslide in South East France, simulated using different GCM scenarios and downscaling methods for local precipitation. *Climate Research*, **15**, 69–81.

Busche, D. (2001). Early Quaternary landslides of the Sahara and their significance for geomorphic and climatic history. *Journal of Arid Environments*, **49**(3), 429–448.

Caine, N. (1980). The rainfall intensity-duration control of shallow landslides and debris flows. *Geografiska Annaler*, **A62**(1–2), 23–27.

Cayan, D. R. and Webb, R. H. (1992). El Niño/Southern oscillation and streamflow in the Western United States. In H. F. Diaz and V. Markgraf (eds.), *El Niño: Historical and Paleoclimatic Aspects of the Southern Oscillation*. Cambridge: Cambridge University Press, pp. 29–68.

Coe, J. A., Godt, J. W. and Wilson, R. C. (1998). Distribution of debris flows in Alameda County, California triggered by 1998 El Niño rainstorms: a repeat of January 1982? *EOS*, **79** (45), 266.

Collison, A., Wade, S., Griffiths, J. and Dehn, M. (2000). Modelling the impact of predicted climate change on landslide frequency and magnitude in SE England. *Engineering Geology*, **55**, 205–218.

Corominas, J. (2001). Landslides and climate. In *Keynote Lectures, VIII ISL, Cardiff, June 2000*, ed. E. N. Bromhead, CD–ROM.

Corominas, J. and Moya, J. (1999). Reconstructing recent landslide activity in relation to rainfall in the Llobregat River basin, Eastern Pyrenees, Spain. *Geomorphology*, **30**(1–2), 79–93.

Corsini, A., Pasuto, A. and Soldati, M. (2000). Landslides and climate change in the Alps since the Late-glacial: evidence of case studies in the Dolomites (Italy). In E. Bromhead, N. Nixon and M.-L. Ibsen (eds.), *Landslides in Research, Theory and Practice*. London: Thomas Telford Publishing, pp. 329–334.

Crozier, M. (1986). *Landslides: Causes, Consequences and Environment*. London: Croom Helm.

Crozier, M. (1997). The climate-landslide couple: a Southern Hemisphere perspective. In J. A. Matthews, D. Brunsden, B. Frenzel, B. Gläser and M. M. Weiß (eds.), *Rapid Mass Movement as a Source of Climatic Evidence for the Holocene*. Palaeoclimate Research, vol. 19. Jena: Gustav Fischer, pp. 1–6.

Crozier, M. J. (1999). Prediction of rainfall-triggered landslides: a test of the Antecedent Water Status Model. *Earth Surface Processes and Landforms*, **24**(9), 825–833.

Crozier, M. J. and Glade, T. (1999). The frequency and magnitude of landslide activity. In M. J. Crozier and R. Mausbacher (eds.), *Magnitude and Frequency in Geomorphology*, Zeitschrift für Geomorphologie Supplementband, **115**, 141–155.

Dapples, F., Lotter, A. F., van Leeuwen, J. F. N. *et al.* (2002). Paleolimnological evidence for increased landslide activity due to forest clearing and land-use since 3600 cal BP in the western Swiss Alps. *Journal of Paleolimnology*, **27**, 239–248.

Dehn, M. (1999). Application of an analog downscaling technique to the assessment of future landslide activity: a case study in the Italian Alps. *Climate Research*, **13**, 103–113.

Dehn, M. and Buma, J. (1999). Modelling future landslide activity based on general circulation models. *Geomorphology*, **30**(1–2), 175–187.

Dehn, M., Bürger, G., Buma, J. and Gasparetto, P. (2000). Impact of climate change on slope stability. *Engineering Geology*, **55**, 193–204.

Dikau, R. and Schrott, L. (1999). The temporal stability and activity of landslides in Europe with respect to climatic change

(TESLEC): main objectives and results. *Geomorphology*, **30**, 1–12.

Dixon, N. and Brook, E. (2007). Impact of predicted climate change on landslide reactivation: case study of Mam Tor, UK. *Landslides*, **4**(2), 137–147.

Ellen, S. D. and Wieczorek, G. F. (eds.) (1988). *Landslides, Floods, and Marine Effects of the Storm of January 3–5, 1982, in the San Francisco Bay Region, California*. U.S. Geological Survey Professional Paper, 1434.

Frenzel, B., Matthews, J. A. and Gläser, B. (1993). *Solifluction and Climatic Variation in the Holocene*. Palaeoclimate Research, vol. 11. Jena: Gustav Fischer, pp. 1–9.

Fukuoka, M. (1980). Landslides associated with rainfall. *Geotechnical Engineering*, **11**, 1–29.

Gamper, M. (1993). Holocene solifluction in the Swiss Alps: dating and climatic implications. In B. Frenzel, J. A. Matthews and B. Gläser (eds.), *Solifluction and Climatic Variation in the Holocene*. Palaeoclimate Research, vol. 11. Jena: Gustav Fischer, pp. 1–9.

Glade, T. (2003). Landslide occurrence as a response to land use change: a review of evidence from New Zealand. *Catena* **51** (3–4), 297–314.

Glantz, M. (1995). *Currents of Change: El Niño's Impact on Climate and Society*. Cambridge: Cambridge University Press.

Godt, J. W. (1999). Maps showing locations of damaging landslides caused by El Niño rainstorms, winter season 1997–98, San Francisco Bay region, California. USGS http://pubs.usgs.gov/mf/1999/mf-2325/.

Godt, J. W., Highland, L. M. and Savage, W. Z. (1997). El Niño and the National Landslide Hazard. Outlook for 1997–1998. *U.S. Geological Survey*, fact sheet 180–97.

González Díez, A., Salas, L., Díaz de Terán, J. R. and Cendrero, A. (1996). Late Quaternary climate changes and mass movement frequency and magnitude in the Cantabrian region, Spain. *Geomorphology*, **15**(3–4), 291–309.

Goudie, A. (1992). *Environmental Change*, 3rd edition. Oxford: Clarendon Press.

Guzzetti, F., Peruccacci, S., Rossi, M. and Stark, C. P. (2008). The rainfall intensity–duration control of shallow landslides and debris flows: an update. *Landslides*, **5**(1), 3–17.

Haas, J. N., Richoz, I., Tinner, W. and Wick, L. (1998). Synchronous Holocene climatic oscillations recorded on the Swiss Plateau and at the timberline in the Alps. *The Holocene*, **8**, 301–304.

Harris, C., Davies, M. C. R. and Etzelmüller, B. (2001). The assessment of potential geotechnical hazards associated with mountain permafrost in a warming global climate. *Permafrost and Periglacial Processes*, **12**, 145–156.

Holm, K., Bovis, M. and Jacob, M. (2004). The landslide response of alpine basins to post-Little Ice Age glacial thinning and retreat in southwestern British Columbia. *Geomorphology*, **57**(3–4), 201–216.

Hormes, A., Muller, B. U. and Schluchter, C. (2001). The Alps with little ice: evidence for eight Holocene phases of reduced glacier extent in the Central Swiss Alps. *The Holocene*, **11**, 255–265.

Hurrell, J. W. (1995). Decadal trends in the North Atlantic oscillation: regional temperatures and precipitation. *Science*, **269**, 676–679.

Ibsen, M. L. and Brunsden, D. (1997). Mass movement and climatic variation on the south coast of Great Britain. In J. A. Matthews, D. Brunsden, B. Frenzel, B. Gläser and M. M., Weiß (eds.), *Rapid Mass Movement as a Source of Climatic Evidence for the Holocene*. Palaeoclimate Research, vol. 19. Jena: Gustav Fischer, pp. 171–182.

IPCC (2007). *Fourth Assessment Report: Climate Change 2007*. www.ipcc.ch (December 2008).

Jibson, R. W. (1989). Debris flows in Southern Puerto Rico. *Geological Society America Special Publication*, **236**, 1–13.

Korup, O., McSaveney, M. J. and Davies, T. R. H. (2004). Sediment generation and delivery from large historic landslides in the Southern Alps, New Zealand. *Geomorphology*, **61**, 189–207.

Lateltin, O., Beer, C., Raetzo, H. and Caron, C. (1997). Landslides in Flysch terranes of Switzerland: causal factors and climate change. *Eclogae Geologicae Helvetiae*, **90**, 401–406.

Magny, M. (1999). Lake-level fluctuations in the Jura and French subalpine ranges associated with ice-rafting events in the North Atlantic and variations in the Polar Atmospheric Circulation. *Quaternaire*, **10**(1), 61–64.

Maisch, M., Wipf, A., Denneler, B., Battaglia, J. and Benz, C. (2000). *Die Gletscher der Schweizer Alpen: Gletscherhochstand 1850, aktuelle Vergletscherung, Gletscherschwund Szenarien*. Zürich: VdF Verlag.

Malet, J.-P., van Asch, Th. W. J., van Beek, R. and Maquaire, O. (2005). Forecasting the behaviour of complex landslides with a spatially distributed hydrological model. *Natural Hazards and Earth System Sciences*, **5**, 71–85.

Margielewski, W. (2001). Late glacial and Holocene climatic changes registered in forms and deposits of the Klaklowo landslide (Beskid Średni range, outer Carpathians). *Studia Geomorphologica Carpatho-Balcanica*, **35**, 63–79.

Marques, R., Zezere, J., Trigo, R., Gaspar, J. and Trigo, I. (2008). Rainfall patterns and critical values associated with landslides in Povoacao County (Sao Miguel Island, Azores): relationships with the North Atlantic Oscillation. *Hydrological Processes*, **22**(4), 478–494.

Matthews, J. A., Brunsden, D., Frenzel, B., Gläser, B. and Weiß, M. M. (eds.) (1997). *Rapid Mass Movement as a Source of Climatic Evidence for Holocene*. Palaeoclimate Research, vol. 19. Jena: Gustav Fischer.

Moreiras, S. M. (2005). Climatic effect of ENSO associated with landslide occurrence in the Central Andes, Mendoza Province, Argentina. *Landslides*, **2**, 53–59.

Ngecu, W. M. and Mathu, E. M. (1999). The El-Niño triggered landslides and their socioeconomic impact in Kenya. *Environmental Geology*, **38**, 277–284.

Noverraz, F., Bonnard, C., Dupraz, H. and Huguenin, L. (1998). *Grands Glissements de versant et Climat*. Rapport final PNR 31. Vdf. Zurich.

Panizza, M., Pasuto, A., Silvano, S. and Soldati, M. (1996). Temporal occurrence and activity of landslides in the area of Cortina d'Ampezzo (Dolomites, Italy). *Geomorphology*, **15** (3–4), 311–326.

Philander, S. G. (1999). El Niño and La Niña predictable climate fluctuations. *Reports on Progress in Physics*, **62**, 123–142.

Reynolds, R., Dettinger, M., Cayan, D., *et al.* (1997). *Effects of El Niño on Streamflow, Lake Level, and Landslide Potential*. U.S. Geological Survey Open File Report.

Schmidt, J. and Dikau, R. (2004). Modelling historical climate variability and slope stability. *Geomorphology*, **60**(3–4), 433–447.

Schoeneich, P., Tercier, J., Hurni, J.-P. and Orcel, C. (1997). Datation par dendrochronologie du glissement des Parchets (Les Diablerets, Alpes vaudoises). *Eclogae Geologicae Helvetiae*, **90**(3), 481–496.

Sidle, R. C., Taylor, D., Lu, X. X. *et al.* (2004). Interactions of natural hazards and society in Austral–Asia: evidence in past and recent records. *Quaternary International*, **118–119**, 181–203.

Sletten, K., Blikra, L. H., Ballantyne, C. K., Nesje, A. and Dahl, S. O. (2003). Holocene debris flows recognized in a lacustrine sedimentary succession: sedimentology, chronostratigraphy and cause of triggering. *The Holocene*, **13**(6), 907–920.

Smith, L. N. (2001). Columbia Mountain landslide: late-glacial emplacement and indications of future failure, Northwestern Montana, USA. *Geomorphology*, **41**, 309–322.

Soldati, M. and Borgatti, L. (2009). Paleoclimatic significance of Holocene slope instability in the Dolomites (Italy). *Geografia Fisica e Dinamica Quaternaria*, **32**, 83–88.

Soldati, M., Corsini, A. and Pasuto, A. (2004). Landslides and climate change in the Italian Dolomites since the Lateglacial. *Catena*, **55**(2), 141–161.

Starkel, L. (1991). Younger Dryas-Preboreal transition and during the early Holocene: some distinctive aspects in central Europe. *The Holocene*, **1**, 234–242.

Thomas, M. F. (1999). Evidence for high energy landforming events of the central African plateau: eastern province, Zambia. *Zeitschrift für Geomorphologie N.F.*, **43**(3), 273–297.

Tinner, W. and Ammann, B. (2001). Timberline paleoecology in the Alps. *Pages News* **9**(3), 9–11.

Tinner, W. and Kalterieder, P. (2005). Rapid response of high-mountain vegetation to early Holocene environmental changes in the Swiss Alps. *Journal of Ecology*, **93**(5), 936–947.

Trauth, M. H., Alonso, R. A., Haselton, K. R., Hermanns, R. L. and Strecker M. R. (2000). Climate change and mass movements in the NW Argentine Andes. *Earth and Planetary Science Letters*, **179**(2), 243–256.

Trauth, M. H., Bookhagen, B., Marwan, N. and Strecker M. R. (2003). Multiple landslide clusters record Quaternary climate changes in the northwestern Argentine Andes. *Palaeogeography, Palaeoclimatology, Palaeoecology*, **194** (1–3), 109–121.

Trigo, R. M., Osborn, T. J. and Corte-Real, J. (2002). The North Atlantic Oscillation influence on Europe: climate impacts and associated physical mechanisms. *Climate Research*, **20**, 9–17.

Trigo, R. M., Zezere, J. L., Rodrigues, M. L. and Trigo, I. F. (2005). The influence of the North Atlantic Oscillation on rainfall triggering of landslides near Lisbon. *Natural Hazards*, **36**, 331–354.

Van Asch, Th. W. J. and Buma, J. T. (1997). Modelling groundwater fluctuations and the frequency of movement of a landslide in the Terres Noires Region of Barcelonnette (France). *Earth Surface Processes and Landforms*, **22**, 131–141.

Van Asch, Th. W. J., Buma J. and Van Beek, L. P. H. (1999). A view on some hydrological triggering systems in landslides. *Geomorphology*, **30**(1–2), 25–32.

Van Beek, L. P. H. (2002). The impact of land use and climatic change on slope stability in the Alcoy region, Spain. PhD Thesis, Utrecht University.

Viles, H. A. and Goudie, A. S. (2003). Interannual, decadal and multidecadal scale climatic variability and geomorphology. *Earth-Science Reviews*, **61**, 105–131.

Walker, G. T. (1923). Correlation in seasonal variations of weather. Part VIII: A preliminary study of world weather. *Memoirs of Indian Meteorological Department*, **24**, 75–131.

Wieczorek, G. F. (1996). Landslides triggering mechanisms. In A. K. Turner and R. L. Schuster (eds.), *Landslides: Investigation and Mitigation*. Transportation Research Board, Special Report 247, Washington, D.C.: National Academy Press, pp. 76–90.

Wieczorek, G. F., Lips, E. W. and Ellen S. D. (1989). Debris flows and hyperconcentrated floods along the Wasatch Front, Utah, 1983 and 1984. *Bulletin of the Association of Engineering Geologists*, **26**, 191–208.

Winter, M. G., MacGregor, F. and Shackman, L. (eds.) (2005). *Scottish Road Network Landslide Study*. Edinburgh: Scottish Executive.

9 The hazardousness of high-magnitude floods

Avijit Gupta

9.1 Introduction

The word flood has a common usage, implying a situation where water in a river spills over its banks. It can be defined as a hydrologic condition where the river discharge exceeds the storage capacity of the channel and the excess water overflows and inundates part of the valley bottom. The height of the overflow and the extent of inundation depend on the size of the flood, which in turn is related to its frequency. In any given location, bigger floods occur rarely and smaller floods are common. Probability of a flood of a given size is frequently expressed in terms of its recurrence interval, i.e., the time period within which the flood of a given size is expected to occur once. Flooding is a normal and expected phenomenon for a river, but as riverbanks are commonly populated, larger floods carry a hazardous component.

All rivers flood, although the magnitude and frequency of floods vary among them. Rivers flowing through certain environments tend to be more flood prone, e.g., those draining hilly basins located across common paths of tropical cyclones. A flood discharge is capable of enlarging the channel, transporting large quantities of sediment, and physically transforming part of the valley flat. Flood pulses also control valley bottom ecology, ranging from rivers with large floodplains, as documented for the Amazon (Junk, 1997), to steep rock-cut canyons, as for the Colorado (Webb *et al.*, 1999). The recurrence of large floods in a river, capable of remodeling the channel and the floodplain, depends on meteorological and physiographical settings. River forms in an arid climate are flood dependent, although these floods may recur at intervals of up to hundreds of years (Schick, 1988; Patton *et al.*, 1993; Bourke and Pickup, 1999). In parts of the seasonal tropics, channel forms are dependent on both large floods

and high flows of the wet season (Gupta, 1975, 1995). The effect of floods on the channel and floodplains in the humid temperate areas could be temporary (Gupta and Fox, 1974), and alluvial channel forms and floodplain evolution may be related not to floods but to bankfull discharges with a recurrence interval between 1 and 2 years (Leopold *et al.*, 1964). Rivers that are flood prone tend to be associated with a characteristic basin physiography that causes intense localized rainfall, rapid surface runoff, and time-synchronization of tributary flood peaks reaching the main-stem. Besides meteorological causes, large floods also result from collapse of anthropogenic dams, ice-jams and moraines or landslide-related barriers. Volcanic eruptions that generate enough heat to melt large volumes of ice and snow also cause flooding. This paper is focused on the nature and function of large floods, their geographical location and hazardousness.

Flooding is a natural and expected phenomenon for a river and periodic flooding modifies its channel and maintains the ecological characteristics of its floodplain. Even high-magnitude floods are expected phenomena for rivers in certain parts of the world. Flooding becomes a hazard when the valley flat of a river is utilized for settlements or economic pursuits. This becomes disastrous when floods and human settlements compete for the same location, e.g., alluvial fans (Larsen, *et al.* 2001a, b), mountain valleys (Stewart and LaMarche, 1967) or low-lying and as yet unfilled alluvial plains (Sarker *et al.*, 2003). Huge efforts are usually needed to deal effectively with such hazards.

A change in climate modifies flood potential. The size and potential of river floods have varied across the Quaternary (O'Connor and Baker, 1992; Baker *et al.*, 1993; Ely *et al.*, 1993; Knox, 1993, 2000; Clarke *et al.*, 2003; Baker, 2007; Blum, 2007). The current floods are probably smaller and less frequent than those in the Late

Geomorphological Hazards and Disaster Prevention, eds. Irasema Alcántara-Ayala and Andrew S. Goudie. Published by Cambridge University Press. © Cambridge University Press 2010.

Pleistocene or Early Holocene. It is possible that flood potential will increase in the future with ongoing climate change, but research in this area is still at a preliminary stage.

9.2 Flood climate

Hayden (1988), in an analysis of flood-producing climates, indicated that barotropic conditions prevail in the tropical latitudes, where a reservoir of atmospheric moisture exists due to high evaporation rates and moisture-holding capacity. The release of intense precipitation from this reservoir requires the presence of tropical storms, easterly waves, local convection, and convergence and uplift of air at the Intertropical Convergence Zone (ITCZ). This advection, convergence and uplift result in high and intense rainfall. Significant wind shear with height is rare in the tropics, and tall convective clouds develop, driven by the latent heat of condensation, as a precursor to thunderstorms or tropical cyclones. Lamb (1972) has mapped an annual distribution of thunderstorms identifying their important role in the hydrology of the Amazon Basin, coastal West Africa, Central Africa, western Madagascar and the islands of Southeast Asia and New Guinea. Tropical storms are initiated in summer when the sea-surface temperature is 27 °C or more and enough Coriolis force exists for generating circulation round a central core. Intense tropical storms therefore usually occur between 10° and 30° latitudes in summer. Tracks of such storms tend to follow a common pattern, moving east to west over the tropical oceans

(Table 9.1), except the South Atlantic, where cooler temperature and vertical wind shear restrict their growth. Hence flood-producing rains from these storm events tend to fall over certain geographical areas, enhanced in places where storm paths cross orographic barriers (Figure 9.1). A proportion of these storms develop into tropical cyclones. Seasonal fluctuations (known as the monsoon) of moisture-bearing winds also give rise to high rainfall and storm events in parts of the humid tropics. In geographical areas where the monsoon system prevails, as in South, Southeast and part of East Asia, flooding commonly happens when a tropical storm invades an area that has already experienced high seasonal rainfall (Gupta et al., 1999; Wood and Ziegler, 2008). This happened in several locations in the

TABLE 9.1. *Estimated annual frequency of tropical cyclones*

Area of origin	Simpson and Riehl (1981)	Reynolds (1985)
Northwest Pacific	22	26
Northeast Pacific	15	13
Australian seas	5	10
Northwest Atlantic	8	9
Southern Indian Ocean	5	8
Northern Indian Ocean	8	6
South Pacific	5	6

The general pattern is the same in both estimates. From Gupta (1988).

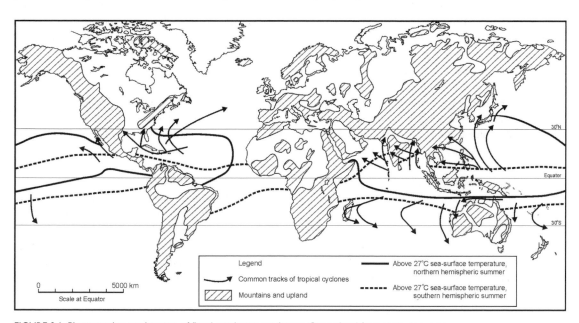

FIGURE 9.1. Physiographic combination of flood-producing conditions. Generalized from various sources.

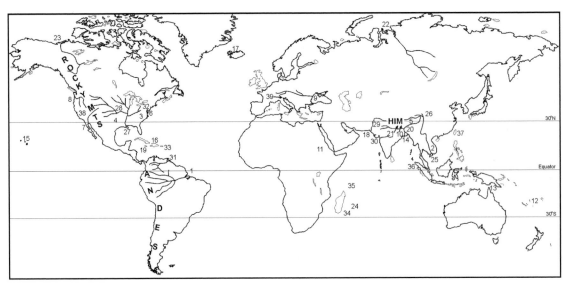

FIGURE 9.2. Location map of places mentioned in the text. Due to scale restriction, only approximate locations shown.

1. Amazon R., 2. Annamite Mts., 3. Appalachian Mts., 4. Balcones Escarpment, 5. Brahmaputra R., 6. Chesapeake Bay, 7. Colorado R., 8. Columbia R., 9. Danube R., 10. Darjeeling, 11. Ethiopian Plateau, 12. Fiji., 13. Fly R., 14. Ganga R., 15. Hawaiian Is., 16. Hispaniola, 17. Iceland, 18. Indus R., 19. Jamaica, 20. Khasi Hills, 21. Kosi R., 22. Lena R., 23. Mackenzie R., 24. Mauritius, 25. Mekong R., 26. Min R., 27. Mississippi R., 28. Missouri R., 29. Nanga Parbat, 30. Narmada R., 31. Orinoco R., 32. Patuxent R., 33. Puerto Rico, 34. Réunion, 35. Seychelles, 36. Sumatra, 37. Taiwan, 38. Teton Dam, 39. Vaiont Dam.

Indian subcontinent in 2008 including the disastrous flooding of the Kosi River Fan (Figure 9.2). Sikka (1977) has referred to eight to ten storms arriving in India in an average rainy season, each with a lifetime of two to five days. These are large storms, 1000–1300 km across with a 7–9 km deep cyclonic circulation with an anticyclonic outflow above. Flooding is obviously rare in arid areas due to the low moisture content of the atmosphere, but incursions of moist air may periodically cause intense rain, leading to rare occasions of flooding (Schick, 1988). Briefly, precipitation capable of producing high-magnitude floods in the tropics arises from cyclonic movements and convergences in tropical storms, easterly waves, and thunderstorms, especially if heightened by orographic uplift. Thus, expectedness of floods is high for certain regions (Figure 9.1).

In higher latitudes, baroclinic conditions prevail and precipitation results from temperate cyclones and frontal convergence. Convergence leading ultimately to precipitation in the higher latitudes is caused by the meeting of air masses of different temperature and moisture content. Commonly such frontal rainfall lacks the deep convective process of the tropics and precipitation falls for a long time period but at a gentle rate. Hayden (1988) has attributed flood-producing precipitation to polar fronts, which separate the barotropic tropical atmosphere from the middle and high latitude baroclinic conditions. High-magnitude floods may occur, but under special circumstances, such as

incursion of remnants of tropical storms to higher latitudes, cumulative rain from periodic passages of a series of large depressions along the polar front, and rainfall on melting snow on the lower slopes of mountains. Little moisture is present close to the poles, and flood-producing meteorological storms are an anomaly.

Of all the flood-related data, rainfall records are the most readily available. Very heavy rainfall has been recorded where depressions and storms have collided against orographic barriers. Over 4,000 mm of rain fell between 11 and 19 March in 1952 in Cilaos, Réunion, which is a volcanic island in the southwestern Indian Ocean (Landsberg, mentioned in Flores and Balagot, 1969). Data on high rainfall for one day and total storm rainfall over several days have been compiled for the humid tropics (Gupta, 1988). One-day rainfall at the scale of 10^2 mm and a total in 10^3 mm accumulated over several days are not unexpected in a tropical storm. For example, 24-hour rainfalls in the Buff Bay and Yallahs river basins of eastern Jamaica have been estimated to have a range of 130–305, 290–584 and 355–735 mm for recurrence intervals of 2, 10 and 25 years respectively (Gupta, 1983). High-magnitude rainfall, however, may occur repeatedly within a short time period. Rainfall from Typhoon Sinlaku (14 September 2008) ranged between 500 and 1000 mm, depending on location in hilly Taiwan. Such rainfall was repeated in the next typhoon, Jangmi (28 September 2008). Four large rainfall

floods occurred in the basin of the Patuxent River in Maryland, USA, between 3 August 1971 and 24 June 1972. Two of these floods had recurrence intervals of about 100 years (Gupta and Fox, 1974). Intense and large amounts of rainfall that translate into high-magnitude floods therefore tend to be associated with the humid tropics between 10° and 30° of latitude. Large floods of meteorological origin occur frequently in the higher humid tropics. Unlike rainfall, data for high-magnitude flood discharges are not available across the world. Sediment fluxes and their chemical quality are very rarely recorded in large floods because of the inherent measurement problems.

9.3 Non-meteorological floods

Floods also occur from causes that are not climatic. These floods may happen in two ways: collapse of a barrier in a valley with an upstream lake, and thermal melting of a large body of accumulated ice and snow. An earthquake or a severe rainstorm or both may produce a landslide to block a river and fill a reservoir upstream in the valley. These natural barriers ultimately give way to produce a catastrophically damaging flood. A narrow steep-sided valley, usually in tectonically unstable mountains with a failure-prone geological structure and lithology, is a prime suspect. Slopes are more likely to fail and blocked reservoirs are filled quickly when seismic disturbances occur in the rainy season. Certain locations meet these requirements. For example, floods from landslide-dam failures occur repeatedly in the Himalaya Mountains, the mountains of Sichuan Province in China, and the hilly Caribbean Islands. An often-used example is the earthquake-caused collapse of a section of Nanga Parbat, northwestern Himalaya, that blocked the Indus River during the winter of 1840–1 forming an upstream lake, 64 km long and over 300 m deep. The dam was breached by the Indus in the following June and a tremendous flood swept down the river. Rising to 30 m in height, this flood washed away part of an army camp on the floodplain at Attock, 400 km downstream of the dam (Mason, 1929). In 1933, following the 7.5 magnitude Diexi earthquake, about 1.5×10^8 m^3 of material fell into the valley of the Min, a northbank tributary of the Changjiang (Yangtze) draining the mountainous Sichuan region, blocked the river, and created a huge lake (Tang et al., 1994). The Min eventually broke through the barrier but a huge lake still exists in its valley. A similar event happened in 2008 in a tributary of the Min, following the 7.9 earthquake in Sichuan.

Floods from collapsed landslide dams are relatively rare, compared to the frequency of meteorological floods, but as most of these failures are associated with high blockage of rivers in gorges (Hewitt, 1982), they are usually hazardous. Such floods also tend to arrive unexpectedly or at short notice. The hydrograph rises very steeply to a high peak, and extreme physical damage happens during the early stage of the flood. The hydrograph is usually short-based as the impounded water tends to empty quickly. O'Connor and Costa (2004) identified at least six historic floods of this type with more than 10^5 m^3 s^{-1} peak discharges. They also refer to the stratigraphic evidence of a landslide-breached flood in the Columbia River about 500 years ago that peaked approximately at 220,000 m^3 s^{-1}. The blocked Columbia rose to a height of 70–80 m behind the landslide, rather idiosyncratically known as the Bridge of the Gods landslide. O'Connor and Costa (2004) listed the current largest dam of this genre as the 550 m-high Usoi landslide dam created by a large earthquake in 1911, with 16 km^3 Lake Sarez behind it in Tadjikistan. The existence of 5 million people downvalley illustrates the hazardousness of such barriers.

Floods also occur due to collapse of barriers of ice. This type of flood usually happens in large north-flowing arctic rivers, such as the Lena or the Mackenzie. The floods happen due to early melting of the southern headwaters while the northern downstream section remains jammed with ice before a sudden collapse and passage of the flood. Some of these floods can be very large. Ice-jam floods are confined to higher latitudes and occur in late spring or early summer. Ice-related floods occur also in temperate latitudes but these are of lesser dimensions. For example, an ice dam on the Missouri in North Dakota eroded in April 1952, increasing the discharge from 2,100 m^3 s^{-1} to 14,000 m^3 s^{-1} within 24 hours (O'Connor and Costa, 2004). Glacial ice and moraine impounded lake outbursts may create large floods. The frequency of this type may increase due to climate change.

Glacial-outburst floods are known as jökulhlaups. In rare instances, volcanic eruptions create floods by melting glacial ice. The ice-melt floods from the heat of volcanic activities could be of extreme dimensions (Tomasson, 1996; Snorrason et al., 2002), but these obviously are location related. A high-magnitude flood event occurred in 1996 in southeastern Iceland, following a volcanic eruption below the Vatnajokull glacier. The glacial meltwater first reached a caldera lake, and subsequently floated the overlying ice cap. About 3.6 km^3 of water was released with a peak discharge of 50,000 m^3 s^{-1} (Snorrason et al., 2002).The largest flood of this kind in historic times occurred in 1918 when a volcanic eruption occurred below glacial ice at Katla, Iceland. The resulting discharge reached 300,000 m^3 s^{-1} at peak. An even bigger jökulhlaup, with a peak discharge of 700,000–1,000,000 m^3 s^{-1} has

been recognized to have occurred about 7,000 years earlier in Jökulsá á Fjöllum, Iceland (Carrivick et al., 2004; O'Connor and Costa, 2004).

Flooding from the failure of a barrier or partial melting of a glacier produces characteristic hydrographs with steeper limbs than those from high rainfall or snowmelt floods. Similar hydrographs also result from the failure of engineered dams, which are not that uncommon. Costa (1988) refers to an inspection of 8,639 dams by the U.S. Army Corps of Engineers in 1981, who found a third of these unsafe. Both dam heights and reservoir volumes determine the size of floods in dam failures. The largest flood from dam failure on record is the 1976 failure of the Teton Dam in Idaho releasing a downstream flood that peaked at $70,000\,\mathrm{m}^3\,\mathrm{s}^{-1}$ (O'Connor and Costa, 2004). A rather unusual flood occurred in 1963 when a large slope failure slid into the reservoir behind Vaiont Dam in Italy. The resulting flood wave spilled over the dam and disastrously inundated the valley downstream. The dam, however, survived. Release of water from reservoirs prior to large rainfall events may also cause hazards downvalley, especially when warnings are not timely or heeded.

9.4 Flood physiography

Certain aspects of basin physiography (alignment of highlands, drainage pattern, types of landforms) accelerate flooding. The windward side of a steeply rising highland (young fold mountain, plateau edge, etc.) facing moisture-bearing winds often experiences heavy rainfall that may, on occasion, translate into floods. As an extreme example, the Khasi Hills at the southern edge of the Meghalaya (Shillong) Plateau, eastern India, that intercept the wet southwest monsoon receive nearly 10,000 mm of rain a year. The southern slopes of the Himalaya, eastern slopes of the northern Andes, the Ethiopian Plateau, and the mountainous islands in the Caribbean and other tropical seas all are good examples of physical barriers against moisture-bearing winds. For example, the windward sides of the Hawaiian Islands, Sumatra, Taiwan and Réunion have all recorded very high rainfall from time to time.

Floods in the mountains usually remain confined within steep and deep valleys (see Carrivick et al., 2004), but tend to become hazardous at the highland–lowland contact where floodwater may spread across alluvial fans. A series of large fans marks the transition from the lower southern slopes of the Himalaya to the Ganges Plains. Such fans carry unstable braided channels that change course in floods. The Kosi River, which has built one of the biggest fans, has changed its course repeatedly in floods, moving 113 km to the west in 228 years, in episodic jumps across its

megafan. The megafan measures 154×147 km (Gole and Chitale, 1966; Wells and Dorr, 1987). In 2008, however, the Kosi changed its course towards the east in a major flood. This flood, although it occurred in the middle of the wet monsoon, may not have been entirely meteorological in origin. Orographic barriers at smaller scales also cause intense local rainfall. Baker has described the role of the Balcones Escarpment in Texas in producing intense rainfall and coarse bedload in flooded streams (Baker, 1977). In a general review of the location of large floods in the United States, O'Connor and Costa (2003) have stated that their distribution is not random but reflects climate and topography. Floods tend to be common in the tropical United States (Hawaii and Puerto Rico) due to the combination of moisture availability, high frequency of storms, and hilly terrain. Within the conterminous United States large floods occur in central Texas, central and northern parts of the hilly terrain and plateaux of the Appalachian Mountains, southern Midwest, basins of large rivers with headwaters in the Rockies that drain into the Pacific, and along the Pacific coast. O'Connor and Costa (2003) identified three key factors for large floods: climate, topography, and basin size. Floods are frequent and hazardous where topographic barriers are oriented orthogonal to moisture-bearing winds, especially storm tracks, and close to the two main sources of moisture: the Pacific and the Gulf of Mexico. Persistent rain on snow in the upper parts of large river basins and convective storms over small basins both cause flood hazards. These generalizations are applicable almost anywhere. Hazards from large floods increase significantly at breaks in slope where rivers are mobile but with a significant drop in velocity. Alluvial fans drained by braided channels at the foot of mountains are especially hazardous.

Drainage basin morphometry partly shapes the flood hydrograph (Strahler, 1964). Other conditions being equal, basin shape and bifurcation ratio (which combines the number and internal arrangement of stream segments) are reflected in the flood hydrograph. In the worst-case scenario, a circular basin and a high ratio, the river would have a steep and peaked hydrograph (Patton, 1988). A basin with a high drainage density floods early as the length of overland flow is reduced, hillslope angles are increased speeding up the runoff, and the path of the floodwater is mostly in channels. Floods from high rainfall occur over longer periods and their peakedness and rate of rise and fall both depend primarily on the amount and intensity of precipitation. River basins that commonly experience high rainfall establish an organized drainage net, increasing the flood potential of the mainstem. The alignment of drainage basins with storm tracks also increases the flood potential of the trunk stream. The Narmada River of central India floods when tracks of tropical cyclones cross its basin in the wet season. Occasionally, the tracks run sub-parallel to basin alignment, and rainfall and discharge peaks travel downstream near-simultaneously (Rajaguru *et al.*, 1995; Gupta *et al.*, 1999).

Infiltration, base flow and surface runoff are controlled to some extent by basin geology and land use, but such controls work only for small floods. With the kind of rainfall mentioned earlier, almost all rivers rise to high-magnitude floods inundating the valley flats to an extraordinary degree.

9.5 Floods and geographical locations

All rivers flood, but rivers in certain parts of the world flood disastrously on a frequent basis. For example, tropical-cyclone-driven floods could be expected to affect river systems in:

North and northeastern Australia (Wohl, 1992).

Parts of Southeast and East Asia where tropical cyclones come onshore against high relief: Myanmar, Cambodia, Vietnam, the Philippines, Taiwan, southern China.

Parts of the Indian subcontinent (Kale *et al.*, 1994; Rajaguru *et al.*, 1995; Gupta *et al.*, 1999).

Madagascar and neighbouring coastal areas of East Africa.

The Caribbean islands, the coastal areas of the Gulf of Mexico, and tropical and subtropical North America (Gupta, 1975; Ahmad *et al.*, 1993; Larsen *et al.*, 2001a, b).

Volcanic islands in the Indian and Pacific Oceans against which tropical cyclones abut: Mauritius, Seychelles, Fiji.

Availability of flood studies for all these locations, however, is limited. Usually, the amount and intensity of rainfall are recorded, the loss of life and property is covered as a news item, but an explanatory account of the causes and consequences of a specific flood event is not always available. Even when they exist, such accounts may be difficult to locate due to the limited distribution of local publications. Flood reports may be limited to a few case studies from a region, which makes it difficult to generalize. Discharge and sediment load are hard to measure in high-magnitude floods and long-term records are available only for certain rivers. For example, although often a single hurricane crosses several islands of the Greater Antilles, flood data or geomorphological reports tend to be available only for Jamaica and Puerto Rico. In contrast, we have very little field information from Hispaniola, although

inundation in hurricane floods is a common hazard on the island. The data from Puerto Rico indicate that hurricane floods can have a recurrence interval of less than 10 years for the region (Gupta, 2000; Table 9.1, compiled from various sources, mostly U.S. Geological Survey data). Floods therefore operate commonly as hazards and the rivers are adjusted to large floods at small intervals. A list of areas with flood hazards exists for the United States (O'Connor and Costa, 2003) but for most areas such lists are not always available outside the region.

9.6 Water and sediment transfer in floods

After rainfall, the best recorded flood parameter probably is flood stage, which is possible to identify after the passage of the flood and is usually recorded in populated areas. Peak stages for large floods can be extremely high (Figure 9.3), inundating large areas of the valley flat. Flood measurement data from a number of case studies are available (Gupta, 1983; Kochel, 1988). Flood depth will be high in gorges, but even elsewhere records indicate depths of 10 m or more in the channel with several metres of water on the floodplain. This, apart from inundating the valley flat, leads to significant stream channel and floodplain response. Stream power data in extreme floods are not routinely computed but two gauged floods (with about 5-year recurrence interval) for the Yallahs River at Llandewey, eastern Jamaica ($A_d = 123 \text{ km}^2$) indicated that in both cases the unit stream power rose sharply with stage and remained high enough to move 500 mm boulders for about half-a-day and cobbles, pebbles and sand for the duration of the entire flood which lasted for about two days (Gupta, 2000). Hurricane floods will have even greater unit power and may last for several days. The Yallahs flows over coarse alluvium on bedrock at

FIGURE 9.3. The Narmada River immediately downstream of the Marble Canyon, Jabalpur, India. The September 1926 flood rose to the eaves of the bungalow on the cliff. (Photograph by Avijit Gupta, from Rajaguru *et al.*, 1995.)

Llandewey. In narrow bedrock rivers, bed shear stresses of $10^3 \, \mathrm{N \, m^{-2}}$ and unit stream powers of $10^4 \, \mathrm{W \, m^{-2}}$ have been calculated (Baker *et al.*, 1988; Wohl, 1992; Kale *et al.*, 1994; Gupta, 1995; Baker and Kale, 1998). The rising stage of a flood hydrograph in an alluvial channel may be associated with net filling, bringing in upstream sediment stored within the channel perimeter. A continued rise in stage leads to scouring of the bed and channel deepening. The falling stage may lead to partial filling of the channel (Leopold and Maddock, 1953). Concentration of suspended sediment is high at the beginning of the rising stage but it drops with increasing flood discharge, and often the water is quite clear towards the end of the flood. As the water level drops, banks in alluvium may collapse due to undercutting and outflow of subsurface water and liquefaction of the bank material, thus widening the channel (Gupta and Fox, 1974). The 1993 flood on the middle Mississippi showed a similar change in the channel. About 4 m of channel bed was scoured from 12 through 20 July at Chester, Illinois. The general pattern was scouring during the rising flood and bed aggradation during the falling stage (Holmes, 1996).

Floodwaters tend to top the bank or the alluvial levee and move onto the floodplain. As the sediment concentration is low at this stage, the water tends to erode, cutting chutes across the floodplain and river bends. A break in the levee results in sediment being splayed across the valley flat. In larger floods, chute-cutting and sediment splays can effectively alter floodplain morphology. In very large floods, where the river effectively scours the floodplain and uses the combined width of the channel and floodplain as the floodway, a low terrace of deposited material is left on the floodplain after the passage of the flood (Gupta, 1975; Baker, 1988).

A rising large flood increases the capacity for erosion and sediment transport in rock-cut channels, as in bedrock gorges (Hancock *et al.*, 1998), but unless sediment is continuously supplied by debris flows down tributary valleys, the excessive energy leads to macroturbulence, as manifested in the formation of large vortices. This ultimately leads to bedrock erosion, inner channel formation, and spectacular channel enlargement (Matthes, 1947; Baker, 1988).

9.7 Source-to-sink passage of a flood

Meade (2007), while analyzing sediment transfer in the Amazon and Orinoco rivers, identified large rivers as massive conveyance systems for moving clastic sediment and dissolved matter across continental distances. This applies to all rivers, the scale of the conveyance being directly related to river dimensions. Huge amounts of sediment tend to move in high-magnitude floods, although one flood is usually not enough to do the complete source-to-sink transfer except for short rivers with high gradient that flood from the headwaters to the sea. Even then a considerable volume of sediment remains stored on fans, floodplains and parts of the channel, awaiting transfer in the next large flood (Ahmad *et al.*, 1993). Starkel (1972) discussed the effect of 700–1100 mm of rainfall from a Bay of Bengal tropical cyclone in the Darjeeling Himalaya. Rock and debris falls, debris slides and avalanches, mudflows and displacement of liquefied soils by piping scarred the mountain slopes in combination, and transferred an enormous amount of material to the swollen rivers, which operated with extreme competence and capacity. Small channels carried boulders 3–5 m in diameter, and up to 12 m in suitable locations. Tributary streams eroded 2–3 m deep in rock and the main valleys were loaded with coarse sediment, building nearly 10 m high terraces and scattered detritus fans at margins. The post-flood valleys were filled with rounded and chipped 2–3 m boulders from which the finer fraction had been removed (Starkel, 1972). Brunsden *et al.* (1981) have described similar valleys in the Nepal Himalaya as floors filled with sediment ranging from silt to coarse boulders, and acting as continuous conveyor belt transportation systems. In flatter areas, the stored sediment is finer but a similar episodic transfer and storage system may operate. Meade (2007) has referred to the millenium-scale storage of the Andean-derived sediment in the floodplain of the Amazon prior to the sediment reaching the ultimate sinks: the lowermost floodplain, the delta and the coastal zone.

It is possible to prepare an indicative model for water and sediment transfer in a flood for a river from the post-flood field studies. It is reasonable to assume that our flood-prone river starts in tectonics-driven mountains, which operate as a barrier to moisture-bearing tropical storms. Most of the sediment of this river system will be derived from the mountains. About 90% of the sediment of the Amazon or the Orinoco is Andean derived. Most of the sediment of the Ganges comes from the Himalaya, that of the Brahmaputra from the Eastern Syntaxis of the Himalaya, that of the Mekong from the eastern extension of the Himalaya and the Annamite Mountains. The sediment will be coarse in nature and is likely to reflect an origin by slope failures and debris flows. If there is enough sediment, the entire valley bottom in the mountains will be covered by coarse material up to boulder size. This material will be transmitted down the main river as progressively diminishing bedload, to form alluvial fans at the contact between the mountains and the plains. Small rivers draining hilly islands or where mountain ranges approach the coast often flow into the sea over fans that are probably better described as fan-deltas (Ahmad *et al.*, 1993; Larsen *et al.*, 2001a, b). Areas of low relief are often limited to these fans, encouraging hazardous settlements on them in the absence of any alternative. The texture of sediment progressively diminishes along the river. Ahmad *et al.* (1993) described a downstream sequence of slope failures, debris flows, boulder-blocked river, small bars of coarse material, and larger bars of mostly sand and pebbles for the Rio Grande, which drains the Blue Mountains of eastern Jamaica.

Beyond the mountains, a long river will flow through a flat valley with a delta at its mouth. The channel will carry coarse bedload in flood but the finer material will be deposited on the floodplain from periodic inundations, with levees of sandy silt on channel banks and silt and clay in the backswamps beyond the levee. Crevasse splays of sand will spread over the silt–clay through levee breaks. High hydraulic heads, built up before levee breaks in the 1993 flood on the middle Mississippi and lower Missouri, produced high-energy flows that carried and deposited huge quantities of sand on floodplains. For example, 4 m of sand and 0.14 m of silt and clay were deposited over more than 20,000 ha of the floodplain, across a bend in the Missouri downstream from Hermann, Mo. (Holmes, 1996). This is a common occurrence. The volume of this transverse deposition can be considerable. In the case of the Amazon, the lateral flux of sediment has been recognized to exceed the downstream transfer (Dunne *et al.*, 1998; Meade, 2007). Similar features have been reported for other rivers, for example, the Fly of Papua New Guinea (Dietrich *et al.*, 1999). The floodplain storage period for a single grain could be long – hundreds or thousands of years. On the other hand, deposited material on the floodplain of mobile rivers can be transferred downstream after a short interval. The coarse material in the channel may have a pulsating flux, being removed during the rising stage of the hydrograph and being replaced by upstream material later in the flood. The ultimate sinks obviously are the deltas, coastal zones and offshore slopes (Figure 9.4).

This is a simple descriptive model of sediment transfer in floods but it provides the background for identifying riverine flood hazards. The overbank inundation is an immediate, and at times, life-threatening problem, but it is a short-term problem. Flood inundation commonly

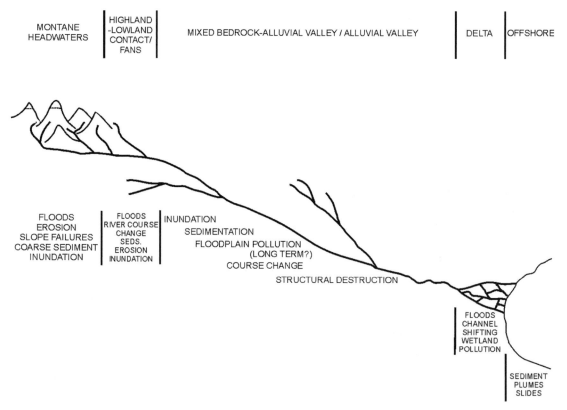

FIGURE 9.4. Sketch indicating source-to-sink flood hazards in a hypothesized large river.

lasts for several days. In exceptional cases, as in the 1993 flood of the Middle Mississippi and Missouri Rivers that broke the existing records, water may remain overbank for months. The hazard from floodplain sedimentation, however, may be long lasting, with extensive burial of farmland soil under sand as the river straightens its course across bends. The grains of silt and clay that are deposited on the floodplain beyond the levee are potential carriers of contaminants such as pesticides and industrial pollution such as heavy metals. Mapping of overbank silt therefore could be a mapping of hazardous elements. Areas where rivers are mobile, such as on fans or where they are able to cut chutes across bends in floods, are particularly hazardous. Such hazardous zones are discussed in the next section.

The flood environment is different in the arid tropics. Floodwater tends to disappear in the coarse alluvium of the channel after flowing for a short distance downstream. Large floods that persist down channel usually happen from rare and unusual meteorological conditions (Schick, 1988). Flooding usually is short term, and originates from desert cloudbursts, or, for rivers that begin in the mountains, from rain on the highlands. Commonly, a runoff-producing event is characterized by a hydrograph with a near-vertical rise and a sharp fall. Such flash floods in small- and medium-sized basins have the ability to considerably modify the channel as they arrive with high velocities, and even supercritical flows are not unknown. Frequently, a metre-high wall of water leads the flood wave. However, if the alluvial fill is present also in the main channels, a downstream diminution of flood sizes and a total disappearance may happen. The relatively rare protracted flows may modify the channel when they occur, but they occur rarely. Desert channels are therefore widened and filled episodically. Channels of different sizes corresponding to floods of varying dimensions and recurrence have been identified, with progressively smaller channels nesting inside bigger ones. Bourke and Pickup (1999) stated that the forms in the Todd River channelway in Central Australia happen episodically and the probability of the form preservation is directly related to the flood size. Flood aggradation builds channel and floodplain insets, deposits overbank veneers of sediment, and causes channels to fill. The sedimentation sequence is interrupted by unconformities and has been described as chaotic (Bourke, 1994). In the arid Central Australia, rivers build low-angle alluvial piedmont fans as they exit mountain ranges. Downslope on flatter ground, the water often spreads out as wide unconfined sheetfloods (locally known as floodouts), finally disappearing into the subsurface (Tooth, 1999).

9.8 Types of flood hazard and their location

Flood hazards are not limited to inundation of populated floodplains. Floods also deposit coarse sediment as crevasse splays and fine sediment in sheets across valley flats, contaminate the soil by depositing pesticides and heavy metals with the clay and silt fraction of the sediment, and cause rivers to shift course across fans, low-gradient alluvial valleys and deltas. The effect of a specific flood hazard can be time dependent. Inundation of floodplains could be disastrous but the floodwater returns to the channel in days, maximally in months, as happened in the 1993 flood of the middle Mississippi River (Holmes, 1996) or that of the Kosi in 2008. In contrast, the effects of floodplain sedimentation, soil contamination and channel change may last for years, centuries, even millennia. Meade (2007) discussed two hypothetical cases with reference to sedimentation on the Amazon floodplain, although one of his examples included the extreme case of radioactive nuclei. The importance of the lateral flux and floodplain storage of sediment is being recognized, especially for major rivers flowing on low gradients along flat lowland, such as the Amazon (Dunne et al., 1998; Meade, 2007), the Fly (Dietrich et al., 1999) or the Ganges (Singh, 2007). Deposition of flood sediment can alter the landscape and become hazardous where a large volume of sediment is delivered from the upper basin with contaminated material locked in with the finer fraction. This happens where rivers drain mining landscapes, agricultural fields and polluted cities. On a floodplain, flood hydraulics and floodplain relief control the distribution of heavy metal pollution that arrives with overbank fine-grained sediment (Lewin and Macklin, 1987; Asselman and Middelkoop, 1995). A large river acquires inputs of contaminated material from a number of sources along its course. The polluted material is stored partly on the floodplain, partly in tributary mouth fans, with some reaching the deltaic wetlands. Lóczy (2007) has summarized the situation for the Danube.

Flood hazards are commonly perceived as site specific or reach specific. That may happen in many cases, as flooding is accentuated by local physiography. The effect of a large flood, however, may extend far down the channel, and different types of hazards may threaten in different places. The 1972 post-tropical-storm-Agnes flooding of rivers flowing into the Chesapeake Bay, eastern United States, damaged the fishing industry. The breaking of submarine cables offshore near the mouths of the Yallahs and Hope Rivers in eastern Jamaica, after the passage of Hurricane Flora in 1963, suggests the nature and volume of sediment pouring off the island shelf (Burke, 1967). Space for urbanization is commonly restricted to river mouth fans on mountainous islands or where a coastal range rises abruptly from the sea. In such locations, floods and debris flows periodically and catastrophically destroy parts of cities (Figures 9.5A and 9.5B), as happened on the northern coast of Venezuela in 1999 (Larsen *et al.*, 2001a, b). In urban areas, flood damage can be unexpected. The Arno has repeatedly flooded Florence with disastrous effects on the art and literary treasures of the city.

Table 9.2 is a generalization of flood hazards along a hypothetical large river organized according to location, time and land use. It is intuitively obvious that all subsets

FIGURE 9.5. Urban location on an alluvial fan (A) in northern Venezuela and city destruction (B) in the debris flows and large floods of 1999. (Photographs by Matthew C. Larsen, U.S. Geological Survey.)

TABLE 9.2. *A list of flood hazards along a hypothetical river system*

1. **Montane headwaters**
   Fast and deep flood discharge
   Floods from collapse of barriers upstream
   Accumulation of coarse sediment from slope failures and debris flows in valley bottoms
   Short-term but intensive flood erosion
   Inundation of settlements on tributary-mouth fans and valley sides
2. **Highland/lowland contact and alluvial fan**
   Low flooding of the entire or a large part of the fan
   Occasional river course change, probability of disastrous floods
   Thick deposition of new sediment at the highland–lowland contact zone
   Sediment transport across fan and deposition, degeneration of agricultural fields and settlements
3. **Mixed bedrock/alluvial valley, alluvial valley**
   Short-term inundation of valley flat with possible loss of life and property
   Coarse sediment deposition as crevasse splays and fine sediment deposition as sheets across floodplains, damage to fertile land and settlements
   Long-term accumulation of pesticides and heavy metals on floodplain, especially in depressions
   Deterioration of floodplain water quality near settlements, causing spread of gastroenteritic diseases
   River course changes leading to loss of life and property
   Destruction of engineered structures, navigational facilities, etc. on major rivers
4. **Deltas**
   Widespread flooding leading to significant loss of life and property, especially if associated with tropical cyclones
   Shifting of river channels disrupting settlements, agricultural fields, and communication systems
   Pollution of deltaic wetlands and degrading ecological effects
5. **Offshore**
   Disruption of submerged engineering structures

Not all hazards are to be expected along a single river. See Figure 9.4.

of flood hazards do not take place in every basin or at every site along a river, but a pattern of hazards does exist along the mainstem (Figure 9.4). If we are to plan amelioration of these hazards, an integrated basin-scale flood management strategy needs to be structured. Hazards that are not on this simplified list because of the rarity of their occurrence may turn out to be significant for certain basins.

9.9 Conclusions: flood hazards and climate change

Major rivers have existed on a geological timescale, adjusting to repeated climate and sea-level changes. It is widely accepted that the Early Holocene was a period of higher precipitation than the present and also a time of repeated large floods. The effect of climate change on extreme flows should be experienced throughout a river system near-instantaneously (Blum, 2007). The impact of climate change as reflected in changes in the magnitude and frequency of floods, however, is not directly proportional to increases or decreases of precipitation, and even a small change in climate may significantly increase or decrease flood sizes over a range of recurrence intervals (Knox, 1993; Blum, 2007). Aalto et al. (2003) showed that larger floods occur in the Beni and Mamore Rivers of northern Bolivia during the La Niña part of the El Niño Southern Oscillation (ENSO). We may therefore anticipate changes in the flood environment of present-day rivers with on-going changes in climate. Blum, however, has shown that the magnitude of such changes is still difficult to model, and as nearly all major rivers are structurally modified, the effect of anthropogenic changes dampens out the climatic change flood signals (Blum, 2007). Instead a number of robust predictions should provide us with qualitative indicators for river flood hazards several decades from now. It is likely that tropical oceans would remain warmer for a longer period each year, and therefore generate stronger, cyclonic circulations. Strategically located tropical and mid-latitudinal rivers therefore may experience both increasing floods and droughts. Secondly, a decrease in stored ice and snow in upper basins of higher latitudes and on high mountains everywhere may lead to earlier flooding in the year. We need to investigate the changing effect of such floods on the rivers, and also their effect on control structures across rivers. It may also be difficult to separate out the effect of a rising sea level from other effects of global warming in the lower reaches of a major river. The problem is accentuated by the restrictive nature of our knowledge regarding floods and hazards arising thereof.

Our understanding of floods in various parts of the world is rather limited. We return to this topic in Chapter 10.

Acknowledgements

This chapter has benefited from an earlier reading by Jonathan Carrivick. Lee Li Kheng drafted Figures 9.1 and 9.2. M. C. Larsen, U.S. Geological Survey kindly provided Figures 9.5A and 9.5B.

References

Aalto, R., Maurice-Bourgoin, L., Dunne, T. et al. (2003). Episodic sediment accumulation on Amazonian floodplains influenced by El Niño/Southern Oscillation. Nature, 425, 493–497.

Ahmad, R., Scatena, F. N. and Gupta, A. (1993). Morphology and sedimentation in Caribbean montane streams: examples from Jamaica and Puerto Rico. Sedimentary Geology, 85, 157–169.

Asselman, N. E. M. and Middelkoop, H. (1995). Floodplain sedimentation: quantities, patterns and processes. Earth Surface Processes and Landforms, 20, 481–499.

Baker, V. R. (1977). Stream-channel response to floods, with examples from central Texas. Bulletin, Geological Society of America, 88, 1057–1081.

Baker, V. R. (1988). Flood erosion. In V. R. Baker, R. C. Kochel and P. C. Patton (eds.), Flood Geomorphology. New York: Wiley, pp. 81–95.

Baker, V. R. (2007). Greatest floods and largest rivers. In A. Gupta, (ed.), Large Rivers: Geomorphology and Management. Chichester: Wiley, pp. 65–74.

Baker, V. R. and Kale, V. S. (1998). The role of extreme floods in shaping bedrock channels. In K. J. Tinkler and E. E. Wohl (eds.), Rivers Over Rocks: Fluvial Processes in Bedrock Channels. Washington, D. C.: American Geophysical Union Monograph 107, pp. 153–165.

Baker, V. R., Kochel, R. C. and Patton, P. C. (1988). Flood Geomorphology. New York: Wiley.

Baker, V. R., Benito, G. and Rudoy, A. N. (1993). Paleohydrology of late Pleistocene superflooding, Altay Mountains, Siberia. Science, 259, 348–350.

Blum, M. D. (2007). Large river systems and climate change. In A. Gupta, (ed.), Large Rivers: Geomorphology and Management. Chichester: Wiley, pp. 627–659.

Bourke, M. (1994). Cyclical construction and destruction of flood dominated flood plains in semiarid Australia. In L. J. Olive, R. J. Loughran and J. A. Kesby, (eds.) Variability in Stream Erosion and Sediment Transport. Wallingford: International Association of Hydrological Sciences Publication 224, pp. 113–123.

Bourke, M. and Pickup, G. (1999). Fluvial form variability in arid central Australia. In A. J. Miller and A. Gupta, (eds.), Varieties of Fluvial Form. Chichester: Wiley, pp. 249–271.

Brunsden, D., Jones, D.K.C., Martin, R.P. and Doornkamp, J.C. (1981). The geomorphological character of part of the Low Himalaya of Eastern Nepal. *Zeitschrift für Geomorphologie*, **37**, 25–72.

Burke, K. (1967). The Yallahs Basin: a sedimentary basin southeast of Kingston, Jamaica. *Marine Geology*, **5**, 45–60.

Carrivick, J.L., Russell, A.J. and Tweed, F.S. (2004). Geomorphological evidence for jökulhlaups from Kverkfjöll volcano, Iceland. *Geomorphology*, **63**, 81–102.

Clarke, G., Leverington, D., Teller, J. and Dyke, A. (2003). Superlakes, megafloods, and abrupt climate change. *Science*, **301**, 922–923.

Costa, J.E. (1988). Floods from dam failures, In V.R. Baker, R.C. Kochel and P.C. Patton, (eds.), *Flood Geomorphology*. New York: Wiley, pp. 439–463.

Dietrich, W.E., Day, G. and Parker, G. (1999). The Fly River, Papua New Guinea: inferences about river dynamics, floodplain sedimentation and fate of sediment. In A.J. Miller and A. Gupta (eds.), *Varieties of Fluvial Form*. Chichester: Wiley, pp. 345–376.

Dunne, T., Mertes, L.A.K., Meade, R.H., Richey, J.E. and Forsberg, B.R. (1998). Exchanges of sediment between the flood plain and channel of the Amazon River in Brazil. *Bulletin, Geological Society of America*, **110**, 450–467.

Ely, L.L., Enzel, Y., Baker, V.R. and Cayan, D.R. (1993). A 5000-year record of extreme floods and climate change in the southwestern United States. *Science*, **262**, 410–412.

Flores, J.F. and Balagot, V.F. (1969). Climate of the Philippines. In H. Arakawa (ed.), *Climates of Northern and Eastern Asia: World Survey of Climatology*. Amsterdam: Elsevier, pp. 159–213.

Gole, C.V. and Chitale, S.V. (1966). Inland-delta building activity of the Kosi River. *Journal of Hydraulic Division, American Society of Civil Engineers*, **HY2**, 111–126.

Gupta, A. (1975) Stream characteristics in eastern Jamaica, an environment of seasonal flow and large floods. *American Journal of Science*, **275**, 825–847.

Gupta, A. (1983). High-magnitude floods and stream channel response. *International Association of Sedimentologists*, Special Publication, 6, 219–227.

Gupta, A. (1988). Large floods as geomorphic events in the humid tropics. In V.R. Baker, R.C. Kochel and P.C. Patton, (eds.), *Flood Geomorphology*. New York: Wiley, pp. 301–315.

Gupta, A. (1995). Magnitude, frequency, and special factors affecting channel form and processes in the seasonal tropics. In J.E. Costa, A.J. Miller, K.W. Potter and P.R. Wilcock (eds.), *Natural and Anthropogenic Influences in Fluvial Geomorphology*. Washington, D.C.: American Geophysical Union Monograph 89, pp. 125–136.

Gupta, A. (2000). Hurricane floods as extreme geomorphic events. In M. Hassan, O. Slaymaker and S. Berkowicz (eds.), *The Hydrology-Geomorphology Interface: Rainfall, Floods, Sedimentation, Land Use*. Wallingford: IAHS Publication 261, pp. 215–228.

Gupta, A. and Fox, H. (1974). Effects of high-magnitude floods on channel form: a case study in Maryland Piedmont. *Water Resources Research*, **10**, 499–509.

Gupta, A., Kale, V.S. and Rajaguru, S.N. (1999). The Narmada River, India, through space and time. In A.J. Miller and A. Gupta (eds.), *Varieties of Fluvial Form*. Chichester: Wiley, pp. 113–143.

Hancock, G.S., Anderson, R.S. and Whipple, K.X. (1998). Beyond power: bedrock river incision process and form. In K.J. Tinkler and E.E Wohl (eds.), *Rivers Over Rock*. Washington, D.C.: American Geophysical Union Monograph 107, pp. 35–60.

Hayden, B.P. (1988). Flood climates. In V.R. Baker, R.C. Kochel and P.C. Patton (eds.), *Flood Geomorphology*. New York: Wiley, pp. 13–26.

Hewitt, K. (1982). Natural dams and outburst floods of the Karakoram Himalaya. *International Association of Hydrological Sciences*, **138**, 259–269.

Holmes, R.R., Jr. (1996). Sediment transport in the Lower Missouri and the central Mississippi rivers, June 26 through September 14, 1993. *U.S. Geological Survey Circular* 1120-I.

Junk, W.J. (ed.) (1997). *The Central Amazon Floodplain: Ecology of a Pulsating System*. Berlin: Springer.

Kale, V.S., Ely, L.L., Enzel, Y. and Baker, V.R. (1994). Geomorphic and hydrologic aspects of monsoon floods on the Narmada and Tapi rivers in central India. *Geomorphology*, **10**, 157–168.

Knox, J.C. (1993). Large increases in flood magnitude to modest changes in climate. *Nature*, **361**, 430–432.

Knox, J.C. (2000). Sensitivity of modern and Holocene floods to climate change. *Quaternary Science Reviews*, **19**, 439–457.

Kochel, R.C. (1988). Geomorphic impacts of large floods: review and new perspectives on magnitude and frequency. In V.R. Baker, R.C. Kochel and P.C. Patton (eds.), *Flood Geomorphology*. New York: Wiley, pp. 169–187.

Lamb, H.H. (1972). *Climate: Past, Present and Future*, Vol. 1. London: Methuen.

Larsen, M.C., Vásquez Conde, M.T. and Clark, R.A. (2001a). Flash-flood related hazards: landslides, with examples from the December 1999 disaster in Venezuela. In E. Gruntfest and J. Handmer, (eds.), *Coping with Flash Floods*. Dordrecht: Kluwer, pp. 259–275.

Larsen, M.C., Wieczorek, G.F., Eaton, L.S., Morgan, B.A. and Torres-Sierra, H. (2001b). Venezuela debris-flow and flash-flood disaster of 1999 studied. *Eos, Transactions, American Geophysical Union*, **82**, 572–573.

Leopold, L.B. and Maddock, T., Jr. (1953). The hydraulic geometry of stream channels and some physiographic implications. *U.S. Geological Survey Professional Paper* **252**, 1–57.

Leopold, L.B., Wolman, M.G. and Miller, J.P. (1964). *Fluvial Processes in Geomorphology*. San Francisco: W.H. Freeman.

Lewin, J. and Macklin, M.G. (1987). Metal mining and flood-plain sedimentation in Britain. In V. Gardner (ed.),

International Geomorphology 1986, Part 1. Chichester: Wiley, pp. 1009–1027.

Lóczy, D. (2007). The Danube: morphology, evolution and environmental issues. In A. Gupta (ed.), *Large Rivers: Geomorphology and Management*. Chichester: Wiley, pp. 235–260.

Mason, K. (1929). Indus floods and Shyok glaciers. *Himalayan Journal*, **1**, 10–29.

Matthes, G. H. (1947). Macroturbulence in natural stream flow. *Transactions, American Geophysical Union*, **28**, 255–262.

Meade, R. H. (2007). Transcontinental moving and storage: the Orinoco and Amazon rivers transfer the Andes to the Atlantic. In A. Gupta (ed.), *Large Rivers: Geomorphology and Management*. Chichester: Wiley, pp. 45–63.

O'Connor, J. E. and Baker, V. R. (1992). Magnitudes and implications of peak discharges from Glacial Lake Missoula. *Bulletin, Geological Society of America*, **104**, 267–279.

O'Connor, J. E. and Costa, J. E. (2003). Large floods in the United States: where they happen and why. *U.S. Geological Survey Circular* 1245, Denver.

O'Connor, J. E. and Costa, J. E. (2004). The world's largest floods, past and present: their causes and magnitudes. *U.S. Geological Survey Circular* 1254, Denver.

Patton, P. C. (1988). Drainage basin morphometry and floods. In V. R. Baker, R. C. Kochel and P. C. Patton (eds.), *Flood Geomorphology*. New York: Wiley, pp. 51–64.

Patton, P. C., Pickup, G. and Price, D. M. (1993). Holocene paleofloods of the Ross River, Central Australia. *Quaternary Research*, **40**, 201–212.

Rajaguru, S. N., Gupta, A., Kale, V. S., *et al.* (1995). Channel form and processes of the flood-dominated Narmada River, India. *Earth Surface Processes and Landforms*, **20**, 407–421.

Reynolds, S. (1985). Tropical meteorology. *Progress in Physical Geography*, **9**, 157–186.

Sarker, M. H., Huque, I., Alam, M. and Koudstaal, R. (2003). Rivers, chars and char dwellers of Bangladesh. *International Journal of River Basin Management*, **1**, 61–80.

Schick, A. P. (1988). Hydrological aspects of floods in extreme arid environments. In V. R. Baker, R. C. Kochel and P. C. Patton (eds.), *Flood Geomorphology*. New York: Wiley, pp. 189–203.

Sikka, D. R. (1977). Some aspects of the life history, structure and movement of monsoon depressions, *Pure and Applied Geophysics*, **115**, 1501–1529.

Simpson, R. H. and Riehl, H. (1981). *The Hurricane and its Impact*. Oxford: Blackwell.

Singh, I. B. (2007). The Ganga River. In A. Gupta (ed.), *Large Rivers: Geomorphology and Management*. Chichester: Wiley, pp. 347–371.

Snorrason, A., Finnsdottir, H. P. and Moss, M. E. (eds.), (2002). *The Extreme of the Extremes: Extraordinary Floods*. Wallingford: IAHS Press.

Starkel, L. (1972). The role of catastrophic rainfall in the shaping of the relief of the Lower Himalaya (Darjeeling hills). *Geografica Polonica*, **21**, 103–147.

Stewart, J. H. and LaMarche, V. C. (1967). Erosion and deposition produced by the floods of December 1964 on Coffee Creek, Trinity County, California. *U.S. Geological Survey Professional Paper*, 422 K Washington, D.C.

Strahler, A. N. (1964). Quantitative geomorphology of drainage basins and channel networks. In V. T. Chow (ed.), *Handbook of Applied Hydrology*. New York: McGraw-Hill, pp. 4.40–4.74.

Tang, B., Liu, S. and Liu, S. (1994). Mountain disaster formation in northwest Sichuan. *GeoJournal*, **34**(1), 41–46.

Tomasson, H. (1996). The jokulhlaup from Katla in 1918. *Annals of Glaciology*, **22**, 249–254.

Tooth, S. (1999). Floodouts in Central Australia. In A. J. Miller and A. Gupta (eds.), *Varieties of Fluvial Form*. Chichester: Wiley, pp. 219–247.

Webb, R. H., Schmidt, J. C., Marzolf, G. R. and Valdez, R. A., (eds.) (1999). *The Controlled Flood in Grand Canyon*, Washington, D.C.: American Geophysical Union Monograph 110.

Wells, N. A. and Dorr, J. A., Jr. (1987). Shifting of the Kosi River, northern India. *Geology*, **15**, 204–207.

Wohl, E. E. (1992). Bedrock benches and boulder bars: floods in the Burdekin Gorge of Australia. *Bulletin, Geological Society of America*, **104**, 770–778.

Wood, S. H. and Ziegler, A. D. (2008). Floodplain sediment from a 100-year-recurrence flood in 2005 of the Ping River in northern Thailand. *Hydrology and Earth System Sciences*, **12**, 959–973.

10 Flood hazards: the context of fluvial geomorphology

Gerardo Benito and Paul F. Hudson

10.1 Introduction

River flooding occurs as high water inundates the adjacent floodplain, and is controlled by a combination of discreet processes operating at local and watershed scales. A floodplain is the relatively flat alluvial landform adjacent to a river that is more or less related to the modern flood regime (Wolman and Leopold, 1957; Nanson and Croke, 1992; Knighton, 1998; Bridge, 2003). Most floods are natural events vital to river and floodplain geomorphological (Leopold et al., 1964) and ecosystem processes (Hupp 1988; Junk et al., 1989; Thoms, 2003). When humans are impacted, however, floods become "natural disasters" (Figure 10.1). For thousands of years floods have been among the most common and severe natural disasters on Earth, in terms of economic damage and loss of life.

Floods in most river basins are caused by excessive rainfall generated by a variety of atmospheric mechanisms (Smith and Ward, 1988; Slade and Patton, 2002). In cold-winter regions, large floods can be generated from snow/ice melt, particularly in combination with rainfall, while along coastal-draining rivers extensive flooding may be associated with storm surge events. Floods are also generated from catastrophic failure of artificial (reservoirs) and natural lakes, a category that includes dams created by ice, glacial moraines, volcanic lava flows, and landslides (Costa, 1988). Flood hazard refers to the potential of a given flood to threaten human life and property (Smith, 1996). Assessment of flood hazards is critical for appropriate flood risk management, which should span the before, during, and post flood event periods to understand, prevent, and mitigate flood hazards and their potential impacts on humans, ecosystems, and natural resources (Smith and Ward, 1998). Flood hazard management includes all planning measures implemented within the upper basin and floodplain to mitigate flooding, and usually includes physical modification of the floodplain and river channel (Goddard, 1976). Flood hazard assessment and management has been dominated by a legacy of "hard" engineering approaches, which in many cases has increased flood risk (White, 1945; Pinter, 2005; Pinter et al., 2008). Most approaches to flood management seek to minimize energy dissipation and increase channel conveyance, but effective flood management should also strive to maintain the "natural" geomorphological functioning of river channels and floodplains to retain lateral and longitudinal connectivity of water, sediment, and nutrients (e.g. Junk et al., 1989; NRC, 2005). Traditional engineering approaches use standardized probabilistic and hydraulic procedures defined by government agencies based on generalized accepted principles (e.g. 100-year flood) reproducible and defensible in a court of law if engineering structures fail (Wolman, 1971; Baker et al., 2002). Because of the lack of extensive instrumental flood data sets, however, modern rigorous hydraulic and hydrologic models cannot actually be validated without the base-line flood data provided by sedimentary and geomorphological approaches (Baker et al., 2002; Lastra et al., 2008).

Fluvial geomorphology has a substantial legacy in analyzing flooding from modern to millennial time-scales, and is increasingly recognized as a vital discipline to rigorously assess flood hazards in response to local and global scale environmental change (House et al., 2002a; Figure 10.1). Over the last two decades increased societal demands for the maintenance and restoration of fluvial ecosystems and dissatisfaction with continued flood devastation, even within heavily managed rivers, has stimulated scientific interest in the application of fundamental geomorphological concepts and methods as a complementary approach to flood mitigation and management

Geomorphological Hazards and Disaster Prevention, eds. Irasema Alcántara-Ayala and Andrew S. Goudie. Published by Cambridge University Press. © Cambridge University Press 2010.

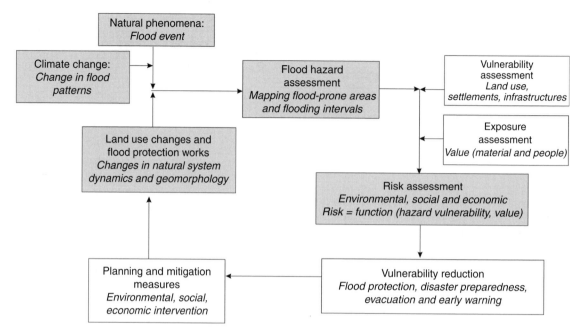

FIGURE 10.1. Basic components for flood risk assessment and management. Geomorphological studies and contributions are mainly directed to fields shown by gray squares.

(Gilvear, 1999; Baker, 2008). Fluvial geomorphology is increasingly contributing to flood management science (Gregory *et al.*, 2008) and the evolution of the discipline parallels complementary advances in computational fluid dynamics, digital remote sensing and geographical information systems (GIS), and geophysical data acquisition and analysis (e.g. Bates *et al.*, 2006), enabling a more comprehensive understanding of how geomorphological approaches are relevant to flood hazard analysis (Lewin, 1989; Baker *et al.*, 2002).

This chapter examines flood hazard assessment and management from the context of the scientific discipline of fluvial geomorphology. The chapter reviews fundamental concepts and methodological approaches commonly utilized within fluvial geomorphology to understand and analyze flooding, spanning from watershed to local spatial scales. The chapter highlights the linkages between flood hazards with different floodplain styles and flood processes, illustrating distinctions between small upland rivers and large lowland rivers. In addition, it provides an overview of approaches to exploit the Quaternary and historical sedimentary flood record, and illustrates its importance for estimating flood risk in the context of global climate change. The chapter concludes by discussing the geomorphological impact of fundamental flood management approaches, and outlines a new paradigm in flood management that strives to enhance and restore "natural" geomorphological and ecological processes.

10.2 Fluvial geomorphology in flood hazard assessment

Fluvial geomorphology has approached flood hazard analysis from several angles. The first approach estimates the hydrological response of small basins ($\leq$50 km^2) utilizing parametric models relating flood hydrograph characteristics (e.g. peak runoff, lag time) to quantitative drainage network and shape indices (catchment area, shape, drainage density, stream network geometry). These concepts were developed by Horton, Strahler, and Schumm in the 1940s and 1950s, and have become increasingly robust (e.g. Rodriguez-Iturbe and Valdés, 1979; Gupta *et al.*, 1980). A second approach delineates flood hazard zones in broad alluvial valleys by mapping flood-related landforms and deposits, soil and plant associations, and flood observations. A third approach involves energy-based inverse hydraulic modeling of discrete paleofloods located in appropriate settings as slackwater deposits (SWD) and other paleostage indicators (Kochel and Baker, 1982; Baker, 2008). This approach is limited to bedrock and confined valleys, but provides accurate discharge estimates of rare floods and subsequent flood frequency analysis, with numerous applications to flood hazard problems (Benito and Thorndycraft, 2005). All three perspectives have tremendously benefited in recent decades with advances in numerical modeling, geospatial methodologies such as global positioning systems (GPS), digital photogrammetry and high resolution remote sensing (e.g. ALS, SAR, LiDAR), and geographical information systems (GIS) (e.g. French, 2003; NRC, 2007), as well as increasing use of computational fluid dynamics (Bates *et al.*, 2005).

10.2.1 Flooding and flood hazards at the drainage basin scale

Because the drainage basin is a fundamental control on stream hydrology (Horton, 1945; Gregory and Walling, 1973), the characteristics of flooding are influenced by a range of factors at the watershed scale. Relevant hydrologic factors controlling runoff generation and flooding are (1) drainage network morphometry, (2) hillslope soil infiltration, (3) geology related to structure, tectonics, and surface erodibility, (4) vegetation and land use, and (5) meteorological-climatic conditions (Patton, 1988). Drainage basin morphology, catchment size, and relief are important controls on flood hydrology such as concentration time, hydrograph shape, and flood peak (Edson, 1951; Rodriguez-Iturbe and Valdés, 1979; Gupta *et al.*, 1980). For many decades these flood–geomorphological relationships have been found to be relatively valid at regional scales where hydroclimatology and geology control stream network development (Horton, 1945; Maxwell, 1960; Morisawa, 1962; Patton and Baker 1976; Patton, 1988; Knighton, 1998; Ward and Trimble, 2004).

The first attempt to combine drainage area and flood magnitude (peak discharge) was conceived by Dickens (1865) in India, and later by Jarvis (1936) in the USA. The equation shows an exponential relation ($Q_T = a. A^b$) between annual peak discharge (Q_T), estimated for a particular return period (T), and drainage area (A). The exponent b varies between 0.5 and 0.8 (Jarvis, 1936), or 0.5 to 0.9 (Thomas and Benson, 1970), depending on the region considered, and generally decreases as the flood return period increases. Although somewhat limited because of not considering the physical processes of runoff generation (Patton, 1988), a set of regional curves for different return periods can be constructed to reasonably approximate annual peak discharges. When fitted with paleoflood data the approach is well suited to questions concerning changes in flooding associated with projected climate change scenarios (Enzel *et al.*, 1993), which may ultimately be utilized to identify an upper hydroclimatic limit in precipitation and peak discharge for a given drainage basin (Wolman and Costa, 1984).

An additional approach at the drainage basin scale examines the influence of drainage density on runoff generation and flood propagation (Horton, 1945; Gregory and Walling, 1973; Baker, 1976), because network geometric parameters condition runoff connectivity and travel distances. Different indices can be constructed from drainage density and morphometric characteristics (basin shape, area, stream length, and relief), and may be used for developing empirical equations to model stream flow (Mosley and McKerchar, 1993). The application is limited, however, because of a lack of extrapolation to different regions (Patton and Baker, 1976), and because of the usage of various techniques to define the extent of drainage networks (see Mosley and McKerchar, 1993). Additionally, drainage density develops over a much longer time period than the relatively short time periods of climate stability, such that relict drainage morphometry formed during older climatic regimes may not produce representative hydrogeomorphic indices (Patton, 1988). Nevertheless, the topologic characteristics of drainage networks have been utilized by modelers to identify a basin-scale transfer function to deal with inherent non-linearity, which represented a problem to the classic unit hydrograph (UH) approach developed by Sherman (1932) as well as Nash's (1957) instantaneous unit hydrograph (IUH). The geomorphological unit hydrograph (GUH) formulated by Rodriguez-Iturbe and Valdés (1979), later generalized by Gupta *et al.*, (1980), attempts to relate the IUH of a catchment to the geometry of the stream network, so that relevant parameters of the IUH such as peak discharge, shape, and time to peak can be related to geomorphological drainage characteristics deduced from the Strahler (1957) stream order and Horton's (1945) "laws of drainage networks".

10.2.2 Flooding and flood hazards at the floodplain (local) scale

The consideration of discreet flood processes in relation to floodplain geomorphology results in several categories of floodplain styles (e.g. Nanson and Croke, 1992) of importance to flood hazard management. Along most fluvial systems floodplain styles range along a continuum upstream to downstream, from confined narrow high-energy floodplains to broad low-energy floodplains. Flood hazards and flooding are characterized by three main valley profiles representative of upstream (Fig.10.2.I), middle (Fig. 10.2.II), and downstream (Fig. 10.2.III) reaches of a typical fluvial system. These valley geometries have significance in terms of discharge-stage relationships, energy dissipation, flood processes, area of inundation, and flood occurrence, and represent varying hazards as related to human settlement and damages (Figure 10.2; Table 10.1).

The inundation of floodplain surfaces within confined narrow river valleys (cross-section I; Figure 10.2), particularly within small mixed bedrock–alluvial valleys, is

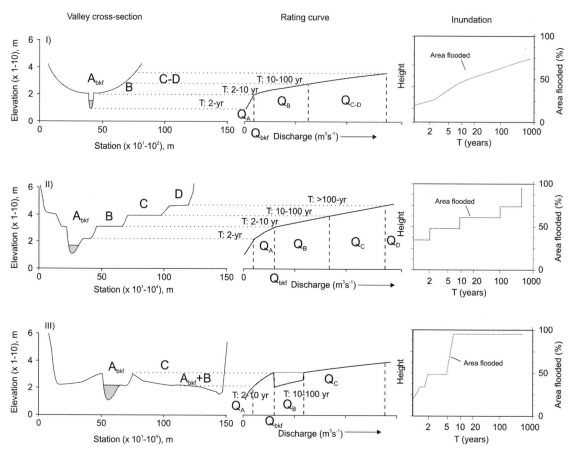

FIGURE 10.2. Models of floodplain morphology and flooding characteristics in the upstream (I), middle (II), and downstream (III) reaches of alluvial river valleys, including the general shape of rating curve and area inundated. The area of flood inundation follows the cross-sectional morphology, and can be linked to flood damages if a curve of exposure and property value is known.

primarily driven by high amounts of precipitation within the upper watershed. This setting may result in "flashy" events (short lag times) with large energy expenditure (unit stream power) on the floodplain surface (Nanson 1986; Grant and Swanson, 1995; Wohl, 2000). The flood process is simple overbank, and is not augmented by groundwater or surface conduits (e.g. crevasses, sloughs, paleochannels). Such settings may be readily identified by considering planform dimensions obtained from aerial photographs or topographic maps, and have ratios of channel width (W_C) to valley width (W_V), generally $< 0.5 W_C/W_V$ (Grant and Swanson, 1995). The floodplain geomorphology of such settings is generally reworked within a cycle of floodplain rejuvenation (Nanson, 1986), and thus provides discreet evidence for understanding flood hazard potential. Specific aspects of the floodplain geomorphology useful to understand the degree or stage of cyclic reworking include soil development (thickness and sequence of soil horizons), floodplain topography (relief and slope), and

flood deposits (texture and thickness). Flood hazards are characterized by physical damage, such as undermining bridges and roads, washing away of human settlements, and destruction of the physical setting by altering the floodplain geomorphology (e.g. floodplain stripping).

Floodplains within middle reaches (cross-section II; Figure 10.2) are commonly composed of coarse-grained lateral accretion sediments (bottom stratum) covered by vertically accreted fine-grained flood sediments (top stratum) (Brakenridge, 1988; Bridge, 2003). Floodplain morphology exhibits different floodplain surfaces directly associated with different flood frequencies. While not all floodplain styles exhibit the same characteristics, similarities often exist in the types of specific channel and floodplain geomorphological units (flood zones A to D in Table 10.1; Ballais *et al.*, 2005). Channel bed unit (zone A; Table 10.1) includes channel lag, high-energy bar deposits, and low-stage slackwater facies (Brakenridge, 1988). Adjacent to the outer channel banks, the lower floodplain

TABLE 10.1 *Geomorphological description of flood hazard zones in relation to geomorphological units of a typical floodplain. Note that these zones may vary for specific fluvial systems, hydroclimatological conditions, and floodplain geometry. Flood zones are referred to Figure 10.2.*

Flood zone	Relative flood magnitude	Return frequency (years)	Geomorphological unit	Sedimentary facies	Dominant geomorphological processes	V m s^{-1}	Vulnerable land use/land cover	Perceived flood risk (relative to a human lifespan)
A	Q_A: minor flood	<2	Channel	Channel lag, point bar, side bar, longitudinal bar, chute, low-stage slackwater facies (e.g. clay drapes)	Seasonal sediment flushing, coarse sediment mobility, incision, erosion–deposition zone	>2	In-channel infrastructure and economic activities (fishing, shipping). Highly dangerous to people	High risk: flooded every year
B	Q_B: moderate flood	2–10	Low floodplain	Lateral accretion, natural levee (sandy ripples and dunes adjacent to channel fining to silt laminations on natural levee flanks), high flow channel (floodway), crevasse splay, oxbow infilling, slackwater sedimentation in low depressions and backswamps	Channel migration and channel bar development, bank erosion, coarse floodplain aggradation. Irregular topography	>1	Riparian vegetation, gravel mining; agriculture and floodplain irrigation (in lesser developed nations). Dangerous to human life	Medium risk: seen 4–6 times
C	Q_C: large flood	10–100	High floodplain	Slackwater flood deposits on low Holocene terraces, significant backswamp sedimentation (fine sediments in slackwater ponded), and sloughs	Dominant vertical accretion, infilling of paleomeanders, activation of recently abandoned channels; channel avulsion; floodplain reworking	usually <1	Agriculture and floodplain irrigation, recreation areas, transportation infrastructure; rural and urban habitation (in lesser developed nations). Only dangerous to people when water depth >1 m	Low risk: 1–2 times
D	Q_D: extreme flood	>100	Low terrace/ highest floodplain	Slackwater flood deposits on older Quaternary terraces and high bedrock ledges; coarse overbank "stringers" on distal floodplains	Colluvial, tributary and alluvial fan accumulation; floodplain "stripping" (e.g. Nanson "cyclic disequilibrium model")	<1	Agriculture and irrigation, flood management infrastructure; urban areas with high population densities (developed nations). Not dangerous to human life	Very low risk: 0–1 times

surface is characterized by an irregular topography formed by frequent flooding (return periods 2–10 years) containing depositional (e.g. crevasse splay) and erosional landforms (high-flow channels) with medium to coarse sands, and occasional gravel (Table 10.1, zone B). This lower floodplain is usually covered by riparian vegetation, which significantly favors flood energy dissipation. The distal (lateral) floodplain contains morphological evidence of former river meanders, episodically inundated by extraordinary floods (10 to 100 year return period). Here, the sedimentology is composed of fine-grained sediments (silt and clay) infilling backswamps, paleomeanders (oxbow and oxbow lake environments), and backwater sloughs along valley margins (Table 10.1, zone C). The low relief and episodic flooding favors human activities (agriculture, roads, and mining), and settlements on the distal (upper) floodplain. The lower river terraces (late glacial) or the highest floodplain surface may be flooded by extreme floods (>100 to 500 year floods; zone D in Table 10.1), indicated by fine sands and silts on older alluvium and soils. On valley sides, the floodplain may contain additional alluvial and colluvial deposits from adjacent high surfaces, including slope and cone deposits, and alluvial fans.

Broad floodplains (cross-section III in Figure 10.2) within wide alluvial valleys or deltas, in contrast to confined narrow valleys, have a more complex floodplain geomorphology that requires consideration of local controls on floodplain inundation. Large floodplains, having low W_C/W_V ratios, are associated with long duration flooding with low unit stream power (Nanson and Croke, 1992; Ferguson and Brierly, 1999). Additionally, these settings generally have fine-grained cohesive floodplains and sufficient space to store older channel belts, as well as high degrees of lateral (hydrologic) connectivity (Mertes, 1997; Burt et al., 2002; Poole et al., 2002). Here, the floodplain sedimentology and topography represents an important control on flooding. In large lowland alluvial valleys (Figure 10.2.III), the meander belt (channel and active floodplain) is perched above the lower-lying floodplain bottoms, increasing the complexity of flood processes (Hudson and Colditz, 2003). During local-scale floods (return periods ~2–10 years), longitudinal flow paths are mainly confined by natural levees (flood zone B; Figure 10.2.III), although lateral flow paths may stagnate distant floodplain bottoms. Flooding of natural levees occurs less frequently than floodplain bottoms (return periods of ~10–100 years; Table 10.1, flood zone C), with a much lower duration, mainly associated with watershed-scale flood mechanisms (Hudson and Colditz, 2003). In combination with fine-grained cohesive top stratums, these settings can remain inundated for weeks and months after river stage has receded (Badji and Dautrebande, 1997; Hudson and Colditz, 2003), a severe limitation to many types of human

activities and hazardous to floodplain agriculture. Older buried coarse-grained channel belts that intersect the active channel represent pathways for rapid groundwater flow (Sharp, 1988; Poole et al., 2002), which can then inundate distant floodplain reaches, even beyond river dikes (levees). High surface connectivity represented by crevasses, sloughs, oxbow lakes, and abandoned channels can be "reoccupied" during a flood event and result in flooding and sediment dispersal in distant floodplain areas (Gomez et al., 1997).

10.2.3 Flood hazard mapping

Flood hazard mapping is a fundamental tool for flood risk assessment and management, and the recognition of different fluvial styles suggests that multiple techniques are required (e.g. Dunne, 1988; Pelletier et al., 2005). Flood hazard maps include the extent of flooding for a given flood recurrence interval, and other fundamental hydraulic information such as flood depth, velocity, and frequency of inundation (Wolman, 1971). Much research in this sector of the discipline is for applied study, funded by government agencies or insurance companies, and is commonly accompanied with flood risk maps that illustrate the area of inundation and the potential damage impact, including human risk, economic value, and impacts on environmental systems (Figure 10.1). The final aims of flood hazard mapping often include (1) support for flood management plans, (2) spatial land use and planning activities, (3) emergency and evacuation plans, and (4) increased public awareness of flood risks.

Traditional hydrological approaches concerning flood hazard analysis are based on the estimation of peak flood discharge (Q_2, Q_{10}, Q_{50}, and Q_{100}) and corresponding flood stage levels for events of various frequencies of occurrence or return periods (where Q_N represents N-year flood), for which associated estimated damages are established for each probability (flood risk map and model). Flood maps based on hydraulic-hydrologic modeling approaches are expensive and require long instrumental discharge or rainfall records, which makes it difficult to cover extensive areas of a state or nation. However, the concept of "acceptable risk" has become widely used, in which delineation of flood-prone areas is based on multicriteria including stream gauge records, geomorphological mapping of flood-related landforms, historical evidence (documentary), and occasional flood events (Figure 10.3). This philosophy underlies the European Council Flood Risk Directive 2007/60/EC (EC, 2007), under which flood hazard maps should distinguish three main flood zones, namely floods with low probability (extreme event scenarios), floods with medium probability (likely return period ≥100 years), and floods with high probability. Some European countries

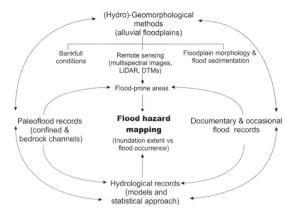

FIGURE 10.3. Integrated methodological approach for flood hazard mapping, based on hydrological, geomorphological, paleoflood, and historical records.

(e.g. France and Spain) recommend that geomorphological criteria be integrated with conventional flood hazard mapping approaches to fulfill the European Directive (DIREN-PACA., 2007; Díez-Herrero *et al.*, 2008).

Flood hazard maps based on a geomorphological approach utilize aerial photography or remote sensing imagery combined with field work to map flood-related landforms, sediments, and high-stage indicators (Baker *et al.*, 1988; Garry and Graszk, 1999). A first approach to floodplain mapping is the identification of two assemblages of deposits and landforms: channel deposits (base flow and high-flow channels), and channel bank and overbank deposits (lower and upper floodplain surfaces). Flood hazard maps also commonly delineate a channel migration zone (CMZ), including recent channel activity and changes (migration) based on aerial photos and historical maps (e.g. Rapp and Abbe, 2003). Additional flood indicators refer to pedogenic conditions, such as soil development, stratification, and drainage (Smith and Boardman, 1989), and biological flood markers, including distinctive floodplain (riparian) vegetation assemblages, and specific vegetation types (e.g. indicator species) related to high-water conditions (e.g. Foxcroft *et al.*, 2008). These physical and biological flood indicators can be combined with hydraulic-hydrological estimates and data from occasional historical flooding to provide relevant hydraulic data directly related to flood stage levels (e.g. water depths, flow velocity).

Recent advances in geospatial methodologies (e.g. French, 2003; NRC, 2007) and absolute dating techniques (Duller, 2004) substantially improve the reliability of geomorphological flood mapping. Common sources of geospatial data used in flood mapping include global positioning systems (GPS) (Hudson and Colditz, 2003), digital photogrammetry, and high-resolution ground and airborne remote

sensing techniques (e.g. ALS, SAR, LiDAR) (Smith, 1997; NRC, 2007) to develop topographic products such as digital elevation models (DEMs). The use of airborne light detection and ranging (LiDAR) has tremendous potential for floodplain mapping because of the high vertical (~10 cm) and horizontal (1,000,000 points per km^2) resolution. LiDAR is an active sensory system mounted on an airborne platform that uses laser light to measure distances between the sensor on the airborne platform and points on the ground (or a building, tree, etc.). The data provide substantial details on flood-related landforms and can easily be integrated with traditional remote sensing (aerial photos and satellite imagery) (NRC, 2007). An additional major technological development to support flood geomorphology studies is numerical age dating of fluvial sediments (sand and silts) and organic materials. Radiocarbon dating of organics (e.g. seeds, charcoal, wood, peat, shells, bones, etc.) is a standard absolute dating tool employed for alluvial sediments (e.g. Baker *et al.*, 1985). New developments in optically stimulated luminescence (OSL) dating (Duller, 2004), and new analytical protocols such as the single-aliquot regenerative-dose (SAR) for determining the equivalent dose (Murray and Wintle, 2000) provide very accurate numerical dating of alluvial sediments, with age uncertainties of 5–10%. Additionally, this approach yields accurate dates for deposits younger than 300 years, a period associated with considerable measurement error in radiocarbon dating (Duller, 2004).

10.2.4 Estimation of rare events using paleoflood hydrology

Significant advances in understanding flood hazards have been provided by using paleoflood hydrology to quantitatively estimate the magnitude and chronology of large floods over the past millennia. Paleoflood studies involve many different techniques relying on regime-based studies of discharge capacity (e.g. bankfull discharge) of alluvial channels (Williams, 1988) and sediment transport-flow competence analysis (Costa, 1983; Williams, 1988; Komar, 1989). The most developed approach, however, is based on flood stage indicators in stable bed-rock channels (Baker and Kochel, 1988; Baker, 2008).

Flow competence evaluations are based on selective-entrainment relationships (empirical and physically based equations), usually based on the largest clasts (Carling, 1983; Costa, 1983) providing mean-flow stress, velocity, and discharges per unit flow width. Deficiencies and constraints of this method are mainly related to inherent problems with entrainment equations, being largely inadequate to predict incipient motion because of issues related to grain sorting, vertical armoring, sediment packing, and mechanisms of grain pivoting versus sliding (Komar, 1989).

The most successful paleoflood techniques are based on field reconnaissance of high water marks (HWM), slackwater flood deposits, and other paleostage indicators (SWD-PSI), which are combined with conventional indirect methods to estimate peak flood discharge magnitudes (Baker, 2008). Paleoflood hydrology has undergone a revolutionary development in the three decades since Kochel and Baker (1982) coined the term (see recent review by Baker, 2008). A major achievement is the establishment of a core array of standardized protocols for the collection of field data and quantitative techniques for estimating paleodischarge. Such procedures can now be included within standard statistical flood frequency analysis. Importantly, the procedures have gained scientific credibility and recognition as an effective tool for numerous applications in understanding flood occurrences and the evaluation of flood hazards (Baker et al., 2002; Saint-Laurent, 2004; Benito and Thorndycraft, 2005; Baker, 2008).

Sources of paleoflood data include geological and botanical indicators such as slackwater flood deposits at high rock ledges, silt lines and erosion lines along the river channel, terraces, and canyon walls (Baker and Kochel, 1988; Greenbaum et al., 2000). The methodological steps to conduct a paleoflood study include: (1) preliminary inventory of potential sites using aerial photographs, (2) field visit and survey for the identification and selection of flood indicators (flood deposits and marks), (3) stratigraphical description with emphasis on identifying flood units, (4) sample collection for age dating, (5) topographic survey of flood sites and river reaches, (6) hydraulic modeling and discharge estimation, (7) comparison with available historical data, and (8) flood frequency analysis.

Sedimentary environments associated with slackwater flood deposition include: (1) channel widening, (2) channel expansions, (3) channel bends, (4) obstacle shadows where flow separation causes eddies, (5) alcoves and caves in bedrock walls, (6) back-flooded tributary mouths and valleys, and (7) on top of high alluvial or bedrock surfaces that flank the channel (Baker and Kochel, 1988; House et al., 2002b; Benito et al., 2003a; Benito and Thorndycraft, 2005). In these environments, depositional landforms include thick, high-standing terraces or "benches", and eddy bars. The flood benches are formed by vertical accretion of slackwater sediment layers deposited by successive floods, which constitute a rising threshold or local censoring level over time (Figure 10.4). Dating of sedimentary flood units (radiocarbon, OSL, Cs-137) provides an understanding of flood frequency, and for recent flooding enables the identification of human recorded events in cases where historical, instrumental, and paleofloods temporally overlap. In Europe, paleoflood hydrology has been combined with information about floods in the pre-instrumental period (last 500 years) from oral and documentary sources (Benito et al., 2003b; Thorndycraft et al., 2003; Werritty et al., 2006).

The stage associated with the different paleoflood units (paleostages) can be readily converted into discharge values using widely accepted hydraulic procedures (e.g. Jarrett, 1987; O'Connor and Webb, 1988). In fact, this conversion is an inverse problem, with the flood discharge obtained by trial and error using a hydraulic model to compare the observed river stage with simulated stage levels. The calculated discharges are minimum discharge values, since the water depth at the site of deposition is unknown (Figure 10.4). These models assume a fixed bed, hence the importance of application on bedrock channels.

Paleoflood hydrology techniques provide an accurate catalogue of flood discharges and their age dating. This paleoflood record can be combined with gauge station data for the estimation of the statistical moments. The basic hypothesis in the statistical modeling of paleoflood information is that all floods exceeding a certain water level or magnitude (threshold of discharge) have been registered through sedimentary records and/or other paleostage indicators (Stedinger and Cohn, 1986). In hydrology, flood observations reported as having occurred above some threshold are known as censored data sets (Leese, 1973). Paleoflood information is considered data censored above a threshold (Figure 10.4) and it is assumed that the number of k observations exceeding an arbitrary discharge threshold (X_T) in M years is known, similar to partial-duration series (Stedinger and Cohn, 1986; Francés et al., 1994). The value of the peak discharge for the paleofloods above X_T may be known or unknown. Paleoflood data are organized according to different fixed threshold levels over particular periods of time exceeded by flood waters. Estimated flood discharges obtained from the minimum high-water paleoflood indicators and maximum bounds (non-exceeded threshold sense, Levish et al., 1997) can be introduced as minimum and maximum discharge values (Figure 10.4). Estimation of statistical parameters of flood distribution functions (e.g. Gumbel, LP3, Generalized Extreme Value) are calculated using maximum likelihood estimators (Leese, 1973; Stedinger and Cohn, 1986), the expected moment algorithm (Cohn et al., 1997), and fully Bayesian approach (O'Connell et al., 2002; Reis and Stedinger, 2005). This provides a practical framework for incorporating imprecise and categorical data as an alternative to the weighted moment method (U.S. Water Resources Council, 1982).

The fields of application of paleoflood hydrology include (Benito and Thorndycraft, 2005): (1) flood risk assessment (Baker et al., 1988; House et al., 2002b; Thorndycraft et al., 2003; Benito and Thorndycraft, 2004), (2) determination of the maximum limit of flood magnitude (Enzel et al., 1993) and non-exceedence as a check of the probable maximum flood

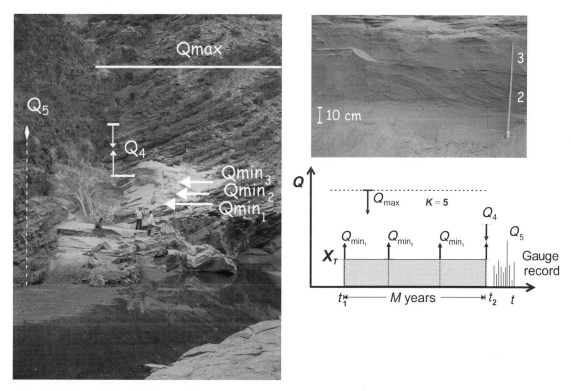

FIGURE 10.4. Slackwater flood deposits of the Kuiseb River (Namibia) with indication of paleostages associated with multiple flood events. The paleoflood records are censored data sets above a threshold and it is assumed that the number of k observations ($k = 5$) exceeding an arbitrary threshold (X_T) in M years is known.

(PMF) and safety risk analysis of critical facilities (e.g. dams and wastewater facilities and power plants; Levish *et al.*, 1997; Benito *et al.*, 2006; Greenbaum, 2007), (3) a better understanding of long-term flood–climate relationships (Ely, 1997; Knox, 2000; Thorndycraft and Benito, 2006; Benito *et al.*, 2008), and (4) assessing sustainability of water resources in dryland environments where floods are an important source of water to alluvial aquifers (Greenbaum *et al.*, 2002; Grodek *et al.*, 2007). An alternative to the analysis of slackwater flood deposition methodology was developed by the U.S. Bureau of Reclamation to be applied to dam safety purposes based on paleostage exceedance information and paleohydrologic bounds (Levish *et al.*, 1997; England *et al.*, 2006). A paleohydrologic bound is physically located on a terrace or abandoned floodplain surface, at a high elevation, on which a paleostage has not been exceeded sufficiently to modify its surface (non-exceedance threshold) during a time interval obtained from age dating of soils and of scarce flood deposits.

10.3 Flood hazards in the context of global climate change

Historical and projected global climate change has raised concerns about flooding and flood hazards (Kundzewicz

and Schellnhuber, 2004). Climate change projections, however, are generally within the range of climate change that occurred over the middle and late Holocene (Knox, 2000). Paleoflood records have revealed the sensitivity of flood magnitude and frequency to subtle alterations in atmospheric circulation (Knox, 1993; Ely, 1997), which have also been observed during the instrumental period because of climatic forcing (Knox, 2000; Redmond *et al.*, 2002). Shifts in climate may have a greater impact on the estimation of large flood quantiles (50-year flood and higher), which have been found to be highly sensitive to climate variability (Knox, 2000). Analysis of gauge records has shown that hydroclimatic homogeneity may have only occurred for 30-year intervals during the twentieth century (Webb and Betancourt, 1992). This must be considered in flood frequency analysis (Baker *et al.*, 2002), and has significant implications for the design of more effective approaches to flood management.

Paleoflood records provide rigorous data for understanding how future climatic variations might influence flood magnitude and frequency (Knox, 2000; Knox and Daniels, 2002; Macklin *et al.*, 2006; Starkel *et al.*, 2006). The aim of these studies is not necessarily to provide analogues of future flood–climate episodes but to analyze flood response to climate shifts. Paleoflood records from slackwater flood deposits of Spanish rivers show distinct flood periods at 1000–600 BC, AD 900–1100, AD 1450–1500 and AD 1600–1900 (Thorndycraft and Benito, 2006; Benito et al., 2008). Some of these flood periods are consistent with those found in other European countries, and in the SW United States (Ely, 1997). The direction of climatic shift is not unequivocal, with flood episodes related to cold (wet) conditions (phases 1000–600 BC, AD 1450–1500 and AD 1600–1900), while others were related to greater hydroclimatic variance (warming-dry periods: phase AD 900–1100). Anomalous magnitudes of recent floods, such as the 2002 flood of the Gardon River (France), the largest on record since 1890, occurred more frequently during the Little Ice Age (Sheffer et al., 2008), a period characterized in the Mediterranean by extreme variability in episodes of flood and drought (Barriendos and Martín Vide, 1998). Floodplain stratigraphy in the upper Mississippi (Knox and Daniels, 2002) has also provided excellent proxy records to characterize long-term changes

in the flood frequency and magnitude, showing a clustering of small floods between 5,000 and 3,300 years BP, and a general increase in flood magnitude after 1,000 years BP, particularly between about 700 and 500 years BP. However, floodplain aggradation requires sufficient sediment yield and overbank flows, hence some recent aggradation episodes may respond more strongly to major environmental impacts in the basin (deforestation, land-use changes) than to specific climate change signals (Benito et al., 2008).

10.4 Geomorphological adjustment to flood management

Flood hazard management most commonly involves a variety of approaches coordinated across a range of government agencies and stakeholders (WMO, 2004; Hudson et al., 2008). Common goals of flood management are (1) reduction in the area of inundation to increase habitable lands, (2)

TABLE 10.2. *Common options associated with flood control*

Approach	Intention (rationale)	Unintended geomorphological response
Floodplain modifications		
Dikes (levees) and flood walls	Reduce area of inundation (creation of embanked floodplain)	Change in floodplain hydrology, generally higher flood stages and rates of flood sedimentation; reduction of floodwater storage capacity; infilling of floodplain water bodies and wetlands; dike breach ponds; enhanced seepage and sand boils behind dikes; floodplain "borrow pits"
Drainage canals and relief wells	Remove waters attributed to rising groundwater and dike seepage	Oxidation of floodplain soils; ground subsidence and change in local drainage
Pumping	Remove waters attributed to rising groundwater and dike seepage	Oxidation of floodplain soils; ground subsidence and change in local drainage
Floodways	Bypass corridors and detention to store floodwaters; lower flood stage	Variable sedimentation and scour within flood diversion corridor and basins
Channel modifications		
Straightening (cutoffs)	Lower flood stage and increase flood conveyance; reduce flood frequency; reduction in channel length, increase in channel gradient	Knickpoint formation, channel incision and bank erosion; downstream channel bed aggradation; creation of new floodplain oxbow lakes
Bank protection (revetment and rip-rap)	Manage bank erosion, protection of dikes	Reduction of channel sediment loads; possible channel bed scour
Groynes (wing dikes)	Align river channel, reduce channel width by selective aggradation of bed material	Increase in roughness at high water, possible increase in flood stages
Dredging	Removal of local bed material associated with shoaling	Localized channel bed change and formation of dredge spoil bars

reduction in flood stage and peak discharge, and (3) reduction in flood duration. Modern flood management strategies generally involve a variety of approaches to physically modify the floodplain and channel (Table 10.2). Fundamental engineering modifications to the floodplain include dikes (levees) to reduce the area of inundation, drainage canals and pumping to remove excessive water, and floodways to reduce downstream flood stages. Channel engineering procedures include straightening to increase flood conveyance, including channel reshaping, canalization, and enlargement, bank protection to reduce erosion, groynes (wing dikes) to reduce channel width, and dredging to manage channel bed aggradation. Although these techniques are extensively utilized all over the world, their effectiveness is highly dependent upon a detailed understanding of modern hydrological and geomorphological processes as well as the Quaternary sedimentological framework of the associated fluvial system (Winkley, 1994; Smith and Winkley, 1996; ASCE, 2007; Hudson et al., 2008). Over the past decade there has been increased attention concerning the unintended geomorphological and environmental consequences associated with conventional flood management options (Smith and Winkley, 1996; Hesslink et al., 2003; Pinter et al., 2006, 2008; Hudson et al., 2008).

The construction of earthen dikes (levees) is the oldest approach to managing flood hazards, having been done along many of the world's great rivers, such as the Nile, Tigris-Euphrates, Yangtze, Danube, and Rhine, for hundreds and even thousands of years (Van Veen, 1962; Butzer, 1976). When properly designed, river dikes can be highly effective in minimizing flood risk, enabling large populations to safely reside adjacent to major river systems (NRC, 1995). Dike construction, however, abruptly alters fundamental floodplain processes and floodplain geomorphology (Middelkoop, 1997; Hesselink et al., 2003; Glynn and Kuszmaul, 2004; Hudson et al., 2008). The embanked floodplain (river side of the dikes) is significantly narrower than the natural floodplain, which

changes flood hydrology and sedimentation processes (Figure 10.5). The narrower floodplain results in higher flood stage levels and greater floodplain shear stress. Floodplain sections landward (behind the dikes) are effectively removed from most active overbank fluvial processes. However, serious flood hazards remain because of groundwater inundation (NRC, 1995; Li et al., 1996). This is because the higher embanked flood stages change the alluvial groundwater hydrology, enhancing dike under-seepage and resulting in an increase in the frequency of emergent sand boils that can destabilize dikes (Li et al., 1996; Glynn and Kuszmaul, 2004). This is particularly problematic when dikes and floodwalls are constructed atop unsuitable sedimentary deposits such as permeable sands and organic rich clays (US-ACE, 1998; Glynn and Kuszmaul, 2004), increasing the risk associated with floodplain economic activities and urban populations (NRC, 1995; Pinter, 2005; ASCE, 2007). Despite relief wells and drainage channels along 150 km of the Lower Mississippi River, for example, the 2008 spring flood resulted in the formation of 40 sand boils and landward inundation of over 40,000 hectares.

The modification of floodplain processes associated with dike construction changes the floodplain geomorphology. The change in flood hydrology and hydraulics alters flood sedimentation processes, disrupting the classical fining sequence (e.g. Kesel et al., 1974; Pizzuto, 1987), and resulting in greater variability in overbank sedimentation rates and texture (Middelkoop, 1997; Hesselink et al., 2003). Over long time periods this results in an anthropogenic style of floodplain architecture (Hesselink et al., 2003). Accelerated floodplain aggradation infills oxbow lakes and floodplain depressions, reducing the capacity of the floodplain to store flood waters, resulting in higher flood stages (Middelkoop and Van Haselen, 1999; Silva et al., 2001; Glynn and Kuszmaul, 2004). Over longer periods the construction of dikes is associated with the formation of unique anthropogenic floodplain water bodies, dike breach ponds and

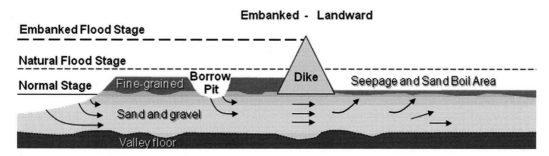

FIGURE 10.5. Embanked floodplain hydrology of a large alluvial river. Groundwater flow depicted at flood stage with high hydraulic head. Note thicker overbank (fine-grained) deposits associated with higher rates of embanked sedimentation because of dikes trapping flood sediments. Borrow pit formation is associated with dike (levee) construction.

borrow pits (US-ACE, 1998; Hudson *et al.*, 2008). Further, the groundwater inundation of landward floodplain reaches requires a network of pumps and canals for floodplain drainage, which fundamentally initiates a sequence of human response and geomorphological adjustment. Over long periods floodplain drainage results in oxidation of hydric floodplain soils and ground subsidence. The lower surface is thereby more susceptible to groundwater inundation and flooding, which requires further land drainage and represents a greater flood hazard in the event of dike breach events (Van Veen, 1962; van de Ven, 1993). In addition to dikes and floodplain drainage, flood management often includes a significant amount of channel engineering to increase flood conveyance and to reduce flood stage and duration. The most significant type of channel modification is straightening by meander neck cutoffs (Hudson *et al.*, 2008), which reduces channel length and increases channel gradient (Winkley, 1994). Channel cutoffs result in the formation of a channel knickpoint, which incises and subsequently migrates upstream (Yodis and Kesel, 1992; Toth *et al.*, 1993; Winkley, 1994; Smith and Winkley, 1996; Shankman and Smith, 2004; Harmar *et al.*, 2005; Hudson and Kesel, 2006; Simon and Rinaldi, 2006). The incision is generally effective at initially lowering flood stages, but over time flood stages may increase in height (Smith and Winkley, 1996; Wasklewicz *et al.*, 2005). Where multiple cutoffs are located along a valley reach, knickpoints initiate a distinctive sequence of geomorphological responses that require further engineering solutions to maintain a stable channel and to minimize flood risk (Hudson *et al.*, 2008).

Flood management infrastructure is designed for discharge–stage relationships having a specific recurrence interval, such as the "100-year flood" (NRC, 2007). A number of allogenic factors, however, can result in changes to the external boundary conditions, thereby increasing flood risk. Specifically, these include ground subsidence, neotectonics, climate change, and, at the coast, sea level rise. Ground subsidence and downwarping along fault zones is important along large alluvial valleys and delta plains, such as the Lower Mississippi and Rhine systems (Schumm, 1986; Dixon *et al.*, 2006). Within several decades vertical displacements of 2–3 mm per year can significantly lower dike levels, such that lower magnitude floods present greater flood hazards (Dixon *et al.*, 2006). Much flood management infrastructure represents a safety concern because of being decades old and in need of maintenance and upgrading, and may be inappropriately designed for our contemporary understanding of flood dynamics and floodplain geomorphology (e.g. Nanson, 1986; Mertes, 1997), ongoing human-induced environmental change (Gregory, 2006), or to the hydrologic

implications of various climate change scenarios (Kundzewicz and Schellnhuber, 2004; IPCC, 2007; NRC, 2007). Indeed, the management of flood hazards is a pressing societal concern, and requires a comprehensive perspective over long time-scales to understand flooding in response to varying climate–land cover scenarios.

10.5 Flood hazard management: an integrated approach

The concentration of human settlements and economic activities along river systems demands that scientists and government agencies develop new effective measures to manage flood hazards. This is required because: (1) the increasing global population and associated economic activities will remain dependent upon floodplain lands and resources; (2) an appreciation exists for documented historical and late Quaternary global hydroclimatic change, and concerns over projections of future change beyond the boundary conditions that flood management infrastructure was designed for; and (3) in many instances the old hard engineering approach triggered socially unacceptable environmental change, and in some instances initiated a sequence of geomorphological responses that increased human vulnerability to flooding (Hudson *et al.*, 2008). The confluence of these concerns represents the stimulus for a paradigm shift in the management of flood hazards, towards an approach increasingly referred to as "integrated flood management" (WMO, 2004).

As with traditional forms of flood management, the primary goal of integrated flood management (IFM) is to minimize loss of life (Silva *et al.*, 2001; WMO, 2004). A major philosophical difference with traditional engineering flood control, however, is that IFM views flooding as a positive attribute of the fluvial system because of its importance to geomorphological and ecological processes (e.g. NRC, 2005), and as such is inherently more environmental. While IFM does not seek to constrain human activities on floodplains that could have adverse economic activities, it does embrace a longer-term perspective of the fluvial system, and prioritizes floodplain land use that does not have adverse environmental impacts (WMO, 2004). Fundamental tenets of IFM commonly include the following: (1) Flood control should be interdisciplinary to understand the myriad of geomorphological, hydrological, ecological, and social processes occurring within a floodplain. (2) Floodplain land use should be ranked by economic measures, and prioritized by its environmental impact. (3) Local stakeholder concerns should be integrated into the decision-making process, such that flood management options are in better accord with community priorities and values. (4) Government agencies should adopt a flexible approach to implementing flood control, such that new knowledge can be integrated into the program. (5) Monitoring is fundamental to IFM, and should be implemented before, during, and after specific flood control measures are implemented, and particularly after large flood events. Monitoring should include an array of ecological, hydrological, and geomorphological processes (WMO, 2004).

IFM represents a collection of ideals more than an exact flood control plan. Because flood control is implemented by large government agencies it is difficult to abruptly change philosophical approaches, or expend the financial resources to replace existing flood control infrastructure. Thus, while IFM represents a paradigm shift in thinking, the actual implementation of IFM is more nuanced, and highly dependent upon financial, political, cultural, and physical conditions within the nation of implementation. Nevertheless, the fundamental tenets outlined above are championed by international organizations, such as the European Union (EC, 2000) and the United Nations World Meteorological Organization (WMO, 2004). IFM is particularly important for developing nations because they are considered more vulnerable to flood hazards associated with climate change than developed nations (e.g. Kundzewicz and Schellnhuber, 2004; IPCC, 2007). In 2000 the European Parliament passed the Water Framework Directive (WFD) (EC, 2000), which explicitly attempts to implement an IFM approach through a series of policies with measurable goals and timetables. An excellent example is the "Room for the River" plan developed for the lower Rhine in the Netherlands (Middelkoop and Van Haselen, 1999; Silva et al., 2001). The Room for the River flood management plan was developed in response to large flood events in 1993 and 1995, public dissatisfaction with the environmental impact of traditional flood management, and a strong awareness of the threat of climate change to lowland floodplains. The Room for the River program is currently being implemented and explicitly takes into account new ideas in floodplain sciences (geomorphological, hydrological, and ecological) and climate change projections.

10.6 Conclusions

Fluvial geomorphology is vital for attaining a comprehensive understanding of flood hazards. For decades fluvial geomorphologists have worked to better understand the timing, controls, and historical changes in flooding and floodplains. With the development of new techniques and advances in digital remote sensing, GIS, and geophysical data, fluvial geomorphology is poised to make substantial contributions to the science of flood hazard management.

This chapter has reviewed a number of important concepts and approaches within fluvial geomorphology relevant to the topic of flood hazard management. The major points include: (1) Fluvial geomorphology has made important contributions to the spatial analysis of flooding, at watershed to local scales, whereby flood processes and flood hazards vary along a continuum with systematic downstream changes in floodplain styles. (2) Traditional hard engineering approaches to flood control have, in many instances, led to adverse geomorphological impacts that triggered unintended geomorphological and environmental changes that undermine existing flood control efforts. (3) Flood management will remain an important societal issue, although individual river basins should strive to attain "integrated flood management" (IFM) to address different types of local-scale environmental change, and for adapting to different environmental change scenarios. (4) Paleoflood data can be quantitatively employed in rigorous models to significantly extend the flood record. Further, the use of paleoflood data implies that, in many instances, different climate change scenarios are inherently integrated into the data set used to design flood control infrastructure. Thus, in an era of global climate change the use of paleoflood data is especially appropriate, and can be integrated into an IFM approach.

Flood hazards have been a mainstay of society since humans first settled upon the banks of the Tigris–Euphrates and Nile Rivers. The crux of the issue is how to assure personal and economic safety from flooding, while maintaining the geomorphological and environmental integrity of the fluvial system. In the context of a changing global environment and a tremendous increase in global population, flood hazard management will long remain an important societal issue and a vital scientific topic, and is best served by explicitly considering fundamental concepts and approaches within the science of fluvial geomorphology.

References

ASCE (American Society of Civil Engineers) (2007). *The New Orleans Hurricane Protection System: What Went Wrong and Why.* American Society of Civil Engineers Hurricane Katrina External Review Panel.

Badji, M. and Dautrebande, S. (1997). Characterization of flood inundated areas and delineation of poor drainage soil using ERS-1 SAR imagery. *Hydrological Processes*, **11**, 1441–1450.

Baker, V. R. (1976). Hydrogeomorphic methods for the regional evaluation of flood hazards. *Environmental Geology*, **1**, 261–281.

Baker, V. R. (2008). Paleoflood hydrology: origin, progress, prospects. *Geomorphology*, **101**(1–2), 1–13.

Baker, V. R. and Kochel, R. C. (1988). Flood sedimentation in bedrock fluvial systems. In V. R Baker, R. C. Kochel and P. C. Patton (eds.), *Flood Geomorphology*. New York: John Wiley Interscience, pp. 123–137.

Baker, V. R., Pickup, G. and Polach H. A. (1985). Radiocarbon dating of flood events, Katherine Gorge, Northern Territory, Australia. *Geology*, **13**, 344–347.

Baker, V. R., Kochel, R. C. and Patton P. C. (eds.) (1988). *Flood Geomorphology*. New York: Wiley Interscience.

Baker, V. R., Webb, R. H. and House, P. K. (2002). The scientific and societal value of paleoflood hydrology. In P. K. House, R. H. Webb, V. R. Baker and D. R. Levish (eds.), *Ancient Floods, Modern Hazards: Principles and Applications of Paleoflood Hydrology*. Water Science and Application Series, 5, pp. 127–146.

Ballais, J. L., Garry, G. and Masson, M. (2005). Contribution of hydrogeomorphological method to flood hazard assessment: the case of French Mediterranean region. *CR Geoscience*, **337**, 1120–1130.

Barriendos, M. and Martín Vide, J. (1998). Secular climatic oscillations as indicated by catastrophic floods in the Spanish Mediterranean coastal area (14th-19th Centuries). *Climatic Change*, **38**, 473–491.

Bates, P. D., Lane, S. N. and Ferguson, R. I. (2005). Computational fluid dynamics for environmental hydraulics. In P. D. Bates, S. N. Lane and R. I. Ferguson (eds.), *Computational Fluid Dynamics*. Applications in Environmental Hydraulics. Chichester: Wiley, pp. 1–15.

Bates, P. D., Wilson, M. D., Horritt, M. S. *et al.* (2006). Reach scale floodplain inundation dynamics observed using airborne synthetic aperture radar imagery: data analysis and modelling, *Journal of Hydrology*, **328**, 306–318.

Benito, G. and Thorndycraft, V. R. (eds.) (2004). *Systematic, Palaeoflood and Historical Data for the Improvement of Flood Risk Estimation: A Methodological Guide*. Madrid: CSIC.

Benito, G. and Thorndycraft, V. R. (2005). Palaeoflood hydrology and its role in applied hydrological sciences, *Journal of Hydrology*, **313**(1–2), 3–15.

Benito, G., Sánchez, Y. and Sopeña, A. (2003a). Sedimentology of high-stage flood deposits of the Tagus River, Central Spain. *Sedimentary Geology*, **157**, 107–132.

Benito, G., Sopeña, A., Sánchez, Y., Machado, M. J. and Pérez González, A. (2003b). Palaeoflood record of the Tagus River (central Spain) during the Late Pleistocene and Holocene. *Quaternary Science Reviews*, **22**, 1737–1756.

Benito, G., Rico, M., Thorndycraft, V. R. *et al.* (2006). Palaeoflood records applied to assess dam safety in SE Spain. In R. Ferreira, E. Alves, J. Leal and A. Cardoso (eds.), *River Flow 2006*. London: Taylor & Francis Group, pp. 2113–2120.

Benito, G., Thorndycraft, V. R., Rico, M., Sánchez-Moya, Y., and Sopeña, A. (2008). Palaeoflood and floodplain records from Spain: evidence for long-term climate variability and environmental changes. *Geomorphology*, **101**, 68–77.

Brakenridge, G. R. (1988). River flood regime and floodplain stratigraphy. In V. Baker, C. Kochel and P. Patton, (eds.), *Flood Geomorphology*. New York: Wiley Interscience, pp. 139–156.

Bridge, J. S. (2003). *Rivers and Floodplains: Forms, Processes, and Sedimentary Record*. Oxford: Blackwell.

Burt, T. P., Bates, P. D., Stewart, M. D. *et al.* (2002). Water table fluctuations within the floodplain of the River Severn, England. *Journal of Hydrology*, **262**, 1–20.

Butzer, K. W. (1976). *Early Hydraulic Civilization in Egypt*. Chicago: University of Chicago Press.

Carling, P. A. (1983). Threshold of coarse sediment transport in broad and narrow natural streams. *Earth Surface Processes and Landforms*, **8**(1), 1–18.

Cohn, T. A., Lane, W. L. and Baier, W. G. (1997). An algorithm for computing moments-based flood quantile estimates when historical flood information is available. *Water Resources Research*, **33**(9), 2089–2096.

Costa, J. E. (1983). Paleohydraulic reconstruction of flash-flood peaks from boulder deposits in the Colorado Front Range. *Geological Society of America Bulletin*, **94**, 986–1004.

Costa, J. E. (1988). Floods from dam failures. In V. Baker, C. Kochel and P. Patton, (eds.), *Flood Geomorphology*. New York: Wiley Interscience, pp. 439–463.

Dickens, C. H. (1865). Flood discharge of rivers. *Prof. Paper India Eng. Roorkee, India, Thomson College Press*, **2**, 133.

Díez-Herrero, A., Laín-Huerta, L. and Llorente-Isidro, M. (2008). *Mapas de peligrosidad por avenidas e inundaciones. Guía metodológica para su elaboración*. Madrid: Instituto Geológico y Minero de España.

DIREN-PACA (Direction Régional de l'Environment-PACA) (2007). L'approache hydrogéomorphologique en mileux mediterranées. Une méthode de détermination des zones inondables. Provence-Alpes-Côte d'Azur. http://www.paca.ecologie.gouv.fr/docHTML/AZIPACA/documents-azi/SITE-DIREN/index.html#ac

Dixon, T. H., Amelung, F., Ferretti A. *et al.* (2006). Subsidence and flooding in New Orleans. *Nature*, **441**, 587–588.

Duller, G. A. T. (2004). Luminescence dating of Quaternary sediments: recent advances. *Journal of Quaternary Science*, **19**, 183–192.

Dunne, T. (1988). Geomorphic contributions to flood-control planning. In V. Baker, C. Y. Kochel, and P. Patton, (eds.). *Flood Geomorphology*. New York: Wiley Interscience, pp. 421–438.

EC (European Council) (2000). Directive 2000/60/EC of the European Parliament and of the Council of 23 October 2000 on the Water, *Official Journal of the European Union*, L 327, 22.12.2000.

EC (European Council) (2007). Directive 2007/60/EC of the European Parliament and of the Council of 23 October 2007. On the assessment and management of flood risks, *Official Journal of the European Union*, L 288, 6.11.2007.

Edson, C. G. (1951). Parameters for relating unit hydrographs to watershed characteristics. *Transactions of the American Geophysical Union*, **32**, 391–396.

Ely, L. L. (1997). Response of extreme floods in the southwestern United States to climatic variations in the late Holocene. *Geomorphology*, **19**, 175–201.

England Jr., J. F., Klawon, J. E., Klinger, R. E. and Bauer, T. R. (2006). *Flood Hazard Study, Pueblo Dam, Colorado, Final Report*. Denver, CO: Bureau of Reclamation, ftp://ftp.usbr.gov/jengland/TREX/pueblo_floodhazard_finalreport.pdf.

Enzel, Y., Ely, L. L., House, P. K., Baker, V. R. and Webb, R. H. (1993). Paleoflood evidence for a natural upper bound to flood magnitudes in the Colorado River basin, *Water Resources Research*, **29**, 2287–2297.

Ferguson, R. J. and Brierly, G. J. (1999). Levee morphology and sedimentology along the lower Tuross River, south-eastern Australia. *Sedimentology*, **46**, 627–648.

Foxcroft, L. C., Parsons, M., McLoughlin, C. A. and Richarson, D. M. (2008). Patterns of alien plant distribution in a river landscape following an extreme flood. *South African Journal of Botany*, **74**, 463–475.

Francés, F., Salas, J. D. and Boes, D. C. (1994). Flood frequency analysis with systematic and historical or palaeoflood data based on the two-parameter general extreme value modes. *Water Resources Research*, **30**, 1653–1664.

French J. R. (2003). Airborne lidar in support of geomorphological and hydraulic modelling. *Earth Surface Processes and Landforms*, **28**, 321–335.

Garry, G. and Graszk, E. (1999). Plans de prevention des risques naturels (PPR). *Risques d'inondation. Guide méthodologique*. Paris: Ministère de l'Environnement, Ministère de l'Equipement, La Documentation Française.

Gilvear D. J. (1999). Fluvial geomorphology and river engineering: future roles utilizing a fluvial hydrosystems framework. *Geomorphology*, **31**(1–4), 229–245.

Glynn, M. E. and Kuszmaul, J. (2004). *Prediction of Piping Erosion Along Middle Mississippi River Levees: An Empirical Model*. U.S. Army Corps of Engineers, ERDC/GSL TR-04–12.

Goddard, J. E. (1976). The nation's increasing vulnerability to flood catastrophe. *Journal of Soil and Water Conservation*, **31** (2), 48–52.

Gomez, B., Phillips, J. D., Magilligan, F. J. and James, A. J. (1997). Floodplain sedimentation and sensitivity: summer 1993 flood, upper Mississippi Valley. *Earth Surface Processes and Landforms*, **22**, 923–936.

Grant, G. E. and Swanson, F. J. (1995). Morphology and processes of valley floors in mountain streams, western Cascades, Oregon. In J. E. Costa, A. J. Miller, K. W. Potter and P. R. Wilcock (eds.), *Natural and Anthropogenic Influences in Fluvial Geomorphology: the Wolman Volume*. Geophysical Monograph 89. Washington, D.C.: American Geophysical Union, pp. 83–101.

Greenbaum, N. (2007). Assessment of dam failure flood and a natural, high-magnitude flood in a hyperarid region using paleoflood hydrology, Nahal Ashalim catchment, Dead Sea, Israel. *Water Resources Research*, **43**, W02401.

Greenbaum, N., Schick, A. P. and Baker, V. R. (2000). The paleoflood record of a hyperarid catchment, Nahal Zin, Negev Desert, Israel. *Earth Surface Processes and Landforms*, **25**, 951–971.

Greenbaum, N., Schwartz, U., Schick, A. P. and Enzel, Y. (2002). Palaeofloods and the estimation of long-term transmission losses and recharge to the Lower Nahal Zin alluvial aquifer, Negev Desert, Israel. In P. K. House, R. H. Webb, V. R. Baker and D. R. Levish (eds.), *Ancient Floods, Modern Hazards: Principles and Applications of Paleoflood Hydrology*. Water Science and Application 5, Washington, D.C.: American Geophysical Union, pp. 311–328.

Gregory, K. J. (2006). The human role in changing river channels. *Geomorphology*, **79**, 172–191.

Gregory, K. J. and Walling, D. E. (1973). *Drainage Basin Form and Process*. New York: Halstead Press, John Wiley & Sons.

Gregory, K. J., Benito, G. and Downs, P. W. (2008). Applying fluvial geomorphology to river channel management: background for progress towards a palaeohydrology protocol. *Geomorphology*, **98**, 153–172.

Grodek T., Enzel Y., Benito G. *et al.* (2007). The contribution of paleoflood hydrology to flood recharge estimations along hyperarid channels, Kuiseb River, Namibia. *Quaternary International, 167–168, Supplement*, p. 148. XVII INQUA Congress, Cairns, Queensland, Australia.

Gupta, V. K., Waymire, E. and Wang, C. T. (1980). A representation of an instantaneous unit hydrograph from geomorphology. *Water Resources Research*, **16**, 855–862.

Harmar, O. P., Clifford, N. J., Thorne, C. R. and Biedenharn, D. S. (2005). Morphological changes of the Lower Mississippi River: geomorphological response to engineering intervention. *River Research and Applications*, **21**, 1107–1131.

Hesselink, A. W., Weerts, H. J. T. and Berendsen, H. J. A. (2003). Alluvial architecture of the human-influenced river Rhine, the Netherlands. *Sedimentary Geology*, **161**, 229–248.

Horton, R. E. (1945). Erosional development of streams and their drainage basins: hydrophysical approach to quantitative morphology. *Geological Society of America Bulletin*, **56**, 275–370.

House, P. K., Webb, R. H., Baker, V. R. and Levish, D. R. (eds.) (2002a). *Ancient Floods, Modern Hazards: Principles and Applications of Paleoflood Hydrology*. Water Science and Application 5, Washington D.C.: American Geophysical Union.

House, P. K., Pearthree, P. A. and Klawon, J. E. (2002b). Historical flood and paleoflood chronology of the lower Verde River, Arizona: stratigraphical evidence and related uncertainties. In P. K. House, R. H. Webb, V. R. Baker and D. R. Levish (eds.), *Ancient Floods, Modern Hazards: Principles and Applications of Paleoflood Hydrology*. Water Science and Application 5, Washington, D.C.: American Geophysical Union, pp. 267–293.

Hudson, P. F. and Colditz, R. R. (2003). Flood delineation in a large and complex alluvial valley, lower Pánuco basin, Mexico, *Journal of Hydrology*, **280**, 229–245.

Hudson, P. F. and Kesel, R. H. (2006). Spatial and temporal adjustment of the lower Mississippi River to major human impacts, *Zeitschrift für Geomorphologie, Supplementband*, **143**, 17–33.

Hudson, P. F., Middelkoop, H. and Stouthamer, E. (2008). Flood management along the Lower Mississippi and Rhine Rivers (the Netherlands) and the continuum of geomorphological adjustment. *Geomorphology*, **101**, 209–236.

Hupp, C. R. (1988). Plant ecological aspects of flood geomorphology and paleoflood history. In V. Baker, C. Kochel and P. Patton (eds.). *Flood Geomorphology*. New York: Wiley Interscience, pp. 335–356.

IPCC (Intergovernmental Panel on Climate Change) (2007). Summary for Policymakers. In M. L. Parry, O. F. Canziani, J. P. Palutikof, P. J. van der Linden and C. E. Hanson (eds.), *Climate Change 2007: Impacts, Adaptation and Vulnerability. Contribution of Working Group II to the Fourth Assessment Report of the Intergovernmental Panel on Climate Change.* Cambridge: Cambridge University Press, pp. 7–22.

Jarrett, R. D. (1987). Errors in slope-area computations of peak discharges in mountain streams. *Journal of Hydrology*, **96**, 53–67.

Jarvis, C. S. (1936). *Floods in the United States*. U.S. Geological Survey, Water-Supply Paper 771.

Junk, W. J., Bayley, P. B. and Sparks, R. E. (1989). The flood pulse concept in river-floodplain systems. In D. P. Dodge (ed.), *Proceedings of the International Large River Symposium (LARS), Canadian Special Publication in Fisheries and Aquatic Science*, **106**, 110–127.

Kesel, R., Dunne, K. C., McDonald, R. C., and Allison, K. R. (1974). Lateral erosion and overbank deposition on the Mississippi River in Louisiana caused by 1973 flooding. *Geology*, **2**, 461–464.

Knighton, A. D. (1998). *Fluvial Forms and Processes.*, Baltimore, MD: Edward Arnold.

Knox, J. C. (1993). Large increases in flood magnitude in response to modest changes in climate. *Nature*, **361**, 430–432.

Knox, J. C. (2000). Sensitivity of modern and Holocene floods to climate change. *Quaternary Science Reviews*, **19**, 439–457.

Knox, J. C. and Daniels, J. M. (2002). Watershed scale and the stratigraphic record of large floods. In P. K. House, R. H. Webb, V. R. Baker and D. R. Levish (eds.) *Ancient Floods, Modern Hazards: Principles and Applications of Paleoflood Hydrology.* Water Science and Application 5, American Geophysical Union, Washington, D.C. pp. 237–255.

Kochel, R. C. and Baker, V. R. (1982). Paleoflood hydrology. *Science*, **215**, 353–361.

Komar, P. D. (1989). Flow-competence evaluations of the hydraulic parameters of floods: an assessment of the technique. In K. Beven and P. Carling (eds.), *Floods: Hydrological, Sedimentological and Geomorphological Implications.* Chichester: John Wiley and Sons Ltd, pp. 185–197.

Kundzewicz, Z. W. and Schellnhuber H. -J. (2004). Floods in the IPCC TAR perspective. *Natural Hazards*, **31**, 111–128.

Lastra, J., Fernández, E., Díez-Herrero, A. and Marquínez, J. (2008). Flood hazard delineation combining geomorphological and hydrological methods: an example in the Northern Iberian Peninsula. *Natural Hazards*, **45**(2), 277–293.

Leese, M. N. (1973). Use of censored data in the estimation of Gumbel distribution parameters for annual maximum flood series, *Water Resources Research*, **9**, 1534–1542.

Leopold, L. B., Wolman M. G. and Miller J. P. (1964). *Fluvial Processes in Geomorphology.* New York: W. H. Freeman and Co.

Levish, D., Ostenaa, D. and O'Connell, D. (1997). Paleoflood hydrology and dam safety. In D. J. Mahoney (ed.), *Waterpower 97: Proceedings of the International Conference on Hydropower.* American Society of Civil Engineering, New York, pp. 2205–2214

Lewin, J. (1989). Floods in fluvial geomorphology. In K. Beven and P. Carling (eds.), *Floods: Hydrological, Sedimentological and Geomorphological Implications*. Chichester: John Wiley and Sons, pp. 265–284.

Li, Y., Craven, J., Schweig, E. S. and Obermeier, S. F. (1996). Sand boils induced by the 1993 Mississippi River flood: could they one day be misinterpreted as earthquake induced liquefaction? *Geology*, **24**, 171–174.

Macklin M. G., Benito G., Gregory K. J. *et al.* (2006). Past hydrological events reflected in the Holocene fluvial history of Europe. *Catena*, **66**, 145–154.

Maxwell, J. C. (1960). *Quantitative Geomorphology of the San Dimas Experimental Forest, California*. Technical Report No. 19, Office of Naval Research, Project NR 389–042, Department of Geology, Columbia University., New York.

Mertes, L. A. K. (1997). Documentation and significance of the perirheic zone on inundated floodplains. *Water Resources Research*, **33**, 1749–1762.

Middelkoop, H. (1997). *Embanked Floodplains in the Netherlands*. Netherlands Geographical Studies 224. KNAG/ Faculteit Ruimtelijke Wetenschappen Universiteit Utrecht.

Middelkoop, H. and Van Haselen, C. O. G. (1999). *Twice a River. Rhine and Meuse in the Netherlands*. RIZA Report 99.003, Arnhem: RIZA.

Morisawa, M. E. (1962). Quantitative geomorphology of some watersheds in the Appalachian Plateau. *Geological Society of America Bulletin*, **73**, 1025–1046.

Mosley, M. P. and McKerchar, A. I. (1993). Streamflow. In D. R. Maidment (ed.), *Handbook of Hydrology*. New York: McGraw-Hill, Inc., pp. 8.1–8.39.

Murray, A. S. and Wintle, A. G. (2000). Luminescence dating of quartz using an improved single-aliquot regenerative-dose protocol. *Radiation Measurements*, **32**, 57–73.

Nanson, G. C. (1986). Episodes of vertical accretion and catastrophic stripping: a model of disequilibrium flood-plain development. *Geological Society of America Bulletin*, **97**, 1467–1475.

Nanson G. C. and Croke J. C. (1992). A genetic classification of floodplains. *Geomorphology*, **4**, 459–486.

Nash J. (1957). The form of the instantaneous unit hydrograph. *International Association of Science and Hydrology*, **45**(3), 114–121.

NRC (National Research Council) (1995). *Wetlands: Characteristics and Boundaries*. Washington, D.C.: National Academy of Sciences Press.

NRC (National Research Council) (2005). *The Science of Instream Flows: A Review of the Texas Instream Flow Program*. Committee on Review of Methods for Establishing Instream Flows for Texas Rivers, Water Science and Technology Board, Division on Earth and Life Studies, Washington D.C.: National Academy of Sciences Press.

NRC (National Research Council) (2007). *Elevation Data for Floodplain Mapping*. Committee on Floodplain Mapping Technologies, Board on Earth Sciences and Resources, Division on Earth and Life Studies, Washington D.C.: National Academy of Sciences Press.

O'Connell, D. R. H., Ostenaa, D. A. Levish, D. R. and Klinger, R. E. (2002). Bayesian flood frequency analysis with paleohydrologic bound data. *Water Resources Research*, **38**, 1058.

O'Connor, J. E. and Webb, R. H. (1988). Hydraulic modeling for palaeoflood analysis. In V. R. Baker, R. C. Kochel and P. C. Patton (eds.), *Flood Geomorphology*. New York: Wiley Interscience, pp. 393–403.

Patton, P. C. (1988). Drainage basin morphometry and floods. In V. Baker, C. Kochel and P. Patton (eds.), *Flood Geomorphology*. New York: Wiley Interscience, pp. 51–64.

Patton, P. C. and Baker, V. R. (1976). Morphometry and floods in small drainage basins subject to diverse hydrogeomorphologic controls. *Water Resources Research*, **12**, 941–952.

Pelletier, J. D., Mayer, L., Pearthree, P. A. *et al.* (2005). An integrated approach to alluvial-fan flood hazard assessment with numerical modeling, field mapping, and remote sensing, *Geological Society of America Bulletin*, **117**, 1167–1180.

Pinter, N. (2005). One step forward, two steps back on U.S. floodplains. *Science*, **308**, 207–208.

Pinter, N., Ickes, B. S., Wlosinski, J. H. and Van der Ploeg, R. R. (2006). Trends in flood stages: contrasting results from the Mississippi and Rhine River systems. *Journal of Hydrology*, **331**, 554–566.

Pinter, N., Jemberie, A. A., Remo, J. W. F., Heine, R. A. and Ickes, B. S. (2008). Flood trends and river engineering on the Mississippi River system. *Geophysical Research Letters*, **35**, L23404, doi:10.1029/2008GL035987.

Pizzuto, J. E. (1987). Sediment diffusion during overbank flows. *Sedimentology*, **34**, 301–317.

Poole, G. C., Stanford, J. A., Frissell, C. A. and Running, S. W. (2002). Three-dimensional mapping of geomorphological controls on floodplain hydrology and connectivity from aerial photos. *Geomorphology*, **48**, 329–347.

Rapp, C. F. and Abbe, T. B. (2003). *A Framework for Delineating Channel Migration Zones*. Washington State Department of Ecology, Report 03–06–027, WS.

Redmond, K. T., Enzel, Y., House, P. K. and Biondi, F. (2002). Climate variability and flood frequency at decadal to millennial time scales. In P. K. House, R. H. Webb, V. R. Baker, and D. R. Levish (eds.), *Ancient Floods, Modern Hazards: Principles and Applications of Paleoflood Hydrology*. Water Science and Application 5, Washington D.C.: American Geophysical Union, pp. 21–45.

Reis, D. S. Jr. and Stedinger J. R. (2005). Bayesian MCMC flood frequency analysis with historical information. *Journal of Hydrology*, **313**, 97–116.

Rodríguez-Iturbe, I. and Valdés, J. B. (1979). The geomorphologic structure of the hydrological response. *Water Resources Research*, **15**, 1409–1420.

Saint-Laurent, D. (2004). Palaeoflood hydrology: an emerging science. *Progress in Physical Geography*, **28**, 531–543.

Schumm, S. A. (1986). *Alluvial River Response to Active Tectonics*. Geophysics Study Committee, Geophysics Research Forum, National Research Council, National Academy of Sciences, Washington D.C.

Shankman, D. and Smith, L. J. (2004). Stream channelization and swamp formation in the US Coastal Plain. *Physical Geography*, **25**, 22–38.

Sharp, J. M. Jr. (1988). Alluvial aquifers along major rivers. In W. Back, J. S. Rosenshein and P. R. Seaber (eds.), *Hydrogeology: The Geology of North America*, vol. 0.2. Boulder, CO: Geological Society of America, pp. 273–282.

Sheffer, N. A., Rico, M., Enzel, Y., Benito, G. and Grodek, T. (2008). The palaeoflood record of the Gardon river, France: a comparison with the extreme 2002 flood event. *Geomorphology*, **98**, 71–83

Sherman, L. (1932). Streamflow from rainfall by unit-graph method. *Engineering News-Record*, 501–505.

Silva, W., Klijn, F. and Dijkman, J. (2001). *Room for the Rhine branches in the Netherlands*. IRMA SPONGE, Rijkswaterstaat, Delft Hydraulics Report.

Simon, A. and Rinaldi, M. (2006). Disturbance, stream incision, and channel evolution: the roles of excess transport capacity and boundary materials in controlling channel response. *Geomorphology*, **79**, 361–383.

Slade, R. M. Jr. and Patton, J. (2002). *Major and Catastrophic Storms and Floods in Texas: 215 Major and 41 Catastrophic Events From 1853 To September 1, 2002*. U.S. Geological Survey Open-File Report 03–193.

Smith, K. (1996). *Environmental Hazards: Assessing Risk and Reducing Disaster*, 2nd edition. London: Routledge.

Smith, L. C. (1997). Satellite remote sensing of river inundation area, stage, and discharge: a review. *Hydrological Processes*, **11**, 1427–1439.

Smith, R. F. and Boardman, J. (1989). The use of soil information in the assessment of the incidence and magnitude of historical flood events in Upland Britain. In K. Beven and P. Carling (eds.), *Floods: Hydrological, Sedimentological and Geomorphological Implications*. Chichester: John Wiley and Sons Ltd, pp. 185–197.

Smith, K. and Ward, R. (1998). *Floods. Physical Processes and Human Impacts*. Chichester: John Wiley.

Smith, L. M and Winkley, B. R. (1996). The response of the Lower Mississippi River to river engineering. *Engineering Geology*, **45**, 433–455.

Starkel, L., Soja, R. and Michczyńska, D. J. (2006). Past hydrological events reflected in Holocene history of Polish rivers. *Catena*, **66**, 24–33.

Stedinger, J. and Cohn, T. A. (1986). Flood frequency analysis with historical and paleoflood information. *Water Resources Research*, **22**, 785–793.

Strahler, A. N. (1957). Quantitative analysis of watershed morphology. *Transactions of the American Geophysical Union*, **38**, 913–920.

Thomas, D. M. and Benson, M. A. (1970). *Generalization of Streamflow Characteristics from Drainage-basin Characteristics*. U.S. Geological Survey, Water-Supply Paper 1975.

Thomas, M. C. (2003). Floodplain–river ecosystems: lateral connections and the implications of human interference. *Geomorphology*, **56**, 335–350.

Thorndycraft, V. R. and Benito, G. (2006). The Holocene fluvial chronology of Spain: evidence from a newly compiled radiocarbon database. *Quaternary Science Reviews*, **25**, 223–234.

Thorndycraft, V. R., Benito, G., Barriendos, M. and Llasat, M. C. (eds.) (2003). *Palaeofloods, Historical Data and Climate Variability: Applications in Flood Risk Assessment*. Madrid: CSIC.

Toth, L. A., Obeysekera, J. T. B., Perkins, W. A. and Loftin, M. K. (1993). Flow regulation and restoration of Florida's Kissimmee river. *Regulated Rivers: Research and Management*, **8**, 155–166.

US-ACE (U.S. Army Corps of Engineers) (1998). *Mississippi River Mainline Levees Enlargement and Seepage Control, Cape Girardeau, Missouri to Head of Passes, Louisiana*, Vol. I.

U.S. Water Resources Council (1982). *Guidelines for Determining Flood Frequency*. Bulletin 17B, Hydrology Committee, Washington D.C.

van de Ven, G. P. (1993). *Man-made Lowlands: History of Water Management and Land Reclamation in the Netherlands*. Utrecht: Matrijs.

Van Veen, J. (1962). *Dredge, Drain, and Reclaim: The Art of a Nation*, 5th edition The Hague: Martinus Nijhoff.

Ward, A. and Trimble, S. W. (2004). *Environmental Hydrology*, Boca Raton FL: CRC-Lewis Press.

Wasklewicz, T., Greulich, S., Franklin, S. and Grubaugh, J. (2005). The 20th century hydrologic regime of the Mississippi River. *Physical Geography*, **25**, 208–224.

Webb, R. H. and Betancourt, J. L. (1992). *Climatic Effects on Flood Frequency of the Santa Cruz River, Pima County, Arizona*. U.S. Geological Survey Water-Supply Paper 2379.

Werritty, A., Paine, J. L., Macdonald, N., Rowan, J. S. and McEwen, L. J. (2006). Use of multi-proxy flood records to improve estimates of flood risk: Lower River Tay, Scotland. *Catena*, **66**, 107–119.

White, G. (1945). *Human Adjustment to Floods. University of Chicago Department of Geography, Research Paper* No. 29 (1942).

Williams, G. P. (1988). Paleofluvial estimates from dimensions of former channels and meanders. In V. Baker, C. Kochel and P. Patton, (eds.), *Flood Geomorphology*. New York: Wiley Interscience, pp. 321–334.

Winkley, B. R. (1994). Response of the Lower Mississippi River to flood control and navigation improvements. In S. A. Schumm and B. R. Winkley (eds.), *The Variability of Large Alluvial Rivers*. New York: American Society of Engineers Press, pp. 45–74.

Wohl, E. (2000). *Mountain Rivers*. Water Resources Monograph 14, Washington, D.C.: American Geophysical Union.

Wolman, M. G. (1971). Evaluating alternative techniques of floodplain mapping. *Water Resources Research*, **7**, 1383–1392.

Wolman, M. G. and Costa, J. E. (1984). Envelope curves for extreme flood events: discussion. *Journal of Hydraulic Engineering*, **110**, 77–78.

Wolman, M. G. and Leopold, L. B. (1957). *River Flood Plains: Some Observations on Their Formation*, U.S. Geological Survey Professional Paper, 282C, 87–109.

WMO (World Meteorological Organization) (2004). *Integrated Flood Management*. The Associated Programme on Flood Management, 1, Geneva.

Yodis, E. G. and Kesel, R. H. (1992). The effects and implications of baselevel changes to Mississippi River tributaries. *Zeitschrift für Geomorphologie*, **37**, 385–402.

11 Geomorphology and coastal hazards

Harley J. Walker and Molly McGraw

11.1 Introduction

Arguably the most conspicuous boundary on Earth – i.e., the relatively narrow zone that separates land from sea – is also the most desirable for human occupation and utilization. However, its apparent discreteness is misleading because it is constantly in flux. Being the juncture between a highly unstable solid and a constantly moving liquid, it is continuously subject to changes in form and position. These changes vary in size and duration and may be random or cyclic and subtle or dramatic. Those who would occupy coastal environments are thus subject to a wide variety of natural variations, some of which can be disastrous. Because, as generally accepted, hazards and the disasters they portend are human related, a proper understanding of those that occur in the coastal zone requires information not only about the geomorphology (materials, forms, and processes) of coastal environments but also the nature of human involvement with and within the coastal zone. After introducing the types of coast present around the world, we look at the processes that prevail within the various coastal zones, and then analyze how humans have adapted to and modified specific forms. Next we consider the hazards that have activated a disaster, or, at least, have the potential to do so. Examples from around the world are used to illustrate the difference in timing, intensity, and complexity of hazards/disasters in the coastal zone.

11.1.1 The natural coastal setting: forms, materials, and processes

As sea plays wife to earth
and molds it to her will
by storm and subtle stroking
shaping the shore with skill

(Walke, 1997, p. 24)

The coastal zone that straddles the shoreline of the world's oceans is finite. The lengths published vary greatly depending upon how much generalization has gone into the measurements. Commonly cited lengths for the world range from 4.4×10^5 km (Inman and Brush, 1973) to 10^6 km (Bird and Schwartz, 1985). Therefore discretion is required when interpreting different published lengths. For example, the various lengths given for the Alaskan coast emphasize how great the contrasts can be and how necessary it is to know the objective of the author being read: the *United States Coast Pilot* (NOAA, 1981) lists the length of the general coastline of Alaska as 10,686 km whereas the U.S. Army Corps of Engineers (1971), by using the tidal shoreline, gives it as being 76,100 km. Spanning most of the Earth from the Arctic Ocean to Antarctica, coasts are subject to a wide variety of geomorphic, oceanographic, atmospheric, cryospheric, and biologic conditions. The CIA *World Factbook* (CIA, 2008) notes that 195 (81.6%) out of 239 countries have an oceanic coastline. Lengths range from 4 km for Monaco to more than 200,000 km (*c.* 30% of the world's total) for Canada.

The basic form of the world's coastal zone has been created by continental drift as postulated in the 1960s paradigm of plate tectonics (Inman and Nordstrom, 1971). Coasts associated with plate movement vary from hills and mountains to low-lying plains and from those that are tectonically active to tectonically passive. Plate movement is so slow that the basic coastal forms have changed little during human history and have only recently been measured (Kerr, 1985). Nonetheless, on a local scale, tectonic activity associated with plate movement has been responsible for major disasters and, of course, must be considered an important hazard.

Plate movement also is mainly responsible for the size and lithology of drainage basins and thus, along with

Geomorphological Hazards and Disaster Prevention, eds. Irasema Alcántara-Ayala and Andrew S. Goudie. Published by Cambridge University Press. © Cambridge University Press 2010.

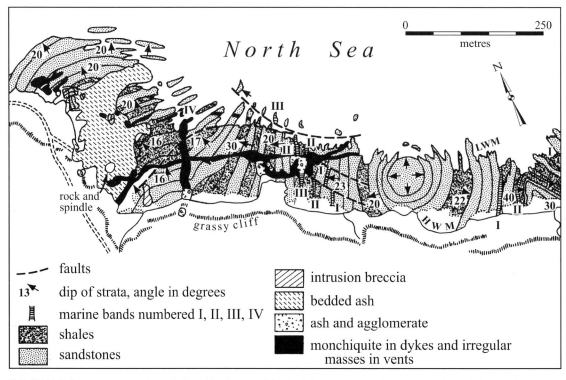

FIGURE 11.1. Contrasts in coastal morphology. North coast of Scotland illustrating material, process and form. (After MacGregor, 1968; from Walker, 1984.)

climate and vegetation, a determinant of the amount and type of sediment delivered to the oceans and partially responsible for the creation of such coastal forms as deltas where some of the world's major disasters occur (Walker, 1990b).

Although the gross characteristics of the coast were established long before humans began to utilize it, numerous other natural modifications have occurred during their occupational history. Most important are the changes associated with variations in sea level. The virtual still-stand of the past 4,000 to 5,000 years has allowed the natural forces that are oceanographic (waves, currents, tides, sea-level change, etc.), geophysical (earthquakes, tsunami, etc.), and meteorological (hurricanes, storm surges, etc.) in origin to operate along the fairly restricted coastal base available to present-day humans.

Many ways of differentiating (classifying) coastal environments have been proposed (see e.g., Inman and Nordstrom, 1971; Dolan *et al.*, 1972; Shepard, 1976; King, 1982); however, from the standpoint of human occupation, utilization, and modification, specific forms – beaches, barriers, cliffs, coral reefs, deltas, dunes, embayments, estuaries, marshes, swamps – are most germane to the present discussion. Each type has its own characteristics, variables, distributions and, therefore, its own unique attributes in so far as humans are concerned. Whereas form is the most conspicuous aspect of the coastal zone upon which the processes listed above operate, important also are the materials of which the various forms are composed (Figure 11.1). Significant from the standpoint of coastal hazards is the fact that there also exists a close correlation between material and form: e.g., solid rock and cliffs; river-derived sediment and deltas; sand and dunes and sand bars; organic matter and marshes; clays and silts and mudflats.

Basically, coastal forms have a random distribution; however, there are many exceptions. For example, ice-bound coasts are found in high latitudes and coral reefs and mangrove swamps are located in tropical and equatorial regions. Although any attempt at giving precise lengths for various coastal forms is imprecise, the following proportions of coastal forms on a world-wide scale will provide an idea of the relative distribution of coastal types available for human utilization. Cliff coasts occupy nearly half of the world's coastline, barrier islands (the most common coastal form in the United States) occupy about 10 to 15%, and deltas about 1.5%. Beaches, some of which are associated with barrier islands, are common throughout the coastal zone and range from the small pocket beaches associated with cliffy coasts to those that stretch for several hundred kilometers. Coastal dunes, present on all continents including Antarctica, are one of the coast's most common and active forms. Estuaries and lagoons, which are highly varied in physical, biological, and chemical characteristics and very attractive to humans, are found along most of the coasts of the world.

11.1.2 Humans and coastal modification: the development of the cultural landscape[1]

> I lie on the beach
> and wait for the wave to come
> the great wave in its turn
>
> (Walke, 1997, p. 1)

Just when humans "discovered" the coastal zone is subject to debate. However, when that happened, they found an ecologic niche that provided usable plants, animals, and materials for tool making and maybe even relaxation, recreation, and expectation as suggested in Roger Walke's poem above. Collecting and scavenging were probably best where cliffs and beaches alternated and where wetlands prevailed. It is possible that these early nomadic coastal dwellers suffered more, proportionally, from disaster than the coastal dwellers of today, given the lack of preparedness, transportation, and warning systems. Without the infrastructures modern humans possess today, their loss from a disaster was mainly a matter of life. It is probable that by the end of the Pleistocene humans had occupied most of the usable coastal zone and, unbeknowingly, began migrating up-slope with sea-level rise – a sea-level rise that buried or destroyed most of the evidence of these early coastal dwellers.

Although throughout most of human history coastal dwellers were nomadic, once boats began to be used for fishing and agriculture spread into floodplain and deltaic areas, permanent settlement, social and political organization, commerce, and over-water transport developed (Walker, 1990b). Ports were often established in embayments and at river mouths especially those that offered protection from the sea. During this period of time coastal infrastructures were being developed and the natural landscape of the coastal zone was beginning to be turned into a cultural landscape, thus adding a new dimension to the concepts of hazards and disasters.

This relatively modest, localized and gradual conversion of the coastal landscape was accelerated dramatically after the Dark Ages with the great explorations and the industrial revolution. The former outlined most of the coastlines of the world and were the beginning of the conversion of the

world into a "migratory network dominated by a single group of technologically advanced and culturally similar states" (Davis, 1974, p. 96). During the exploratory period many of the world's coastal features, such as the major bays, peninsulas, straits, and river mouths were mapped and named, setting the stage for the major developments that accompanied the industrial revolution. Although most of the present-day uses and modifications of the coast – e.g., harvesting coastal resources, construction of harbors, and reclamation – existed prior to the industrial revolution, it gave impetus to the rapid increase in the developments that followed.

There has long been a close relationship between harbors and human settlements. Small ships could find safe harbor in river mouths or shallow bays; however, they became inadequate as larger vessels were developed. Coastal cities followed apace. Although in the pre-industrial revolution days coastal cities were only occasionally among the world's largest, they soon began to dominate. In 1500, only 20% of the world's largest cities were coastal; by 1800, the proportion had increased to 50% and by 1900, 70% (Chandler, 1977). Today, 14 of the 20 (70%) most populous cities in the world are coastal (Table 11.1) and stretch long distances along much of the coastline of their respective countries (United Nations, 2003). The number of people, along with the cultural landscapes they are creating (Figure 11.2), continues to expand in the coastal zone.

Humans, the *sine qua non* of hazards and disasters, are continuing to increase in number in the coastal zone not only in actual number but also in proportion to the overall increase in total world population. Although more than 80% of the countries of the world have coastlines, the distribution of humans along them is highly varied. In some countries, especially insular nations, virtually all people are coastal dwellers. In Japan, for example, it is about 80%; for the world as a whole, it is about half that or 41%. However, their proximity to the actual coast and therefore the potential impact on them of coastal hazards/disasters is difficult to quantify because of the various definitions of the term "coastal" (see e.g., Cohen *et al.*, 1997).

The increased utilization and modification of the coastal zone is being accompanied by a rapid shrinking of the natural landscape. Equally as important, humans are rapidly becoming a dominant geomorphological agent in some coastal countries.

Some of the most conspicuous modifications include the lengthening of the shoreline with groins and marinas, shortening

---

[1] We use the term "cultural landscape" in this paper to include all changes made by humans to and in the coastal zone whether intentional or unintentional.

TABLE 11.1. *The world's 20 largest metropolitan areas: coastal cities in bold*

Rank	Location	Pop. $\times 10^6$	Rank	Location	Pop. $\times 10^6$
1	**Tokyo, Japan**	35.0	11	**Los Angeles, USA**	12.0
2	Mexico City, Mexico	18.7	12	**Dhaka, Bangladesh**	11.6
3	**New York City, USA**	18.3	13	**Osaka-Kobe, Japan**	11.2
4	Sao Paulo, Brazil	17.9	14	**Rio de Janeiro, Brazil**	11.2
5	**Bombay, India**	17.4	15	**Karachi, Pakistan**	11.1
6	Delhi, India	14.1	16	Beijing, China	10.8
7	**Calcutta, India**	13.8	17	Cairo, Egypt	10.8
8	**Buenos Aires, Argentina**	13.0	18	Moscow, Russia	10.5
9	**Shanghai, China**	12.8	19	**Manila, Philippines**	10.4
10	**Jakarta, Indonesia**	12.8	20	**Lagos, Nigeria**	10.1

Source: United Nations, 2003

FIGURE 11.2. The anthropogenic cliff coast of Hong Kong. (Photograph by H. Walker.)

it with dikes, converting lowlands into runways, making beaches where none existed before, changing proportions of land and water ... and even making islands and artificial reefs.
(Walker, 1981, p. 99)

Some of the most obvious modifications are at the shoreline itself, often for the purpose of stabilization but also for disaster prevention. Sea walls, groins, breakwaters, jetties, and dikes now are found along thousands of kilometers of shoreline. More than 80% of Belgium is bordered by artificial structures; for Japan it is 51%, for England, 38%, and for South Korea, 21% (Walker, 1988). Included in the artificialization of the coastline are a number of structures designed to protect people and their constructs from disastrous surges from storms and tsunami. Although modified coastlines are still only a small percentage of the world's total, they nonetheless occur mainly where people are concentrated and therefore

where potential disasters (in the sense used in this essay) prevail.

11.1.3 Coastal hazards and disasters

> The sea that I sailed has known a wave
> Immense as a tsunami and winds that sing
> In rigging shredded now to ribbons, string
> Typhoons and hurricanes, no port to save
> (Walke, 1997, p. 2)

There are few, if any, places on Earth that are safe from disaster. Nonetheless, many places are more prone to disaster than others. The coastal zone is no exception. Along some coasts, because of population density and position in relation to natural and/or anthropogenic hazards, disasters may occur relatively frequently. Along others, disasters are rare. Even though all coastlines have a potential for disaster, the types of hazardous situations that exist along any particular coastline are varied. Within the coastal zone, their domain can be lithospheric, hydrospheric, atmospheric, cryospheric, biospheric, anthrospheric, or some combination thereof (Figure 11.3). The geomorphologic element of coastal disasters is important directly or indirectly in those disasters occurring in all six of the domains listed above. Although geomorphology is most significant in lithospheric-related disasters, it nonetheless has an impact on the nature of the disasters in the other five domains as well.

The hazardous situations that occur in the six domains are essentially passive; it takes some triggering mechanism to activate them and convert them into disasters (Figure 11.3). It is the triggering mechanisms (especially earthquakes, volcanic eruptions, and cyclones) that form the basis for the following discussion. These events and the impacts from

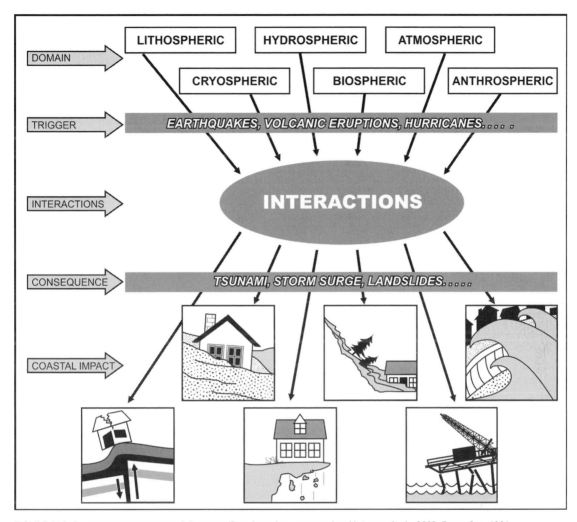

FIGURE 11.3. Factors germane to coastal disasters. (Based on ideas contained in Alcántara-Ayala, 2002, Figure 9, p. 120.)

them vary in a number of ways including magnitude, frequency, duration, areal extent, spatial dispersion, and temporal spacing (Burton *et al.*, 1978; Gares *et al.*, 1994).

11.2 Earthquakes, volcanic eruptions, and tsunami

Earthquakes and volcanic eruptions are two of the most important geophysical phenomena resulting in disaster. They are especially active along lengthy sections of the Earth's coastline, mainly because of plate tectonics. The Earth, composed of numerous relatively rigid yet mobile plates, is constantly in flux. Although there are different types of coasts within plates as well as at some plate boundaries, it is especially at those boundaries where plates are converging that earthquakes and volcanic eruptions are most common (Figure 11.4) and

disasters often occur. The zone that surrounds most of the Pacific Ocean, known as the circum-Pacific seismic belt (often called the "Pacific Ring of Fire"), extends from near the southern tip of South America counter-clockwise around to New Zealand. This belt, which contains such densely populated coastlines as the western United States, Japan, and the Philippines, is responsible for some 90% of the world's earthquakes and supports some three-quarters of its volcanoes. The second most active seismic belt, known as the Alpide belt, extends from Java westward through the Himalayas past the Mediterranean Sea. Although much of it is non-coastal, both the eastern (e.g., Java and Sumatra) and the western ends are seismically active. Although only about 6% of the world's earthquakes occur along it, some of them have been responsible for some of the world's greatest disasters.

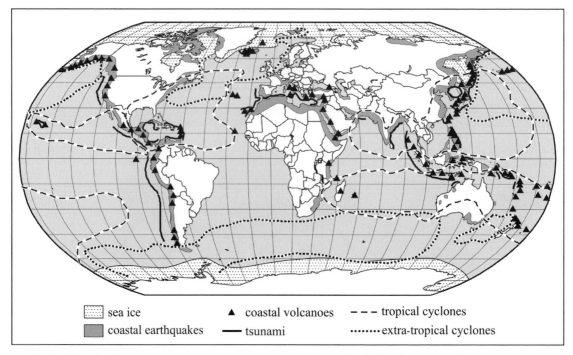

FIGURE 11.4. World-wide distribution of coastal hazards. (From a slide prepared for the 2003 Regional Conference on Geomorphology in Mexico City.)

Because earthquakes are powerful redistributors of energy and often cause massive losses of life and much damage to cultural landscapes, historical records are numerous and extend back centuries (Table 11.2). For example, an earthquake at Corinth, Greece, in 856 killed an estimated 45,000 people; one in Lisbon, Portugal, in 1755 was responsible for 34,000 deaths and the one in Calcutta, India, in 1837 killed 300,000. More recent earthquakes have been just as deadly: the 1923 Tokyo, Japan, earthquake killed 143,000 and the massive earthquake in Sumatra in 2004 was responsible for more than 230,000 deaths, with two million displaced, some as far away as Africa (USGS, 2008). Although all of these were major earthquakes, some equally as powerful have occurred at locations with relatively low population densities; for example, the 9.2 magnitude earthquake in 1960 in Chile and the 9.4 earthquake near Anchorage, Alaska, in 1964. Although not many people (relatively speaking) lost their lives near those earthquakes' centers, there was loss of life as far away as Japan from the tsunami they spawned. Incidentally, some of the major shoreline structures built along the coast of Japan are designed to mitigate the impact of tsunami (Figure 11.5) (see e.g., Chen *et al.*, 2002).

From the standpoint of geomorphology there are many other factors and responses accompanying and following earthquakes that are relevant to the landscape and to associated human structures. Earthquake hazards include ground shaking, soil liquefaction, uplift, subsidence, landslides, and tsunami. The various types of waves set up in the ground, by ground shaking can travel great distances and result in both vertical and horizontal movements in the ground, both of which cause changes in topography and may result in damage to any structures that may be present. The amount and kind of ground motion are affected by lithology and topographic setting: unconsolidated materials are more prone to modification than those that are more indurated. Soil liquefaction, the loss of strength caused by shaking in loose sediments, can result in lateral displacement, ground oscillation, and loss of bearing strength (Smith, 2001). Earthquakes often cause large areas on either side of the fault line to be lifted up or dropped down. Of course, along coastlines, such differential movements expose former sea beds to sub-aerial processes or submerge former shoreline features such as tidal flats, beaches, dunes, wetlands, and the like. Landslides and rockfalls often accompany earthquakes. Because cliffs are common along coasts and because most are subjected to erosion due to normal and storm-generated wave action, earthquake-generated coastal landslides are frequent and in some cases major modifiers of the coastal landscape.

TABLE 11.2. *Deaths associated with major coastal earthquakes*

Date	Location	Magnitude	Deaths
856	Greece, Corinth		45,000
1290	China, Hebi		100,000
1755	Portugal, Lisbon	9	34,000*
1837	India, Calcutta		300,000
1906	USA, San Francisco	8.25	700
1908	Italy, Messina		75,000
1923	Japan, Tokyo	8.2	143,000
1960	Chile	9.2	5,000*
1964	USA, Alaska	9.4	2,000
1995	Japan, Kobe	7.2	5,400*
2004	Indonesia, Sumatra	9.1	230,000

Primary source: Nott, 2006;* Gunn, 2008

FIGURE 11.5. Artificial cliff in Japan constructed for tsunami protection. (Photograph by H. Walker.)

TABLE 11.3. *Major tsunami ranked by number of deaths*

Date	Location	Trigger	Max. Ht. (m)	Deaths
2004	Sumatra, Indonesia	Earthquake	50	230,000
1755	Lisbon, Portugal	Earthquake	30	60,000
1883	Krakatau, Indonesia	Volcanic eruption	35	36,000
1896	Sanriku, Japan	Earthquake	29	27,212
1869	Chile	Earthquake	18	25,000
1923	Sagami Bay, Japan	Earthquake	12	2,144
1690	Port Royal, Jamaica	Earthquake	1.8	2,000
1946	Dominican Republic	Earthquake	5	1,790
1946	Honshu Coast, Japan	Earthquake	6.6	1,362
1960	Central Chile	Earthquake	25	1,263

Source: NOAA, 2008.

Although an earthquake itself tends to be localized, its ramifications are felt virtually world-wide in the case of seismic earth waves and nearly as extensively in the case of tsunami or seismic sea waves. Tsunami, like earthquakes, typhoons, and volcanic eruptions, have been recorded for centuries (Table 11.3) and have been partially blamed by some historians for the collapse of cultures such as the Minoan in the Mediterranean Sea. Crete, its major stronghold, was destroyed by the tsunami generated with the eruption of Thera (Santorini).

Although near-shore earthquakes are the major generator of tsunami they can also be caused by volcanic eruptions, as in the case of the Thera eruption in 3,600 BP, and by submarine slides. Major historic tsunami include: one in 1755 that destroyed Lisbon, Portugal; the Krakatau, East Indies, tsunami in 1883 that was the most famous until 2004 when the Sumatran tsunami impacted most Indian Ocean coastlines. The run-up from tsunami are major morphologic modifiers as well as killers of people and destroyers of property. Run-up can extend far inland along coastlines and submerge islands that lie in a tsunami's path. The 1883 Krakatau-generated tsunami in the Sunda Straits was as much as 40 m high (SDSU, 2008). Coastal vegetation was destroyed and coastal configuration altered. As a major erosional, transportational, and depositional agent, tsunami are one of the Earth's most potent processes. At some locations, the surge from the 2004 tsunami ran more than 2 km inland and deposited sheets of sand on ridges and in swales that were 5 to 20 m thick. Recent research has shown that such tsunami deposits, although rare, are not unique for some were spread inland hundreds of years ago. The infrequency of such events, lost to history on a human time scale, remain out of the conscious thinking of residents in such areas (Monecke *et al.*, 2008). Although earthquakes and volcanic eruptions are the most common causes of tsunami, submarine landslides, as mentioned above, asteroid and meteorite impacts, and even weather systems can also cause them. The most disastrous of all tsunami would certainly be those originating from the impact of asteroids from outer space as has happened in the geologic past.

The 1964 "Good Friday" earthquake in Alaska, one of the world's strongest, is used here to illustrate the various

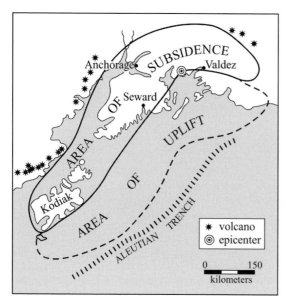

FIGURE 11.6. Uplift and subsidence from the 1964 Alaskan earthquake. (Modified from Bolt et al., 1977.)

impacts and ramifications of earthquakes discussed above. Its epicenter was in the coastal Chuguch Mountain range, about 130 km south of Anchorage, Alaska. Although the earthquake killed relatively few people (about 130, most in Alaska), it devastated a number of coastal communities throughout south central Alaska, e.g., Valdez, the terminal for the Alaskan Pipeline, lost its ocean frontage and at Seward, a major port in Alaska, a large section of the waterfront slid into Resurrection Bay (Lemke, 1967). Ground shaking in Anchorage caused many buildings to collapse, liquefaction of the clay deposits upon which Turnagain Heights sat resulted in flowage across the bluffs toward Cook Inlet, areas of Prince William Sound moved horizontally 20 m, some areas were uplifted more than 11 m, and subsidence exceeded 3.5 m in some locations. The size of the area affected morphologically by either subsidence or uplift exceeded 200,000 km² (Figure 11.6). Another impact that has had long-lasting ramifications is at the head of Turnagain Arm, southeast of Anchorage. The town at its head (Portage) suffered subsidence of about 2 m partly because of the consolidation of sediment that occurred during some 15 minutes of seismic shaking. In a high tidal range location, subsidence exposed the area to saltwater incursion which killed the small forests that existed before the earthquake. Further, rapid sedimentation of the flooded areas buried some of the buildings and within 10 years had deposited 20×10^6 m³ of silt (Bartsch-Winkler and Ovenshine, 1977).

Whereas the direct modification and the sedimentation that occurred as a result of the 1964 earthquake were localized, the waves the earthquake generated were not. For example, its seismic earth wave caused a seiche in Lake Pontchartrain north of New Orleans, USA (5,600 km from Anchorage, Alaska) that tore pleasure boats from their moorings. Much more significant, however, was the seismic sea wave (tsunami) that not only worked its havoc along the Gulf of Alaska shoreline but also along the coasts of Oregon and Hawaii, USA, and Japan. The tsunami killed more than 100 people in Alaska and a sizeable number on the west coast of Canada and the United States. However, it caused only material damage in Hawaii. Nonetheless, because of major loss of life and property at Hilo, Hawaii, from the Aleutian-generated tsunami of 1946 and the massive Chilean tsunami of 1960, the city decided to leave the low-lying coastal section of Hilo Bay as a so-called "green belt", one of the techniques used for mitigating the damages from tsunami and, in the process, a basic return toward a natural shore zone.

As in the case of earthquakes, coastal disasters resulting from volcanic eruptions have been newsworthy since humans began to occupy coastal sites in the vicinity of volcanoes. Famous examples include the eruption of Thera (Santorini) in the Aegean Sea in 3,600 BP mentioned above because of the tsunami it generated. The eruption not only destroyed the settlements on Thera, such as Akrotiri, but also completely transformed its shoreline from one of circular configuration surrounding the former volcano to a crescent-shaped tephra-lined rim bordering a large caldera, the form that is present today. Although somewhat later in time, the eruption of Vesuvius in AD 79 destroyed not only the famous Pompeii but also Herculaneum, a fishing and resort city of about 5,000 people. It was completely buried by pyroclastic flows from the erupting Vesuvius, which changed the lithology and morphology of the sea shore as well as destroying the cultural landscape. Sir Charles Lyell, in his classic *Principles of Geology* first published in 1838, using Pliny the Younger's *Letters* as a source, describes the Vesuvius eruption as follows:

Ashes fell even upon the ships at Misenum, and caused a shoal in one part of the sea – the ground rocked, and the sea receded from the shores, so that many marine animals were seen on the dry sand. (Lyell, 1854, p. 364).

Although Pliny the Younger did not mention Herculaneum or even Pompeii (his home town), excavations, by the time of Sir Charles Lyell, showed that both cities had been "overwhelmed" and the landscape drastically changed.

The two volcanic eruptions discussed above occurred in the Mediterranean portion of the Alpide belt. However, most volcanic eruptions on Earth are associated with the Pacific seismic belt (Figure 11.4). In addition to the volcanoes present in these two belts there are others known as "hot spots", some of which are remote from plate boundaries. Hot spots within oceanic plates have coastal expressions such as on Hawaii (the Big Island) and in the Society Islands. Other hot spot volcanoes are present within diverging plate boundaries such as the Galapagos Islands in the eastern Pacific and Iceland on the Mid-Atlantic Ridge. One of the Icelandic volcanoes recently (1978) erupted and virtually destroyed the harbor of Heimaey. Although the continuing eruptions in Hawaii have resulted in minimal loss of life, they have destroyed a sizeable amount of its cultural landscape and are continually creating a new shoreline.

Volcanic eruptions are of many types ranging from the highly explosive (acidic) type, typified by the almost continuously erupting Stromboli, Italy, to the non-explosive (basic) type like the nearly equally as active volcano in Hawaii. The coastal hazards resulting from volcanic eruptions include lava flows, volcanic ballistics, pyroclastic flows, lahars, and acid rain (Nott, 2006). All of these volcanic products not only impact any humans within their vicinity but also have geomorphic consequences.

Mount Pelée, one of the most destructive volcanic eruptions in history, is a large composite volcano in northern Martinique in the Lesser Antilles. Its eruption in 1902 illustrates well the hazardous nature of establishing a settlement in the vicinity of a volcano, as was done in the seventeenth century when the town of Saint Pierre was founded. Although there had been some eruptions before (e.g., 1792 and 1851), it was the 1902 eruption that destroyed the town and changed the coastal landscape. The devastation was preceded by a series of earthquakes and lahars that flowed into the town. These early events resulted in some evacuation but, nonetheless, some 29,000 people remained behind and were killed (with one or two exceptions) when the major eruption occurred, most of them by hot noxious gases. Following the eruption, heavy rains mixed with pyroclastic material covered the entire area leaving behind thick layers of mud (Hill, 1902). The area of destruction extended some 15 km out from Saint Pierre. The eruption of Mount Pelée shows how rapidly such events can transpire and how vast and conclusive the damage can be to both natural and cultural landscapes.

11.3 Landslides

Rocky coasts stretch along some 80% of the world's shoreline (Sunamura, 1992), with cliffy types making up more than half of the total. During much of history, such coasts have not been so attractive for urban, industrial, or recreational development as other types. However, the rapidly increasing use of coasts around the world includes expansion into formerly neglected rocky and cliffy coastal situations and thus subjection of more and more people to greater risk from disaster.

Landslides and rockfalls are geomorphological occurrences that can be triggered by earthquakes (as noted above), rainfall, de-vegetation, overloading, oversteepening and undercutting, most of which can be enhanced by human activity; even such a simple (naturally human) activity as over-watering gardens at the top of coastal bluffs. The factors affecting the instability of slopes and thus their potential for slides to develop include their bedrock and soil characteristics, slope, configuration, and groundwater characteristics. These factors also affect the type (falling, sliding, flowing) and rapidity of movement of the material down slope.

Most coastal cliffs, because of their juxtaposition to the sea, are subjected to wave action, some continuously, some seasonally, and some episodically, resulting in erosion. It is, as Sunamura (1992) points out, an irreversible process unlike the destruction of beaches, which can be re-nourished, naturally (some seasonally) or artificially.

Most coastal landslides do not become major disasters. Many, especially those occurring along rugged coastlines such as the cliffy coasts of California, USA, frequently damage highways and small settlements that have been constructed in precarious positions. For example, the highway along the Big Sur section of the California coastline suffered a major landslide in 1983, which resulted in the closure of the coast highway for a year. Some of the chalk cliff coasts of northwest Europe are subject to falls that turn into flows that spread seaward for distances four to five times cliff heights. Occurring in soft chalk cliffs, they are potentially dangerous because of the rapidity of their flow. However, they tend to be winter phenomena, when beaches are deserted, and therefore less of a threat than they might be otherwise (Hutchinson, 2002).

Lyme Regis, a small town on the cliffy coast of southern England that dates back centuries, has been subjected to major landslides and extensive shoreline erosion from its beginning. A recent (6 May 2008) landslide delivered stone, sand, and clay to the shoreline from along a 95 m high, 400 m long section of the World Heritage Jurassic Coast at the east end of Lyme Regis. One of the major results of the slide is the supply of numerous fossils that will be winnowed from the basal materials during future storms. However, from the standpoint of this paper about disasters, it should be added that the 2008 slip

… has also started to dissect the old Lyme Regis tip. Glass, tyres, machinery and a lot of battery acid, toxics etc. are now descending to the beach … the first attitude seems to be to let the sea deal with it but this will prove to be a mistake because the material will quickly spread to the tourist beaches. (D. Brunsden, 2008, personal communication).

In contrast, there have been major slides that have proved very disastrous; often not directly by the slides themselves but rather by the tsunami they have generated. In the Mediterranean Sea, Papadopoulos *et al.* (2007) noted that, for the 32 landslide-generated tsunami they studied, triggering was by both sub-aerial and submarine landslides as well as earthquakes and volcanic eruptions. One of the major disasters in history occurred in 1998 in Papua New Guinea, when the tsunami triggered by the undersea landslide caused by an earthquake sent 18 m high waves across shorelines killing more than 2,000 people and destroying numerous villages. It may well be that the most disastrous slides will be those of submarine origin, especially if the Storegga submarine slide that occurred in the North Atlantic Ocean between Iceland and Norway about 8,150 years ago is an example. The slide, which covered an area of about 3,500 km^2, resulted in a tsunami that rose to heights of more than 12 m on the coast of Norway and 20 m in the Shetland Islands (Dawson, 1994).

11.4 Meteorological events and coastal disaster

Whereas above we concentrated primarily on the endogenic events that are responsible for coastal disasters, there is another group that is primarily exogenic, many parts of which are atmospheric phenomena. According to Smith (2001, p. 208) "… most environmental hazards are atmospheric in origin"; a large proportion of which have to do with tropical and extra-tropical cyclones and have coastal relevance. Tropical cyclones (going by other names, such as hurricanes, typhoons, willy-willies among others, in other parts of the world) are especially destructive. They are seasonal and occur in tropical ocean areas between about 5° and 30° latitude. Coastal areas commonly impacted by tropical cyclones include those in the western Pacific Ocean, Indian Ocean, Australasia, western North Atlantic Ocean, and eastern North Pacific Ocean (Figure 11.4). Destruction to both natural and cultural landscapes by tropical cyclones can be by wind, wave, storm surge, rainfall, and flood, separately but usually in some combination. The coastal zone, because it is where tropical cyclones come onshore, receives the brunt of the force.

Some of the hazardous phenomena of tropical cyclones (such as wave action and storm surge) are not in play inland after the storm has passed the coast. Other factors (wind, rain, flooding, tornadoes), however, can continue to have disastrous results at great distances from where the storms made landfall.

Although the tropical cyclone is not a geomorphic phenomenon per se, geomorphology is of major importance to the nature and extent of the disaster that occurs from any given storm. All coastal morphologies are impacted when tropical cyclones strike them. However, it is coastal basins whose geometries

… range from those of broad geomorphic features such as river deltas, barrier islands, and bays, down to detailed features – coastal rivers and streams, [and] distributary channels in wet lands that suffer most from them. (Resio and Westerink, 2008, p. 34)

In turn, the impact of a storm depends on a number of geomorphic factors such as off-shore bathymetry and the relief and configuration of the coast itself. Barrier islands and coastal dunes (which are the least durable of coastal forms) and ridges (e.g., cheniers) affect the impact of wind and serve as barriers to storm surges, which are one of the major causes of death and destruction of the cultural landscape. Storm surges and tsunami run-up are sudden rises in the sea surface that on occasion reach tens of meters above normal sea level. The steepness of the shelf offshore affects the height of both storm surges and wind-generated waves. Generally, the shallower the water offshore and the more gentle the slope of the shelf the larger the surge and run-up. Oceanographic and other meteorological factors also affect the height of surge, especially the stage of the tide and atmospheric pressure.

Coastal wetlands (as well as coastal forests and coral reefs) can be heavily damaged during tropical cyclones. They also play a role in coastal protection by attenuating the storm surge as it progresses inland. The amount of attenuation depends on such details as

… the storm's track, forward speed, duration, size, and associated waves; the regional bathymetry and topography, including shelf width and barrier islands; the local geometry; levee and raised-feature elevations, inland bathymetry and topography, including the channels that interconnect water bodies; and local surface roughness. (Resio and Westerink, 2008, p. 36)

In addition to the often complete destruction of human constructs (Figure 11.7), tropical cyclones and tsunami can alter the morphology of the coastal zone through erosion, transportation, and deposition: channels can be deepened or filled; barrier islands and coastal dunes breached, shifted,

FIGURE 11.7. Destruction from the storm surge generated by Hurricane Katrina in 2005 in Slidell, Louisiana, USA. (Photograph by H. Walker.)

and destroyed; and levees (both natural and artificial) modified. Much modification is also caused by the backwash of the floodwaters that surges, run-ups, and waves produce.

Hurricanes, typhoons, and tropical cyclones have resulted in disasters in so many parts of the world that histories of their destructive nature are numerous. Although they have many characteristics in common, the actual destructiveness of each is unique. An example that illustrates many of the geomorphic, as well as the atmospheric and oceanographic, factors involved is the unnamed cyclone that struck Bangladesh on 12–13 November 1970. Bangladesh, one of the most populous places on Earth, is located at the head of the funnel-shaped Bay of Bengal where the Ganges, Brahmaputra, and Meghna rivers have created a vast low-lying deltaic plain, much of which is covered with mangroves (Figure 11.8). The cyclone, which killed an estimated half-million people, left another million homeless, and affected as many as 50 million others, had winds of up to 225 km/h and a surge of at least 15 m (Bryant, 2005). Exacerbating factors include the reduction of sediment to the delta by dams on the upstream portions of the rivers, mangrove forest cutting, artificial diversions of deltaic drainage systems, coastal submergence, delta-front erosion, salinity intrusion, and sea-level rise (Khalequzzaman, 1988).

More recent tropical cyclones include Katrina (2005) that submerged much of New Orleans and caused approximately 1,500 deaths, Rita (2005) that completely destroyed Holly Beach, Louisiana, Ike (2008) that virtually destroyed Galveston, Texas, but without the loss of life that had occurred during the infamous 1900 Galveston hurricane, and the 2008 Myanmar (Burma) tropical cyclone

Nargis with probably more than one hundred thousand deaths (Table 11.4).

Although the severest storms striking coastlines occur in the lower latitudes, mid-latitude (both west and east) coasts are not exempt from storm-related disaster (Figure 11.4). Most of the factors discussed above that were related to the coastal impact from tropical storms also apply to extra-tropical storms. Severe storms in the westerly wind systems of both the Pacific and Atlantic Oceans occasionally cause powerful wave action and resultant coastal damage. An example of the extra-tropical nature along east coasts is the so-called "nor'easter". It is a cyclonic system that occurs in winter and, in the United States and Canada, can affect more than 1,000 km of coastal property as occurred during the Ash Wednesday, 1962, storm. The configuration of the coast accentuated the storm's severity.

Equally as relevant to the geomorphic enhancement of storm surges as the setting of the northeast coast of North America is the configuration and bathymetry of the southern part of the North Sea (Figure 11.4). Storms moving south in the North Sea progress into a narrowing (funnel-shaped) environment and over a shallowing sea bottom both of which impact water levels. When a storm's movement coincides with a lowering atmospheric pressure and a rising tide, the storm surges and associated waves can be extremely high. During recorded history such occasions have occurred numerous times along North Sea coastlines. For example, in 1492, some 10,000 people died and 72 villages were destroyed in the Netherlands. Although records of North Sea surges extend back nearly 1,000 years, the most famous one occurred in 1953 when a storm surge exceeded 3 m in the Netherlands and killed nearly 2,000 people. Eastern England was also impacted heavily. Although additional storm surges have occurred in the North Sea since 1953, extensive construction such as the Delta Plan of the Dutch and the Thames floodgates of the British have proven to be effective barriers to disaster.

11.5 Other coastal hazards/disasters

For the most part, the disasters above resulted from virtually instantaneous occurrences in the case of those caused by earthquakes and volcanic eruptions or from longer but still of relatively short progressions in the case of cyclones. Most were high-magnitude events that impacted limited sections of the coastal zone (excepting, of course, the tsunami that some of the events triggered) and caused immediate loss of life and property. In addition, the initial impact of each such event is usually only the beginning of the disaster because of the many side-effects spawned, some of which may never be rectified.

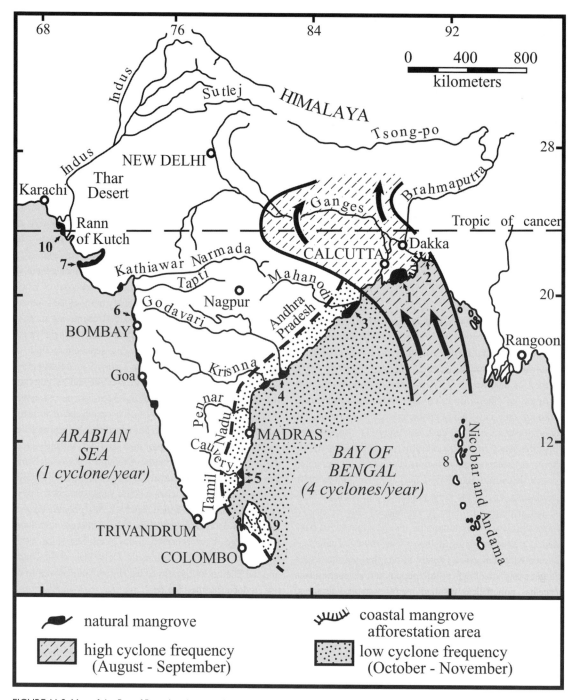

FIGURE 11.8. Map of the Bay of Bengal with tropical cyclone tracks. (Adapted from Blasco et al., 1994.)

In contrast to the above types of disaster, most of which are relatively common even on a human time scale, are those that are infrequent, slow in materializing and continuous in their impact, even if at a low level on a hazard/disaster scale. Most of them have a geomorphic component. Included are such events as human-induced subsidence, sea-level change, sea-ice depletion, glacial volume changes, biological and chemical changes, and war.

Subsidence is a naturally occurring phenomenon that may be instantaneous, as exampled in the discussion above about the 1964 Alaskan earthquake, or may be slow, as during the formation of deltas. However, it also is

TABLE 11.4. *Major tropical cyclones ranked by number of deaths*

Date	Location	Event	Deaths
1970	Bangladesh	Bangladesh Cyclone	500,000
1881	Vietnam	Typhoon Haiphong	300,000
1975	China	Typhoon Nina	171,000
1991	Bangladesh	Cyclone Bangladesh	138,883
2008	Myanmar	Cyclone Nargis	138,000
1974	Honduras	Hurricane Fifi	10,000
1971	India	Cyclone Orissa	10,000
1900	USA	Galveston Hurricane	8,000
2005	USA	Hurricane Katrina	1,500
1899	Australia	Cyclone Mahina	400

Data primarily from Longshore, 2008 and
Wisner *et al.*, 2003

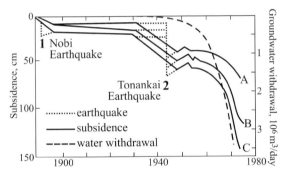

FIGURE 11.9. Subsidence resulting from earthquakes and groundwater withdrawal at Nobi, Japan. (After Iida *et al.*, 1976; from Walker and Mossa, 1986.)

one of those geological processes that can be caused and accelerated by humans, mostly unintentionally. Although not generally causing loss of life, human-induced subsidence has, nonetheless, resulted in environmental disasters, some of quite large scale. Most major human-induced subsidence occurs when substrata are tapped for water or petroleum products. The compaction of the sediments from which fluids are removed results in sinking at the surface. Venice, Italy; Long Beach, California; Niigata, Tokyo and Nagoya areas, Japan; and Houston, Texas are major examples. Subsidence at Nobi (near Nagoya), Japan, resulted from a combination of natural events (earthquakes) and the exploitation of water from subterranean aquifers (Figure 11.9). Much of the pumping was in support of unagi (freshwater eels) aquaculture and coincided with over-pumping throughout Japan in support of expanding industry during the middle of the last century. The area is also frequented by typhoons, the flooding from which is

deepened because of the area's subsidence history (Walker and Mossa, 1986). Subsidence in the Nagoya area, as elsewhere in Japan, has been reduced drastically by restricting pumping and by recharging aquifers. One of the first areas to be seriously studied because of subsidence was the Galveston-Houston, Texas, low-lying coastal plain. Over-pumping of both water and petroleum products resulted in sufficient subsidence to flood large sections of a number of coastal communities leaving many residents stranded. It also increased the area that would be covered by a 5 m storm surge (not uncommon in the area) by more than 65 km^2 (Kreitler, 1977).

At the beginning of the twenty-first century, the major concern, insofar as potential coastal disasters are concerned, is a rise in sea level; considered to be one of the main results of a "global/climate change" scenario. Sea-level change is nothing new in history; humans have experienced it often during their tenure on Earth, usually without realizing what was happening, as discussed above. However, during the last few thousand years humans have become accustomed to a relatively stable sea level and have adapted their activities accordingly. Thus a rising sea level, because of melting global ice and the ocean-water expansion caused by rising water temperatures, will accentuate coastal disasters. Low-lying cities will be flooded; tropical cyclone storm surges and tsunami run-ups will become more effective agents of destruction; beaches, barrier islands, and wetlands will be reduced in size or destroyed; cliff erosion will be accelerated; and salt water will intrude into wetlands and coastal aquifers. The potential modifications of a sea-level rise will not be limited to the cultural landscape but will also affect the physical, chemical, and biological elements in coastal environments.

Another change associated with global warming is the reduction of the sea-ice cover in the Arctic Ocean (Serreze *et al.*, 2007). Although its melting does not add to sea-level rise, it does lengthen the fetch over the seasonally ice-free portion of the ocean, which can result in an increased height of storm waves along arctic coasts and, in turn, can enhance the erosion of permafrost-bound coastal shorelines and increase the destruction of coastal villages. Such a condition involves all of the domains (lithosphere, hydrosphere, atmosphere, cryosphere, biosphere, and anthrosphere) displayed in Figure 11.3.

Many coasts have been subject to long periods of erosion since first being settled; some with almost continuous impact. An example is the Suffolk coast of England, which fronts the stormy North Sea and has had a long history of occupation. Dunwich, a settlement on the coast, was, during its early history, a thriving port that rivaled London in medieval England. However, subject to North

FIGURE 11.10. The ultimate human-induced disaster potential as exampled by the detonation of a thermonuclear bomb at Bikini Atoll, 1954. (Photo courtesy of U.S. Department of Energy, National Nuclear Security Administration / Nevada Site Office.)

Sea storms, it was frequently devastated. Not only were houses, windmills, and churches destroyed but the shoreline was eroded and the harbor filled with sand and mud. Total shoreline retreat has been as much as 500 m since 1587 when one of the first maps of the area appeared (Pye and Blott, 2006). Today, it is little more than a handful of houses, but the residents, at least since Roman times when it was founded, have suffered periodic disasters.

In contrast to the more "conventional" types of hazards/disasters discussed above are those that might be classified as "abnormal", "inconceivable", or rare. Included might be those caused by war, by nuclear detonations and by asteroids. Such events as the beach landings at Normandy in 1945, the 1991 torching of the oil wells in Kuwait during Desert Storm, and the destruction of corals and other coastal ecosystems on many of the islands of the Pacific during WWII had their geomorphic impacts. The nuclear bombs dropped during WWII and at Bikini atoll in 1954 demonstrated the disastrous nature of such detonations to humans and landscapes (Figure 11.10). However, even more disastrous to the coastlines of the world (and to other environments as well) would be the impact of an asteroid or meteorite. From the standpoint of coastal environments, tsunami produced by such an impact in an ocean would have world-wide ramifications.

11.6 Conclusions

One of the major conclusions that stems from the discussion above is that living and working in a hazardous coastal area is not without risk. Nearly two-fifths of the world's citizens live within 100 km of the shoreline, including the many millions crowded into large cities that directly face the sea. Given the structure of the coastline and the dynamic nature of the processes that operate there, disasters are not only inevitable but they are also increasing in complexity and destructiveness, possibly partly naturally, but mainly because of the increased presence of humans. These developments have led to an increase in the development of preventive and mitigation procedures. Although prevention is impossible for most natural disasters, reduction of their impacts can be achieved by increased knowledge about the hazards themselves and improved predictive techniques. Because geomorphological components are present directly (as in the case of earthquakes, volcanic eruptions, and landslides) and indirectly (as in the case of tropical cyclones and tsunami) in virtually all coastal disasters (including those humanly induced), geomorphologists play a major role in advancing the understanding that is needed for improved prediction and mitigation of the hazards/disasters that will occur in the coastal zone as well as elsewhere in the world.

Acknowledgments

We thank M. Eggart and C. Duplechin, Jr. for the drafting of all figures, W. Grabau for constructive criticism during the preparation of the manuscript, Maryann Walke for permission to utilize portions of Roger Walke's poetry, and I. Alcántara-Ayala for enthusiastic promotion of all things geomorphologic.

References

Alcántara-Ayala, I. (2002). Geomorphology, natural hazards, vulnerability and prevention of natural disasters in developing countries. *Geomorphology*, **47**, 107–124.

Bartsch-Winkler, S. and Ovenshine, A. T. (1977). Coastal modification at Portage Alaska, resulting from the Alaskan Earthquake of March 17, 1964. In H. J. Walker (ed.), *Geoscience and Man: Research Techniques in Coastal Environments*, Vol. 18, pp. 21–28.

Bird, E. C. F. and Schwartz, M. L. (1985). *The World's Coastline*. New York: Van Nostrand Reinhold Company.

Blasco, F., Janodet, E. and Bellan, M. (1994). Natural hazards and mangroves in the Bay of Bengal. In C. Finkl, Jr. (eds.), *Coastal Hazards: Perception, Susceptibility and Mitigation*. Charlottesville. The Coastal Education & Research Foundation, pp. 277–288.

Bolt, B., Horn, W., Macdonald, G. and Scott, R. (1977). *Geological Hazards: Earthquakes, Tsunamis, Volcanoes, Avalanches, Landslides, Floods*. New York: Springer-Verlag.

Bryant, E. (2005). *Natural Hazards*, 2nd edition. Cambridge: Cambridge University Press.

Burton, I., Kates, R. and White, G. (1978). *The Environment as Hazard*. New York: The Guilford Press.

Chandler, T. (1977). The forty largest cities: a statistical note. *Historical Geography*, **7**(182), 21–23.

Chen, J., Eisma, D., Hotta, K. and Walker, H. (2002). *Engineered Coasts*. Dordrecht: Kluwer Academic Publishers.

CIA (2008). *The World Factbook – Field Listing – Coastline*. Retrieved 26 October 2008, from https://www.cia.gov/library/publications/the-world-factbook/fields/2060.html.

Cohen, J., Small, C., Mellinger, A., Gallup, J. and Sachs, J. (1997). Estimates of coastal populations. *Science*, **278**, 1211–1212.

Davis, K. (1974). The migrations of human populations. *Scientific American*, **231**(3), 93–105.

Dawson, A. G. (1994). Geomorphological effects of tsunami run-up and backwash. *Geomorphology*, **10**, 83–94.

Dolan, R., Hayden, B. P., Hornberger, G., Zieman, J. and Vincent, M. K. (1972). *Classification of the Coastal Environments of the World*. Part 1, *The Americas*. University of Virginia, Technical Report 1.

Gares, P., Sherman, D. and Nordstrom, K. (1994). Geomorphology and natural resources. *Geomorphology*, **10**, 1–18.

Gunn, A. (2008). *Encyclopedia of Disasters: Environmental Catastrophes and Human Tragedies*. Westport, Connecticut: Greenwood Publishing Group.

Hill, R. T. (1902). Reports on the volcanic disturbance in the West Indies. *National Geographic*, **13**(7), 223.

Hutchinson, J. (2002). Chalk flows from the coastal cliffs of northwest Europe. In S. Evans and J. DeGraff, (eds.), *Catastrophic Landslides: Effects, Occurrence, and Mechanisms*. Boulder, Colorado: The Geological Society of America, pp. 257–302.

Iida, K., Sazanami, K., Kuwahara, T. and Ueshita, K. (1976). Subsidence of the Nobi Plain. *Proceedings Second, International Symposium on Land Subsidence*. Anaheim, pp. 1–8.

Inman, D. L. and Brush, B. M. (1973). The coastal challenge. *Science*, **181**, 20–32.

Inman, D. L. and Nordstrom, C. E. (1971). On the tectonic and morphologic classifications of coasts. *Geology*, **79**, 1–21.

Kerr, R. A. (1985). Continental drift nearing certain detection. *Science*, **239**, 953–955.

Khalequzzaman, M. (1988). Environmental hazards in the coastal areas of Bangladesh: geologic approach. In

S. Ferraras and G. Pararas-Carayannis (eds.), *Natural and Man-made Coastal Hazards*. Ensenada, Mexico: Third International Conference on Natural and Man-made Coastal Hazards, pp. 37–42.

King, C. A. M. (1982). Classification. In M. Schwartz, (ed.), *The Encyclopedia of Beaches and Coastal Environments*. Stroudsburg: Hutchinson Ross Publishing Company, pp. 210–222.

Kreitler, C. (1977). Fault control of subsidence, Houston-Galveston, Texas. In H. J. Walker (ed.), *Geoscience and Man: Research Techniques in Coastal Environments*, vol. 18, pp. 7–20.

Lemke, R. (1967). *Effects of the Earthquake of March 27, 1964, at Seward, Alaska*. Washington: U.S. Geological Survey Professional Paper 542-E.

Longshore, D. (2008). *Encyclopedia of Hurricanes, Typhoons, and Cyclones*. New York: Checkmark Books.

Lyell, C. (1854). *Principles of Geology: or, The Modern Changes of the Earth and its Inhabitants*, 9th edition. New York: D. Appleton & Co.

MacGregor, A. (1968). *Fife and Angus Geology*. London: Blackwood.

Monecke, K., Finger, W., Klarer, D. *et al.* (2008). A 1,000-year sediment record of tsunami recurrence in northern Sumatra. *Nature*, **455**, 1232–1234.

NOAA (1981). *United States Coast Pilot. 9, Pacific and Arctic Coasts, Alaska, Cape Spencer to Beaufort Sea*. Washington, D.C.: U.S. Department of Commerce, National Ocean Survey.

NOAA (2008). *Historical Tsunami Database*. National Geophysical Data Center. Retrieved 23 October 2008, from http://www.ngdc.noaa.gov/hazard/tsu_db.shtml.

Nott, J. (2006). *Extreme Events: A Physical Reconstruction and Risk Assessment*. Cambridge: Cambridge University Press.

Papadopoulos, G., Daskalaki, E. and Fokaefs, A. (2007). Tsunamis generated by coastal and submarine landslides in the Mediterranean Sea. In *Submarine Mass Movements and Their Consequences*. London: Springer-Verlag, pp. 415–422.

Pye, K. and Blott, S. (2006). Coastal processes and morphological change in the Dunwich-Sizewell area, Suffolk, UK. *Journal of Coastal Research*, **22**(3), 453–473.

Resio, D. T. and Westerink, J. J. (2008). Modeling the physics of storm surges. *Physics Today*, **61**(9), pp. 33–38.

SDSU (2008). *Krakatau, Indonesia (1883)*. Retrieved 16 October 2008, from http://geology.sdsu.edu/how_volcanoes_work/Krakatau.html.

Serreze, M., Holland M. and Stroeve, J. (2007). Perspectives on the Arctic's shrinking sea-ice cover. *Science*, **315**(5818), 1533–1536.

Shepard, F. P. (1976). Coastal classification and changing coastlines. *Coastal Research, Geoscience and Man*, **xiv**, 53–64.

Smith, K. (2001). *Environmental Hazards: Assessing Risk and Reducing Disaster*, 3rd edition. London: Routledge.

Sunamura, T. (1992). *Geomorphology of Rocky Coasts*. Chichester: John Wiley & Sons.

United Nations (2003). *Urban Agglomerations 2003*. Retrieved 3 November 2008, U.N. Dept. of Economic and Social Affairs, http//www.un.org/esa/population/publications/wup 2003/2003urban_agglo.htm

U.S. Army Corps of Engineers (1971). *National Shoreline Study. Inventory Report, Alaska Region*. Anchorage: U.S. Army Corps of Engineers, Alaska District.

U.S. Department of Energy, National Nuclear Security Administration / Nevada Site Office. (2008). DOE Library. Retrieved 1 November 2008, http://www.nv.doe.gov/library/photos/photodetails.aspx?ID=1047

USGS (2008). *Most Destructive Known Earthquakes on Record in the World*. Retrieved 2 October 2008, from http://earthquake.usgs.gov/regional/world/most_destructive.pho.

Walke, R. (1997). *The Beach and Other Poems*. Blacksburg, VA: Pocahontas Press.

Walker, H. (1981). The peopling of the coast. In L. J. C. Ma and A. G. Noble (eds.), *The Environment: Chinese and American Views*. New York: Methuen and Co. Ltd, pp. 91–105.

Walker, H. (1984). Man's impact on shorelines and nearshore environments: a geomorphological perspective. *Geoforum* **15** (3), 395–417.

Walker, H. (1988). *Artificial Structures and Shorelines*. Dordrecht: Kluwer Academic Publishers.

Walker, H. (1990a). Nature, humans, and the coastal zone. *The International Journal of Social Science*, **5**(2), 50–62.

Walker, H. (1990b). The coastal zone. In B. L. Turner II (ed.), *The Earth as Transformed by Human Action*. Cambridge: Cambridge University Press, pp. 271–294.

Walker, H. and Mossa, J. (1986). Human modification of the shoreline of Japan. *Physical Geography*, **7**(2), 116–139.

Wisner, B., Wisner, P., Blaikie, P., Cannon, T. and Davis, I. (2003). *At Risk: Natural Hazards, People's Vulnerability and Disasters*. New York: Routledge.

12 Weathering hazards

Andrew S. Goudie and Heather Viles

12.1 Introduction

Like many geomorphological processes, weathering can pose a hazard if it affects humans, their property and livelihoods. Unlike many geomorphological processes that are known to cause a hazard, such as landslides, avalanches and coastal erosion, weathering is generally slow and small scale, though once thresholds have been breached or humans have accelerated the processes involved, weathering can have dramatic results (such as the rapid, catastrophic weathering of limestone facades recognised by Smith and Viles, 2006). The slow rate and small-scale nature of weathering means that it is often overlooked in hazard research, but it can have serious economic effects as well as being capable of irreparably damaging items of our cultural heritage. For example, relatively minor (in geomorphological terms) rates of weathering can disfigure valuable carving and statuary or induce fragments of stonework to fall from a facade, which are capable of causing injury or death. It can also contribute to more catastrophic events such as rockfalls (Figure 12.1) and rockslides (Matsuoka and Sakai, 1999; Jaboyedoff et al., 2004; Borrelli et al., 2007) and promote landslides in deeply weathered materials (Durgin, 1977) and in clays (Gullà et al., 2006). Chemical weathering may also pose hazards to human health by liberating toxic chemicals (e.g. excessive amounts of arsenic, fluoride, heavy metals etc. from bedrock and mine waste) (Islam et al., 2000; Saxena and Ahmed, 2001; Dang et al., 2002).

This chapter focuses on how weathering contributes to the deterioration of buildings and structures. Like most environmental hazards, weathering is at least partially produced by human activity, through placing vulnerable materials in polluted environments, or through increasing the presence and destructive capabilities of pollutants such as salts, or a combination of both. Another similarity with most other environmental hazards is the way in which weathering can interact with other processes and events to produce a larger hazard. Weathering of buildings and structures can often be highly damaging, complex and difficult to diagnose. For example, the Angkor monuments and temples in Cambodia were built between the ninth and thirteenth centuries and became entombed in tropical forest for six centuries after the sites were abandoned. Now listed as a World Heritage Site, there is much concern about the severity of deterioration of the ruins here. André et al. (2008), for example, note that between 1963 and 2008 the amount of deteriorated areas on one portion of the Ta Keo temple has more than tripled. However, it is hard to disentangle the processes responsible for the decay, with some authors suggesting that salt weathering is a key process (perhaps induced by bat guano according to Uchida et al., 1999), and others suggesting that the forest clearance that exposed the monuments in the 1920s might have had an important impact on subsequent weathering (André et al., 2008).

Many different weathering processes are involved with the deterioration of building materials, and many can be viewed as hazardous. For example, frost weathering has been reported to be producing rapid damage to the brickwork of a historic railway tunnel 100 km north of Tokyo, Japan (Thomachot et al., 2005), and some plants and microorganisms have been observed to be capable of producing rapid biodeterioration (Seaward, 1997). However, in this chapter we will assess the hazards associated with three key types of weathering, i.e. weathering associated with salts, air pollution and the role of fire, lightning and thermal fatigue. In reality, of course, most buildings and monuments are affected by a range of weathering and other hazardous processes, sometimes operating synergistically, and it can be difficult to untangle the exact causality. Hazards connected with the dissolution of soluble rocks are discussed in Chapter 13.

Geomorphological Hazards and Disaster Prevention, eds. Irasema Alcántara-Ayala and Andrew S. Goudie. Published by Cambridge University Press. © Cambridge University Press 2010.

FIGURE 12.1. A rockfall in sandstone at Wadi Rum (Jordan) caused by the slope being undermined by cavernous weathering. The area has numerous slope failures created by salt sapping.

12.2 Salt weathering

Some of the world's great cultural treasures are afflicted by the crystallisation and hydration of salts in rock pores and chemical reactions between salts and building materials (as extensively reviewed by Goudie and Viles, 1997). Many important UNESCO World Heritage Sites are recorded to have problems of salt weathering. Mohenjo Daro in Pakistan, for example, is known to be affected by increasing groundwater levels and salinity, with sodium sulphate producing particularly aggressive deterioration (Master Plan, 1972; Lohuizen-de Leeuw, 1973; Fodde, 2007). Similarly, in Uzbekistan the ancient towns of Kiva, Bukhara and Samarkand (all World Heritage Sites) have been recorded by Akiner *et al.* (1992) to be suffering irrigation-induced groundwater rise and salinisation, causing deterioration to the lower courses of the buildings. The same applies to the fabric of some of the great new cities of the Middle East, including those of Bahrain (Figure 12.2), Egypt and the United Arab Emirates (Figure 12.3). The problem is particularly serious where groundwater levels are high and the upward movement of salts takes place into buildings and their foundations through capillary rise (Figure 12.4) – a process termed 'the wick effect'. It is not only buildings that are affected, as bridges, roads and runways have also been recorded to suffer serious salt problems in Australia, Southern Africa, Algerian Sahara, India and USA (see Januszke and Booth, 1984; Horta, 1985; and the global review produced by the Botswana Roads Department, 2001). Salts affecting bridges, roads and runways may come from groundwater but also may be directly applied as deicing agents (Wang 2006). Indeed, deicing salts have been implicated as the most serious cause of bridge corrosion problems in the UK (Mallett, 1994, p. 145), and are

FIGURE 12.2. A nurses' home in Bahrain, showing the severe deterioration that has taken place in the foundations and also in the pillars as a consequence of salt attack.

FIGURE 12.3. A four-year-old wall in Ras Al Khaimah, United Arab Emirates, showing the consequences of upward movement and subsequent crystallisation of salt-rich water.

estimated to have affected *c.* 19% of bridges in North America (Vassie, 1984, p. 713). Let us now consider two case studies which illustrate the problems that salt weathering hazards can create for valuable cultural heritage.

FIGURE 12.4. A wall in the city of Catalyud, NE Spain, showing severe brick decay because of migration upwards (rising damp) of salt-rich solutions derived from the gypsum deposits that underlie the town.

FIGURE 12.5. Rising damp on a wall in Karnak, near Luxor, Egypt. This problem has been created by the spread of irrigation in the vicinity and also by such factors as increases in groundwater levels created by leaking pipes and sewers.

12.2.1 The Sphinx and other Ancient Egyptian archaeological sites

Egypt contains some of the most remarkable sites of ancient cultural heritage, including the world famous pyramids at Giza, the Sphinx, and the many temples and tombs surrounding Thebes (modern-day Luxor). There are many threats to these monuments, with salt damage and other problems relating to rising groundwater being perhaps the most serious, as identified by Keatings *et al.* (2007) at the mudbrick ruins of Hawara Pyramid, some 90 km south of Cairo. Studies by Smith (1986) and Wüst and Schlüchter (2000) illustrate the widespread threat of groundwater rising and salt crystallisation processes around Thebes, with damage caused by sodium chloride the major problem (Figure 12.5). In some cases a range of processes may be at work, linking salt weathering to other hazards. The World Heritage Site of Abu Mena, for example, was put on the 'danger list' by UNESCO in 2001 because of rising groundwater from nearby land reclamation schemes causing liquefaction of sensitive clay soils, resulting in destabilisation and collapse. The problems caused by weathering are particularly notable in Cairo, as summarised by Fitzner and Heinrichs (2002, p. 217) who comment: 'Weathering damage to many historical stone monuments in Cairo is alarming'. Fitzner and Heinrichs (2002) note the multiple threats posed by rising groundwater over the last few decades and associated salt weathering (affecting the lower parts of monuments) and increasing air pollution in recent years linked to gypsum crust damage to the upper parts. Kamh *et al.* (2008), in a study of Islamic archaeological sites in Cairo, note the synergy between salt weathering and the 1992 earthquake. Some of the sites most badly damaged by the earthquake were those previously weakened by intense salt weathering.

The complex nature of deterioration and conservation of the Sphinx of Giza reflects well the general problems of salt weathering of ancient Egyptian sites. The Sphinx is a huge monument, carved into layers of limestone bedrock of variable durability and has deteriorated notably since it was first photographed in 1850, and by one estimate, loss of stone is occurring at the rate of about 30 cm per century (Selwitz, 1990, p. 854). Some of the bands of the Upper Mokattam limestones that make up the Sphinx are very prone to deterioration. Salt weathering (by sodium chloride and calcium sulphate) has been seen to be a dominant process affecting the Sphinx especially in the lower parts where water can build up (Livingston, 1989; Gauri *et al.*, 1995), but aeolian abrasion has been hypothesised to be causing damage to the windward, upper parts of the Sphinx (Camuffo, 1993). Numerous attempts have been made to remediate the problems, and intensive microclimatic monitoring was carried out by the Getty Conservation Institute in the early 1990s. In some cases it has been observed that Roman and modern replacement stone has decayed much faster than the original fabric of the Sphinx. Overall, it is difficult to diagnose exactly what is causing the damage when dealing with such a large, ancient artefact with a long and complex history – as is the case with most historic Egyptian monuments.

12.2.2 Petra, Jordan

At Petra in Jordan the lower portions of many of the Nabatean rock-hewn facades and other monuments show a substantial degree of decay (Figure 12.6). According to Wedekind and Ruedrich (2006) more than 50% of the monument surfaces are now damaged by weathering phenomena, with almost 12% totally destroyed by salt-induced

FIGURE 12.6. Weathering at the base of a Nabatean tomb in Petra, Jordan.

cavernous weathering. Such cavernous weathering features have also been widely observed and studied by Heinrichs (2008). There is also abundant evidence in the sandstone cliffs of Petra of the development of large numbers of cavernous weathering forms, ranging in size from small honeycombs to huge tafoni.

The serious and multiple hazards posed by weathering at Petra are well summarised by Wedekind and Ruedrich (2006, p. 261) who note: 'Today, the existence of the unique rock architecture of these monuments is in danger due to decomposition, poor maintenance and lack of conservation'. Salt weathering, especially by sodium chloride and calcium sulphate, plays a key role in the deterioration as discussed by Albouy *et al.* (1993) and Fitzner and Heinrichs (1991, 1994). Extensive desalinisation trials have been carried out on monument no. 826 to try and reduce the future threats, and there have also been calls to reactivate ancient drainage systems around the base of the facades to provide long-term sustainable management of the ruins (Wedekind and Ruedrich, 2006). Paradise (2005) also notes the direct role of tourists in accelerating the

weathering-related hazards at Petra – with relative humidity increases and touching by visitors both seen to be associated with high surface recession rates inside some chambers, as well as trampling observed to damage the sandstones of the Roman Theatre. Links between salt weathering and other hazards have also been noted at Petra, with Bani-Hani and Barakat (2006) illustrating the role of salt weathering in reducing compressive strength and increasing the susceptibility of the Qasr al-Bint monument to earthquake shaking. Thus, whilst salt weathering is a major hazard here, tourism, earthquakes and removal of ancient drainage systems are undoubtedly compounding the severity of deterioration.

12.3 Changing dimensions of the salt weathering hazard

In essence, in order for salt weathering to be a problem it is necessary to have an abundant supply of salts and water, suitable environmental conditions and vulnerable stone or other materials. Whilst the dryland environments of the world are particularly prone to salt weathering problems (because of the high rates of evaporation that encourage salts to precipitate out of groundwater etc.), salts can also be derived from air pollutants, coastal salt spray, deicing salts and the local geology and thus salt weathering can potentially affect a very wide range of environments.

There is much evidence to suggest that the salt weathering hazard is increasing, as a result of four factors, i.e. changing climatic conditions, human interference, expansion of settlements into salt-prone environments or enhanced vulnerability of ageing monuments. Looking at the vulnerability issue first, it is clear that once monuments have suffered from salt weathering and other problems over a number of years they can reach a critical state where even modest rates of weathering can effect great change. For example, where salt damage has caused hollowing out of the lower parts of walls, whole structures may become weakened and vulnerable to earthquakes, flooding or other hazards. Detailed survey and diagnosis is required to elucidate the condition of old buildings and structures in order to aid prediction and management of vulnerability.

Within recent decades there has been a marked expansion of settlement into salt-prone areas, especially in places such as the Gulf States of the Middle East. The city of Abu Dhabi had a population of around 8,000 in 1960. By 2005 it had reached 1.35 million, a 169-fold increase. The city of Dubai had a population of around 45,000 in 1960. By 2005 this had reached 650,000, representing a 14-fold increase. As settlements expand in many dryland areas, more building occurs on areas such as sabkhas, or on reclaimed land in the coastal zone, both of which are likely to suffer from enhanced salt weathering. Such environments are also likely to suffer from a range of other geomorphological and geotechnical hazards, including subsidence, flash floods, sand and dust movements (Shehata and Amin, 1997).

Human interference is enhancing the salt weathering hazard in many areas through three main processes, i.e. increasing groundwater levels and/or salinity, enhanced use of deicing salts, and alterations to the microclimate in and around buildings and structures. Increasing groundwater levels become a problem when the water table becomes near enough to the surface for capillary rise and evaporative concentration of salts to occur (Rhoades, 1990). Rising groundwater may be caused by irrigation, vegetation clearance/land use change and urbanisation. Irrigation has experienced very rapid expansions in recent decades (Rozanov *et al.*, 1991). In 1961 the global area of irrigated land was *c*. 139 million ha; by 2003 this had almost doubled to 277 million ha. Salinised irrigated land in general ranges between 10 and 50% of the total, depending on how the figures are estimated and how salinisation is defined. Vegetation clearance and land use change can cause groundwater levels to rise by allowing deeper penetration of rainfall into soils and removing transpiration losses. Western Australia has been particularly affected by this process with an estimated 1.8 million ha damaged, and Goudie (2006, p. 98) notes that around 6.1 million ha in the area have the potential to be affected. Similar problems have been noted in other parts of Australia as well as the Canadian Prairies and Niger (Leduc *et al.*, 2001). Urbanisation can also produce rising groundwater, notably through leaking water and sewage pipes, although the spread of impermeable surfaces in urban areas can also reduce the amount of water lost through evapotranspiration (as found in sabkhas along the Arabian Gulf coast by Shehata and Lotfi, 1993).

Human interference can also result in increased salinity of groundwater, such as in coastal areas where overpumping of groundwater encourages saltwater intrusion into the aquifer, and salinity problems. Along the Nile Delta similar problems have arisen, in this case as a result of the building of the Aswan High Dam, which has reduced freshwater recharge in the aquifer. Surface waters can also become more saline as a result of human interference, such as the development of inter-basin water transfers from the rivers draining the Aral Sea, which have led to desiccation of the Aral Sea and increased salinity of the surfaces.

Direct application of salts to roads, pavements and runways, as deicing agents, can also enhance the salinity of groundwater and surface water draining these sites and over

the late twentieth and early twenty-first centuries as traffic levels have risen dramatically, so there have been concomitant rises in the use of deicing salts in many colder countries (Scott and Wylie, 1980; Howard and Beck, 1993). Within and around buildings and structures, humans can also interfere with microclimatic conditions, enhancing the threat of salt weathering damage. For example, in many historic churches attempts to heat the building while in use in winter can lead to low internal relative humidities and encourage the crystallisation of salts and associated stone deterioration (Arnold and Zehnder, 1988; Laue *et al.*, 1996).

Climatic change at present and in the near future may have serious impacts on the severity of the salt weathering hazard in many areas both directly and indirectly. As Brimblecombe and Grossi (2008, Figure 2) found in a simple modelling exercise, predicted climatic conditions for London should dramatically increase the number of damaging thenardite–mirabilite phase transitions as a result of higher numbers of dry days. This can be seen as a direct impact of climate change, whereby changes in climatic variables themselves produce alterations in the rate of salt weathering. Indirect impacts may also be very important, as climatic change will itself influence the availability of the water and salts required for salt weathering, through impacts on groundwater levels and salinities. In some areas the effects of climate change may be quite complex, as for example in North America and the UK where warming is likely to decrease the need for deicing salts, whilst increasing the number of dry days.

12.4 Atmospheric pollution and weathering

Air pollution is an important cause of hazardous weathering in many urban and roadside environments (Davidson *et al.*, 2000; Grossi and Brimblecombe, 2002). Many important cathedrals within the UK and Europe, for example, have been dirtied and damaged by the polluted urban atmosphere that surrounds them, and much money has gone into cleaning and repairing iconic buildings such as St Paul's Cathedral and Westminster Abbey in London. Buildings such as York Minster in northern England have been recorded to suffer from air pollution damage for a very long time with records showing that soot and gypsum crusts were removed from the stonework in the early 1700s (Brimblecombe and Camuffo, 2003, p. 19). Understanding the hazard associated with air pollution-induced weathering is complex, as there is a wide range of pollutants involved, and different materials have varying vulnerabilities to them. Gaseous pollutants, notably sulphur dioxide and nitrogen oxides, can cause weathering of building materials in association with particulates (such as carbonaceous spheres produced by fossil fuel combustion) and oxidants (such as ozone produced by photochemical reactions) which act as important catalysts. Sulphur dioxide, for example, reacts with dust and other particles, water and reactive minerals (such as calcite found in limestone and marble) to produce unsightly and damaging gypsum crusts (Charola *et al.*, 2007). The production of such crusts requires the oxidation of sulphur dioxide either in the gas phase, or in moisture films upon the building surface. Atmospheric oxidants, such as ozone and hydrogen peroxide, and catalysts such as soot and smoke assist this oxidation. Sulphur dioxide can also promote kaolinisation of granites in urban environments (Schiavon, 2007). Organic carbon compounds may promote the growth of black crusts and play a catalytic role in gypsum formation (Sabbioni *et al.*, 1999; Sabbioni, 2003) and Simao *et al.* (2006) have assessed the role of diesel and gasoline vehicle exhausts in causing sulphation of silicate stones. The weathering of calcareous building stones can also be attributed to surface dissolution processes (Hoke and Turcotte, 2004). The following two case studies exemplify the nature and complexities of the air pollution weathering hazard, showing how air pollution-induced deterioration can operate synergistically with other processes or make stonework more susceptible to subsequent weathering processes.

12.4.1 The historic centre of Oxford, UK

The university town of Oxford contains a wealth of historic buildings, largely constructed in limestone, ranging widely in date, architectural style and function. Today there is much evidence of extensive deterioration of the stonework – ranging from the exfoliation of thick, black gypsum-rich crusts to the granular disintegration and flaking of replacement stones (probably caused by salt weathering). In some cases large areas of facades are decaying rapidly and catastrophically as seen in Figure 12.7. The severity of Oxford's stone decay problems relates to two key factors: the use of vulnerable stone types and a long history of air pollution. The early buildings in the city were constructed from local Jurassic oolitic limestones. By the late seventeenth century the quality of Headington stone (mainly the softer beds used for ashlar, and known as Headington freestone) had declined and the stone quickly proved to be extremely sensitive to polluted air. During the nineteenth century, air pollution in Oxford rose dramatically and by the earliest photographs taken in the mid to late nineteenth century it is clear that many facades were

FIGURE 12.7. Deterioration of parts of the facade of Worcester College, Oxford, England.

blackened and experiencing rapid decay in the form of exfoliation (Viles, 1994).

Extensive repair and refacing of buildings took place from the late nineteenth century onwards, and by the 1970s many of Oxford's historic buildings exhibited clean and unweathered facades. However, despite the subsequent improvements in Oxford's air quality, and the use of generally better-quality, more durable replacement stone, deterioration has continued to pose a problem for many buildings. Traffic is now a major nuisance producing localised blackening and decay of stone surfaces near busy roads (Viles and Gorbushina, 2003; Thornbush and Viles, 2006a). Much old, encrusted and blistering stone still remains in Oxford, and it is very sensitive to ongoing deterioration by salts and other agents of weathering. Research is currently under way to try to elucidate how and why complex associations of past and present processes mean that some parts of limestone facades in Oxford are suffering such serious deterioration (Smith and Viles, 2006; Gomez-Heras *et al.*, 2008). This research is finding much evidence that recent reduction in air pollution within the city is enhancing the creation of rapidly forming hollows, because of the slower rate of formation of black crusts which can act to 'heal' blistering stonework.

Thus in summary, air pollution-induced weathering remains a localised hazard for many roadside walls in Oxford, and damage caused by past air pollution predisposes the stonework to further hazardous weathering.

12.4.2 The historic centre of Budapest

Budapest, the historic centre of which was listed as a World Heritage Site in 1987, contains a wide array of limestone buildings that show extensive and severe damage by air pollutants, including the Citadella Fortress and Houses of Parliament. A commonly used stone type in many of these buildings in Budapest is a local Miocene oolitic limestone, which is a very soft and porous stone. According to Zador (1992) the deterioration rate of limestones in Budapest has increased by five to ten times over the last 50–60 years because of the increase in air pollution and the frost sensitivity of the limestones. Torok (2002, p. 376) gives some vivid examples of the rapid recent progress of deterioration here, illustrating rapid removal of patches of black crust through exfoliation between 1993 and 2001. The detachment, through blistering and flaking, of the gypsum-enriched crusts on limestones in Budapest appears to result from the development of microcracking under the crusts, and to be controlled mainly by air pollution and freeze–thaw cycles (Torok et al., 2007). Whilst air quality has improved since 1980, air pollution levels in Budapest are still very high (current sulphur dioxide levels are double those of London and Paris, according to Torok et al., 2007, pp. 263–4), and air pollution-induced deterioration still appears to be widespread and showing no signs of slowing down. The combination of sensitive stone, high levels of air pollution and a climate that encourages frequent freeze–thaw cycles all contribute to a serious weathering hazard here.

12.5 Changing dimensions of the air pollution-induced weathering hazard

For air pollution to cause a weathering hazard, a combination of bad air quality (usually measured in terms of sulphur dioxide, nitrogen oxides and particulates) and sensitive materials are all that are required. Severe climatic conditions can enhance the problems further. Air pollution conditions are very dynamic across the world, varying hugely across time and space. The seriousness of acid rain caused by sulphur dioxide emissions in the western industrialised nations peaked in the mid 1970s or early 1980s. Changes in industrial technology, in the nature of economic activity, and in legislation caused the output of sulphur dioxide in

Britain to decrease by 35% between 1974 and 1990. This was also the case in many industrialised countries, including the United States (Malm et al., 2002). However, there has been a shift in the geographical sources of sulphate emissions, so that whereas in 1980 60% of global emissions were from the United States, Canada and Europe, by 1995 only 38% of world emissions originated from this region (Smith et al., 2001). There are also increasing controls on the emission of NO_x in vehicle exhaust emissions, although emissions of nitrogen oxides in Europe have not declined. In many UK cities such as Oxford, for example, roadside nitrogen dioxide levels regularly exceed the EU guideline amounts. A similar picture emerges from Japan (Seto et al., 2002) where sulphate emissions have fallen because of emission controls, whereas nitrate emissions have increased with an increase in vehicular traffic. This means that the geography of air pollution is changing, becoming less serious in the developed world, but increasing in locations like China, where economic development will continue to be fuelled by the burning of low-quality sulphur-rich coal in enormous quantities. Brimblecombe and Grossi (2008, Figure 1) provide a useful historical modelling exercise illustrating the long-term trend in limestone recession rates in London. Pollution damage, according to their model, increased rapidly in the late 1600s when high-sulphur coal was imported into the rapidly growing city. Rates of recession peaked around 1900, and dropped dramatically over the second half of the twentieth century, as coal consumption and sulphur dioxide levels both fell rapidly.

Polluted areas near coasts and within drylands can suffer from particularly potent cocktails of gases and particulates, including corrosive fogs and dust. In many cases such places experience a combination of salt and air pollution-induced weathering hazards. In Venice, for example, sulphate and chloride levels are often very high during fog events as a combined result of pollution and coastal influences; whilst nitrate levels are generally low because of the lack of road vehicles (Fassina et al., 2002). The Carrara marble building facades, such as that of the Basilica of St Mark, within the historic area of Venice reflect these multiple effects, with blackened gypsum crusts, spalling and granular disintegration.

Climate change may have direct and indirect impacts on the severity of air pollution-induced weathering. Increases in temperature and rainfall may both enhance chemical reactions and thus speed up air pollution-induced weathering. As Brimblecombe and Grossi (2008) note, increased winter rainfall and higher carbon dioxide concentrations will both enhance the recession of limestone facades in London. However, climate change may also affect the rate of other processes that act synergistically with air pollution-induced weathering. In Budapest, for example, future warming may lower the frequency of freeze–thaw cycles thus reducing the potency of frost as an agent of deterioration there. Conversely, in Oxford increasing summer droughts will enhance the likelihood of salt weathering, thus increasing the likelihood of deterioration.

12.6 Fire, lightning and thermal fatigue

One cause of hazardous building-stone decay is the breakdown of material as a result of fire and lightning effects, which cause abrupt and extreme heating resulting in thermal shock damage. It is known from field studies of natural rock outcrops that wildfires can cause spalling and cracking (Thoms, 2007), and there are now various experimental simulations that indicate how fire can weaken rock and cause it to disintegrate (e.g. Goudie et al., 1992; Allison and Bristow, 1999; McCabe et al., 2007) and to discolour (Hajpál and Török, 2004). In addition, fire can cause damage to stone buildings, tunnels (Smith and Pells, 2008, 2009) and to rock art (Tratebas et al., 2004). As Gomez-Heras et al. (2006, p. 513) say, 'Fire can cause a rapid and irreversible decay of building stones', and fire damage has been the subject of a European COST Action (COST-17) which has noted that about one historic building is lost to fire in the EU each day. Fire damage may be particularly intense as a result of wars, as recorded in Vienna after World War I (Sippel et al., 2007). Several major historic buildings have had stonework damaged as a result of fires, e.g. Windsor Castle, Lisbon Cathedral and Heidelberg Castle (Chakrabati et al., 1996; Dionisio, 2007). Perhaps most importantly of all, the high temperatures attained in building and tunnel fires can cause severe damage to concrete, including explosive spalling, reduction of compressive and tensile strength, and failure of bonds between steel and concrete (Khoury, 2000). Explosive spalling of concrete involves a combination of pore pressure and thermal stress effects (Khoury, 2006). There is also a small literature which indicates that lightning strikes can cause fracture of masonry and rock (Farmer, 1939; Wilson, 2003). Such fractures, as well as causing direct damage to buildings, may also predispose the stonework to further deterioration.

Whilst fire and lightning can cause abrupt and severe temperature anomalies across building surfaces, smaller-scale temperature fluctuations (often related to diurnal cycles) can create thermal fatigue. Whether or not thermal fatigue caused by heating and cooling of building surfaces by daily temperature cycles causes major deterioration of buildings is the subject of debate, but it seems to be a particularly likely cause of problems with marble cladding because of the particular characteristics of marble that render it prone to this mechanism (Goudie and Viles, 2000). Some modern buildings, such as the Amoco Building in Chicago and the Finlandia Hall in Helsinki have experienced marble cladding failure (Widhalm et al., 1996). While it is possible that various mechanisms could contribute to this degradation, Rayleigh (1934) suggested that marbles that had been baked at temperatures of 100 °C or even less had their rigidity diminished. He suggested that this was due to the uneven way in which calcite crystals expand on heating. More recently, Royer-Carfagni (1999) has used scanning electron microscopy to identify changes in marble structure on heating, and has suggested why it is that certain types of marble may be prone to the effects of quite modest heating treatments (p. 119):

Calcite is known to expand on heating much more in the direction of its optical axis than perpendicular to it. The grains' shapes change with temperature and a grain which fits snugly into the mosaic at a given temperature is no longer able to do so when the temperature is varied; this is because the anisotropy directions of the individual grains are oriented randomly. The result is a springing apart of contiguous grains, giving rise to a non-zero residual stress state inside the material.

Widhalm et al. (1997) have shown that calcite expands by 26.10^{-6} K^{-1} parallel to the crystallographic c-axis, and contracts by 6.10^{-6} K^{-1} normal to it, while Rosenholtz and Smith (1949) used dilatometry to show the high sensitivity of calcite to heating effects.

12.7 Mitigation

Having discussed the causes and consequences of hazardous weathering we now consider strategies to mitigate the problems. The first important stage in such mitigation is to diagnose the problem(s). Following diagnosis, four basic types of mitigation methods can be applied, i.e.:

1. Reduce the threat of hazardous weathering by:
   locating new structures away from zones with the environmental ingredients that can cause weathering;
   moving vulnerable objects to safer ground.
2. Reduce the severity of environmental factors that can cause hazardous weathering by:
   reducing groundwater levels;
   reducing air pollution;
   reducing likelihood of fire/lightning;
   modifying microclimate conditions, e.g. reducing diurnal humidity and temperature cycles.

3. Employ conservation and preventative solutions to repair existing damage to buildings and structures and reduce vulnerability to weathering hazards in future by:

using treatments such as poulticing to remove potentially damaging salts;

strengthening damaged stonework through the use of consolidants/repairs;

applying protective coatings to prevent future pollution, salts, moisture etc. damaging the surface;

bioremediation;

developing better design and materials specification to improve the resilience of buildings and structures in the face of future weathering hazards.

4. Do nothing.

Diagnosing the major weathering hazards affecting buildings and structures can be difficult, especially when they are ancient and composed of complex associations of materials. Three main diagnostic methods can be applied, i.e. visual assessment, *in situ* non-destructive testing and detailed laboratory testing. In most cases a combination of these methods is needed in order to provide clear diagnosis. Visual assessment methods, sometimes called 'condition surveys', are largely based on techniques for classifying and mapping the visible signs of deterioration, along with adjunct factors such as materials types, architectural elements etc. The most detailed damage diagnosis scheme has been put forward by Bernd Fitzner and colleagues (Fitzner *et al.*, 1995; Fitzner and Heinrichs, 2002). Their scheme allows the production of GIS-based damage maps which can be interrogated in order to establish correlations with factors such as materials type, position, presence of former treatments, aspect etc. Such schemes can be very time consuming, and a degree of subjectivity can creep in when identifying particular decay features.

A range of non-destructive *in situ* test methods have been developed to find out more about the non-visible progress of stone decay, which can provide useful insights. A range of techniques, for example, have been developed to investigate factors such as surface hardness (e.g. Schmidt hammer, Duroscope and Equotip methods as reviewed by Török, 2002; Aoki and Matsukura, 2007). *In situ*, non-destructive techniques are also available to investigate the chemical make-up of weathering crusts and the near-surface zone of deteriorating facades, such as portable XRF (Thornbush and Viles, 2006b). A range of techniques is also being developed to provide better information on the subsurface moisture characteristics of deteriorating walls, such as 2D resistivity surveys (Sass

and Viles, 2006). Many investigations also commonly use a range of probes to monitor the external microclimate affecting deteriorating buildings and structures, as well as surface and subsurface conditions. Where it is possible and ethically acceptable to remove samples from a deteriorating building or structure, laboratory investigations can reveal much about the state of the stonework and the potential causes of decay. Bityukova (2006), for example, shows the use of multiple analysis tools (XRD, ICPMS, sulphur isotope analysis) on samples of stone and black crust from medieval buildings in Tallinn, Estonia, to elucidate the causes of deterioration.

Once diagnosis of weathering hazards has been undertaken, choice of management strategy is necessary. Reducing the threat of hazardous weathering through avoiding building in susceptible locations is a particularly suitable strategy where conditions are very patchy, and weathering 'hot spots' can be relatively easily avoided. This is particularly true for salt weathering situations where groundwater level and salinity are the crucial controls, as in the Middle East's low-lying coastal cities. Jones (1980) illustrates the production and use of hazard zonation maps for such areas, based on groundwater height and salinity measurements. Knowing the spatial distribution of weathering hazards can also help managing problems for movable objects, such as carvings and sculptures, which can be moved away from particularly damaging environments.

The strategy of reducing the hazard by reducing the severity of the factors that are causing the weathering can be very useful in circumstances where relocation of highly valuable artefacts and structures is not possible. For example, manipulation of groundwater levels or the reinstigation of drainage systems can help mitigate problems of groundwater-based salt weathering. Reduction of traffic levels in front of valuable items of cultural heritage can help to reduce air pollution and associated weathering damage. However, in many situations reducing the weathering hazard is only one issue that may need solving, and concerns over economic growth and vitality may mean that irrigation-induced groundwater problems or traffic-based air pollution cannot be easily removed as that would damage agricultural and urban development. In the immediate surroundings of a building or structure, modification of microclimatic conditions can be attempted in order to prevent further damage from salt weathering and other hazards. There has been much research done, for example, on the manipulation of relative humidities in order to reduce the threat of damage from salt crystallisation (e.g. Price and Brimblecombe, 1994; Steiger and Zeunert, 1996).

The third category of approach to dealing with the weathering hazards is to employ conservation and preventative solutions to the materials that are under threat. Where materials have already become damaged, consolidants and water repellents can be used to strengthen and protect them (e.g. Littmann *et al.*, 1993). New techniques continue to be developed, including a range of bioremediation strategies that utilise microbial processes as reviewed by Webster and May (2006). Where damage is severe and hard to mitigate through conservation of the existing materials, replacement of materials is necessary. For example, where salts are damaging bituminous roads and runways the use of salt-free aggregates, thick and dense asphalt and semi-permeable membranes can all contribute to ensuring that, when surfaces have to be replaced because of damage, the same problem does not arise again. Conservation techniques use the best available knowledge and skills, but are not always successful on the short or longer term and, in some instances, can accelerate the weathering hazard.

The final mitigation strategy, 'do nothing', is of course a valid and acceptable response to hazardous weathering in many circumstances, especially when other environmental problems are more pressing, or when economic conditions are unfavourable.

12.8 Conclusions

We have demonstrated in this chapter that weathering poses a significant threat to the built environment, whether composed of ancient monuments or modern structures (Figures 12.8 and 12.9). We have also shown that a range of human activities, such as atmospheric pollution or changes to hydrological conditions brought about by land-use and land-cover changes, has caused rates of material decay to be catastrophically accelerated. Still further changes may occur as a result of future climate change. There are various actions that can be taken, however, to mitigate the worst consequences of aggressive weathering attack, although sometimes these can themselves cause further damage if not properly applied or if conditions change. Some of the complexity and interlinkages involved with weathering processes, other environmental hazards and the development of hazardous or damaging weathering are illustrated in Figure 12.10. Combinations of weathering processes can create a hazard over time (especially when they affect highly vulnerable and valuable carvings etc., for which even small amounts of damage can prove catastrophic, or when some thresholds of stability are overtopped). Furthermore, weathering can predispose many rock slopes,

FIGURE 12.8. Disintegrated water pipelines in the Namib Desert, Namibia, produced by salt corrosion of the iron reinforcements and of the concrete.

FIGURE 12.9. The deterioration of these Bath stone columns at the entrance to an hotel on the sea front in Weymouth, southern England, has taken place since they were replaced in the early 1990s. The stone selected was inadequate in the face of the salty environment to which they have been subjected.

buildings and structures to damage from further hazards such as earthquakes or flooding, or even weathering under future changed environmental conditions. Whilst conservation and management strategies can reduce the threat of weathering hazards, they can also accelerate problems. Thus, as Figure 12.10 depicts, there are many pathways

TABLE 12.1. *Possible consequences of future climate change for building stone decay in the UK*

	Dominant building stone type	Process responses	Other threats	Overall response
Northwest: warmer, wetter winters	Siliceous sandstones and granites	Increased chemical weathering. Less freeze–thaw weathering. More organic growths contributing to soiling.	Increased storm activity may cause episodic damage. Increased wave heights may encourage faster weathering in coastal areas. Increased flooding may encourage decay.	Enhanced chemical decay processes and biological soiling, reduced physical decay processes.
Southeast: warmer, drier summers	Limestones	Less freeze–thaw weathering. Reduced chemical weathering as a result of less available water. Increased salt crystallization in summer. More deteriorating organic growths.	Increased drying of soils (especially clay-rich soils) will encourage subsidence and building damage. Low-lying coastal areas will be particularly prone to marine encroachment. Increased drought frequency may encourage decay.	Enhanced physical and biological weathering, more dust for soiling, reduced chemical weathering.

Source: Viles (2002, Table 3)

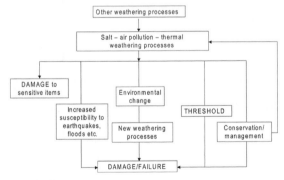

FIGURE 12.10. Conceptual model of how salt, air pollution and thermally induced weathering processes can result in hazardous outcomes.

by which a range of seemingly small-scale and slow-acting weathering processes can have hazardous outcomes.

It is possible that the nature of weathering hazards may change as a result of future climate changes, because of their effects on the relative importance of freeze–thaw cycles, rates of solution, biological activity, stone dampness, fire frequency and phase changes in contained salts,

etc. (Hall and Walton, 1992; Viles, 2002; Smith *et al.*, 2004; Brimblecombe *et al.*, 2006). Viles (2002) considered future stone deterioration in the UK. There she contrasted the likely impacts of climate change on building stones between the northwest and the southeast of the country (Table 12.1). She argued that in the northwest, which would have warmer and wetter winters, chemical weathering's importance might be increased, whereas in the southeast, which would have warmer, drier summers, processes like salt weathering might become more important. Furthermore, as the climate warms, the number of freeze–thaw cycles required to produce frost weathering in areas like Central England are likely to decrease, as they did during the warming of the nineteenth and twentieth centuries (Brimblecombe and Camuffo, 2003). Already, many localities are suffering far less frost damage than, for example, during Little Ice Age conditions as exemplified by the changing nature of weathering of soapstone artefacts in Norway (Stohrmeyer, 2004). Thus, the nature of weathering hazards remains in flux – and whilst in this chapter we have separated out individual weathering hazards such as salt, air pollution and fire effects, in reality multiple weathering processes act in synergy to create the hazard, and the contribution of different processes is likely to change as a result of climate change and changing human impacts.

Acknowledgements

Heather Viles acknowledges the support of EU Life Programme and EPSRC (Grant EP/D008689/1) for funding research in Oxford from which some of the ideas and observations in this chapter were developed.

References

Akiner, S., Cooke, R. U. and French, R. A. (1992). Salt damage to Islamic monuments in Uzbekistan. *Geographical Journal*, **158**, 257–272.

Albouy, M., Deletie, P., Haguenauer, B. *et al.* (1993). Petra, la cité rose des sables. La pathologie des grès et leur traitement dans la perspective d'une préservation et d'une restauration des monuments. In M. J. Thiel (ed.), *Conservation of Stone and Other Materials*. London: E. & F. Spon, pp. 376–385.

Allison, R. J. and Bristow, G. E. (1999). The effects of fire on rock weathering: some further considerations of laboratory experimental simulation. *Earth Surface Processes and Landforms*, **24**, 707–713.

André, M.-F., Etienne, S., Mercier, D., Vautier, F. and Voldoire, O. (2008). Assessment of sandstone deterioration at Ta Keo temple (Angkor): first results and future prospects. *Environmental Geology Special Issue*, doi 10.10007/s00254-008-1408-8.

Aoki, H. and Matsukura, N. (2007). A new technique for non-destructive field measurement of rock-surface strength: an application of the Equotip hardness tester to weathering studies. *Earth Surface Processes and Landforms*, **32**, 1759–1769.

Arnold, A. and Zehnder, K. (1988). Decay of stony materials by salts in humid atmosphere. In *Proceedings 7th International Congress on the Deterioration and Conservation of Stone*, pp. 138–148.

Bani-Hani, K. and Barakat, S. (2006). Analytical evaluation of repair and strengthening measures of Qasr al-Bint historical monument. Petra, Jordan. *Engineering Structures*, **28**, 1355–1366.

Bityukova, L. (2006). Air pollution effect on the decay of carbonate building stones in old town of Tallinn. *Water, Air and Soil Pollution*, **172**, 239–271.

Borrelli, L., Greco, R. and Gullà, G. (2007). Weathering grade of rock masses as a predisposing factor to slope instabilities: reconnaissance and control procedures. *Geomorphology*, **87**, 158–175.

Botswana Roads Department (2001). *The Prevention and Repair of Salt Damage to Roads and Runways*. Ministry of Works, Transport and Communications, Gaborone, Botswana, ISBN 99912-0-380-X.

Brimblecombe, P. and Camuffo, D. (2003). Long term damage to the built environment. In P. Brimblecombe (ed.), *The Effects of Air Pollution on the Built Environment*. London: Imperial College Press, pp. 1–30.

Brimblecombe, P. and Grossi, C. M. (2008). Millennium-long recession of limestone facades in London. *Environmental Geology Special Issue*, doi 10.10007/s00254-008-1465-z.

Brimblecombe, P., Grossi, C. M. and Harris, I. (2006). The effect of long-term trends in dampness on historic buildings. *Weather*, **61**, 278–281.

Camuffo, D. (1993). Controlling the aeolian erosion of the Great Sphinx. *Studies in Conservation*, **38**, 198–205.

Chakrabati, B., Yates, T. and Lewry, A. (1996). Effect of fire damage on natural stonework in buildings. *Construction and Building Materials*, **10**, 539–544.

Charola, A. E., Pühringer, J. and Steiger, M. (2007). Gypsum: a review of its role in the deterioration of building materials. *Environmental Geology*, **52**, 339–352.

Dang, Z., Liu, C. and Haigh, M. J. (2002). Mobility of heavy metals associated with the natural weathering of coal mine spoils. *Environmental Pollution*, **118**, 419–426.

Davidson, C. I., Tang, W., Finger, S. *et al.* (2000). Soiling patterns on a tall limestone building: changes over 60 years. *Environmental Science and Technology*, **34**, 560–565.

Dionisio, A. (2007). Stone decay induced by fire on historic buildings: the case of the cloister of Lisbon Cathedral, Portugal. In R. Prikryl and B. J. Smith (eds.), *Building Stone Decay: From Diagnosis to Conservation*. Geological Society of London Special Publication 271, pp. 87–98.

Durgin, P. B. (1977). Landslides and the weathering of granitic rocks. *Geological Society of America Reviews in Engineering Geology*, **III**, 127–131.

Farmer, M. F. (1939). Lightning spalling. *American Antiquity*, **4**, 346–348.

Fassina, V., Favaro, M. and Naccari, A. (2002) Principal decay patterns on Venetian monuments. In S. Siegesmund, T. Weiss and A. Vollbrecht (eds.), *Natural Stone, Weathering Phenomena, Conservation Strategies and Case Studies*. Geological Society of London Special Publication 205, pp. 381–391.

Fitzner, B. and Heinrichs, K. (1991). Weathering forms and rock characteristics of historical monuments carved from bedrock in Petra/Jordan. In N. S. Baer, C. Sabbioni and A. I. Sors (eds.), *Science, Technology and European Cultural Heritage*. Oxford: Butterworth-Heinemann, pp. 908–911.

Fitzner, B. and Heinrichs, K. (1994). Damage diagnosis at monuments carved from bedrock in Petra, Jordan. In V. Fassina, H. Ott and F. Zezza (eds.), *Proceedings Third International Conference on the Conservation of Monuments in the Mediterranean Basin, Venice, 22–25 June, 1994*, pp. 663–671.

Fitzner, B. and Heinrichs, K. (2002). Damage diagnosis at stone monuments: weathering forms, damage categories and damage indices. In R. Prikryl and H. A. Viles (eds.), *Understanding and Managing Stone Decay*. Prague: The Karolinum Press, pp. 11–56.

Fitzner, B., Heinrichs, K. and Kownatzki, R. (1995). Weathering forms: classification and mapping. In R. Snethlage (ed.), *Fortdruck ans Denkmalpflege und Naturwissenschaft Natursteinkonservierung* 1. Berlin: Ernst und Sohn, pp. 41–88 (in German and English).

Fodde, E. (2007). Fired brick and sulphate attack: the case of Moenjodaro, Pakistan. *Journal of Architectural Conservation*, **13**, 69–80.

Gauri, K. L., Sinai, J. J. and Bandyopadhyay, J. K. (1995). Geologic weathering and its implications on the age of the Sphinx. *Geoarchaeology*, **10**, 119–133.

Gomez-Heras, M., Alvarez de Buergo, M., Fort, R. *et al.* (2006). Evolution of porosity in Hungarian building stones after simulated burning. In R. Fort, M. Alvarez de Buergo, M. Gomez-Heras and C. Vazquez-Calvo (eds.), *Heritage, Weathering and Conservation*. London: Taylor & Francis, pp. 513–519.

Gomez-Heras, M., Smith, B. J. and Viles, H. A. (2008). Laboratory modelling of gypsum crust growth on limestone related to soot pollution and gaseous sulphur: implications of 'cleaner' environments for stone decay. In J. W. Lucaszewicz and P. Niemcewicz (eds.), *Proceedings, 11th International Congress on Deterioration and Conservation of Stone, Torun*. Torun: Nicolaus Copernicus University Press, pp. 105–112.

Goudie, A. S. (2006). *The Human Impact on the Natural Environment*, 6th edition. Oxford: Blackwell.

Goudie, A. S. and Viles, H. A. (1997). *Salt Weathering Hazard*. Chichester: Wiley.

Goudie, A. S. and Viles, H. A. (2000). The thermal degradation of marble. *Acta Universitatis Carolinae 2000 Geographica*, **35**, Supplement, 7–16.

Goudie A. S., Allison, R. J. and McLaren, S. J. (1992). The relations between modulus of elasticity and temperature in the context of the experimental simulation of rock weathering by fire. *Earth Surface Processes and Landforms*, **17**, 605–615.

Grossi, C. M. and Brimblecombe, P. (2002). The effect of atmospheric pollution on building materials. *Journal of Physics IV France*, **12**, 197–210.

Gullà G., Mandaglio, M. C. and Moraci, N. (2006). Effect of weathering on the compressibility and shear strength of a natural clay. *Canadian Geotechnical Journal*, **43**, 618–625.

Hajpál, M. and Török, A. (2004). Mineralogical and colour changes of quartz sandstones by heat. *Environmental Geology*, **46**, 311–322.

Hall, K. J. and Walton, D. W. H. (1992). Rock weathering, soil development and colonization under a changing climate. *Philosophical Transactions Royal Society of London, B*, **338**, 269–277.

Heinrichs, K. (2008). Diagnosis of weathering damage on rock-cut monuments in Petra, Jordan. *Environmental Geology*, **56**, 643–675.

Hoke, G. D. and Turcotte D. L. (2004). The weathering of stones due to dissolution. *Environmental Geology*, **46**, 305–10.

Horta, J. C. de O. S. (1985). Salt heaving in the Sahara. *Géotechnique*, **35**, 329–337.

Howard, K. W. F. and Beck, P. J. (1993). Hydrogeochemical implications of groundwater contamination by road de-icing chemicals. *Journal of Contaminant Hydrology*, **12**, 245–268.

Islam, M. R., Lahermo, P., Salminen, R., Rojstaczer, S. and Peuraniemi, V. (2000). Lake and reservoir water quality affected by metals leaching from tropical soils, Bangladesh. *Environmental Geology*, **39**, 1083–1089.

Jaboyedoff, M. Baillifard, F., Bardou, E. and Girod, F. (2004). The effect of weathering on Alpine rock instability. *Quarterly Journal of Engineering Geology and Hydrogeology*, **37**, 95–103.

Januszke, R. M. and Booth, E. H. S. (1984). Soluble salt damage to sprayed seals on the Stuart Highway. *Australian Road Research Board Proceedings*, **12**(3), 18–31.

Jones, D. K. C. (1980). British applied geomorphology: an appraisal. *Zeitschrift für Geomorphologie. N. F.*, Suppl.-Bd., **36**, 48–73.

Kamh, G. M. E., Kallash, A. and Azzam, R. (2008). Factors controlling building susceptibility to earthquakes: 14-year recordings of Islamic archaeological sites in Old Cairo, Egypt: a case study. *Environmental Geology Special Issue*, doi 10.10007/s00254-007-1162-3.

Keatings, K., Tassie, G. J., Flower, R. J. *et al.* (2007). An examination of groundwater within the Hawara Pyramid, Egypt. *Geoarchaeology*, **22**, 533–554.

Khoury, G. A. (2000). Effect of fire on concrete and concrete structures. *Progress in Structural and Engineering Materials*, **2**, 429–447.

Khoury, G. A. (2006). Tunnel concretes under fire: Part 1, Explosive spalling. *Concrete*, November 2006, 62–65.

Laue, S., Bläuer Böhm, C. and Jeannette, D. (1996). Salt weathering and porosity: examples from the crypt of St. Maria im Kapitol, Cologne. *Proceedings 8th International Congress on Deterioration and Conservation of Stone*, pp. 513–522.

Leduc, C., Favreau, G. and Schoreter, P. (2001). Long term rise in a Sahelian water-table: the continental terminal in southwest Niger. *Journal of Hydrology*, **243**, 43–54.

Littmann, K., Sasse, H. R., Wagener, S. and Hocker, H. (1993). Development of polymers for the consolidation of natural stone. In M. J. Theil (ed.), *Conservation of Stone and Other Materials*. London: Spon, pp. 689–696.

Livingston, R. A. (1989). The geological origins of the Great Sphinx and implications for its preservation. Abstract of paper presented to International Geological Congress, Washington D.C., July 9–19, 1989.

Lohuizen-de Leeuw, J. E. van (1973). Moenjo Daro: a cause of common concern. *South Asian Archaeology 1973*, Leiden: Brill, pp. 1–11.

Mallett, G. P. (1994). *Repair of Concrete Bridges*. London: Thomas Telford.

Malm, W. C., Schichtel, B. A., Ames, R. B. and Gebhart, K. A. (2002). A 10-year spatial and temporal trend of sulfate across the United States. *Journal of Geophysical Research*, **107**(D22), article no. 4627.

Master Plan (1972). *Master Plan for the Preservation of Moenjodaro*. Department of Archaeology and Museums, Ministry of Education and Provincial Coordination, Government of Pakistan.

Matsuoka, N. and Sakai, H. (1999). Rockfall activity from an alpine cliff during thawing periods. *Geomorphology*, **28**, 309–328.

McCabe S., Smith, B. J. and Warke, P. A. (2007). Sandstone response to salt weathering following simulated fire damage: a comparison of the effects of furnace heating and fire. *Earth Surface Processes and Landforms*, **32**, 1874–1883.

Paradise, T. (2005). Sandstone weathering and aspect in Petra, Jordan. *Zeitschrift für Geomorphologie*, **46**, 1–17.

Price, C. and Brimblecombe, P. (1994). Preventing salt damage in porous materials. In *Preventive Conservation*. London: International Institute for Conservation of Historic and Artistic Works, pp. 90–93.

Rayleigh, A. (1934). The bending of marble. *Proceedings of the Royal Society of London*, **144A**, 266–279.

Rhoades, J. D. (1990). Soil salinity: causes and controls. In A. S. Goudie (ed.), *Techniques for Desert Reclamation*. Chichester: Wiley, pp. 109–134.

Rosenholtz, J. L. and Smith, D. T. (1949). Linear thermal expansion of calcite, var. Iceland Spar and Yule Marble. *American Mineralogist*, **34**, 846–854.

Royer-Carfagni, G. F. (1999). On the thermal degradation of marble. *International Journal of Rock Mechanics and Mining Sciences*, **36**, 119–126.

Rozanov, B. G., Targulian, V. and Orlov, D. S. (1991). Soils. In B. L. Turner (ed.), *The Earth as Transformed by Human Action*. Cambridge: Cambridge University Press.

Sabbioni, C. (2003). Mechanisms of air pollution damage to stone. In P. Brimblecombe (ed.), *The Effects of Air Pollution on the Built Environment*. London: Imperial College Press, pp. 63–106.

Sabbioni C., Ghedini N. and Bonazza A. (1999). Organic anions in damage layers on monuments and buildings. *Atmospheric Environment*, **37**, 1261–1269.

Sass, O. and Viles, H. A. (2006). How wet are these walls? Testing a novel technique for measuring moisture in ruined walls. *Journal of Cultural Heritage*, **7**, 257–263.

Saxena, V. K. and Ahmed, S. (2001). Dissolution of fluoride in groundwater: water-rock interaction study. *Environmental Geology*, **40**, 1084–1087.

Schiavon, N. (2007). Kaolinisation of granite in an urban environment. *Environmental Geology*, **52**, 399–407.

Scott, W. S. and Wylie, N. P. (1980). The environmental effects of snow dumping: a literature review. *Journal of Environmental Management*, **10**, 219–240.

Seaward, M. R. D. (1997). Major impacts made by lichens in biodeterioration processes. *International Journal of Biodeterioration and Biodegradation*, **40**, 269–273.

Selwitz, C. (1990). Deterioration of the Great Sphinx: an assessment of the literature. *Antiquity*, **64**, 853–859.

Seto, S., Nakamura, A., Noguchi, I. *et al.* (2002). Annual and seasonal trends in chemical composition of precipitation in Japan during 1989–1998. *Atmospheric Environment*, **31**, 3505–3517.

Shehata, W. M. and Amin, A. A. (1997). Geotechnical hazards associated with desert environment. *Natural Hazards*, **16**, 81–95.

Shehata, W. and Lotfi, H. (1993). Preconstruction solution for groundwater rise in Sabkha. *Bulletin of the International Association of Engineering Geology*, **47**, 145–150.

Simao, J., Ruiz-Agudo, E. and Rodriguez-Navarro, C. (2006). Effects of particulate matter from gasoline and diesel vehicle exhaust systems on silicate stones sulfation. *Atmospheric Environment*, **40**, 6905–6917.

Sippel, J., Siegesmund, S., Weiss, T., Nitsch, K.-H. and Korzen, M. (2007). Decay of natural stones caused by fire damage. In R. Prikryl and B. J. Smith (eds.), *Building Stone Decay: From Diagnosis to Conservation*. Geological Society of London Special Publication 271, pp. 139–151.

Smith, A. G. and Pells, P. J. N. (2008). Impact of fire on tunnels in Hawkesbury Sandstone. *Tunnelling and Underground Space Technology*, **23**, 65–74.

Smith, A. G. and Pells, P. J. N. (2009). Discussion of the paper 'Impact of fire on tunnels in Hawkesbury Sandstone' by Smith and Pells. *Tunnelling and Underground Space Technology*, **24**, 112–114.

Smith, B. J. and Viles, H. A. (2006). Rapid, catastrophic decay of building limestones: an overview of causes, effects and consequences. In R. Fort, M. Alvarez de Buergo, M. Gomez-Heras and C. Vazquez-Calvo (eds.), *Heritage, Weathering and Conservation*, London: Taylor & Francis, pp. 191–197.

Smith, B. J., Warke, P. A. and Curran, J. M. (2004). Implications of climate change and increased 'time of wetness' for the soiling and decay of sandstone structures in Belfast, Northern Ireland. In R. Prikryl, (ed.), *Dimension Stone 2004*. Leiden: A. A. Balkema, pp. 9–14.

Smith, S. E. (1986). An assessment of structural deterioration on ancient Egyptian monuments and tombs in Thebes. *Journal of Field Archaeology*, **13**, 503–510.

Smith, S. J., Pitcher, H. and Wigley, T. M. L. (2001). Global and regional anthropogenic sulfur dioxide emissions. *Global and Planetary Change*, **29**, 99–119.

Steiger, M. and Zeunert, A. (1996). Crystallization properties of salt mixtures: comparison of experimental results and models calculations. *Proceedings 8th International Congress on Deterioration and Conservation of Stone*, pp. 535–544.

Stohrmeyer, P. (2004). Weathering of soapstone in a historical perspective. *Materials Characterization*, **53**, 191–207.

Thomachot, C., Matsuoka, N., Kuchitsu, N. and Morii, M. (2005). Frost damage of bricks composing a railway tunnel monument in Central Japan: field monitoring and laboratory simulation. *Natural Hazards and Earth System Sciences*, **5**, 465–476.

Thoms, A. V. (2007). Fire-cracked rock features on sandy landforms in the Northern Rocky Mountains: toward establishing reliable frames of reference for assessing site integrity. *Geoarchaeology*, **22**, 477–510.

Thornbush, M. J. and Viles, H. A. (2006a). Changing patterns of soiling and microbial growth on building stone in Oxford, England after implementation of a major traffic scheme. *Science of the Total Environment*, **367**, 203–211.

Thornbush, M. J. and Viles, H. A. (2006b). Use of portable X-ray fluorescence for monitoring elemental concentrations in surface units on roadside stone at Worcester College, Oxford. In R. Fort, M. Alvarez de Buergo, M. Gomez-Heras and C. Vazquez-Calvo (eds.), *Heritage, Weathering and Conservation*. London: Taylor & Francis, pp. 613–620.

Torok, A. (2002). Oolitic limestone in a polluted atmospheric environment in Budapest: weathering phenomena and alterations in physical properties. In S. Siegesmund, T. Weiss and A. Vollbrecht (eds.), *Natural Stone, Weathering Phenomena, Conservation Strategies and Case Studies*. Geological Society of London Special Publication 205, pp. 363–379.

Torok, A., Siegesmund, S., Muller, C. *et al.* (2007). Differences in texture, physical properties and microbiology of weathering crust and host rock: a case study of the porous limestone of Budapest (Hungary). In R. Prikryl and B. J. Smith (eds.), *Building Stone Decay: From Diagnosis to Conservation*. Geological Society of London Special Publication 271, pp. 261–276.

Tratebas, A. M., Cerveny, N. V. and Dorn, R. I. (2004). The effects of fire on rock art: microscopic evidence reveals the importance of weathering rinds. *Physical Geography*, **25**, 313–333.

Uchida, E., Ogawa, Y., Maeda, N. and Nakagawa, T. (1999). Deterioration of stone materials in the Angkor monuments, Cambodia. *Engineering Geology*, **55**, 101–112.

Vassie, P. (1984). Reinforcement corrosion and the durability of concrete bridges. *Proceedings of the Institution of Civil Engineers*, **76**, 713–723.

Viles, H. A. (1994). *Time and Grime*. School of Geography, University of Oxford Research Paper No. 50.

Viles, H. A. (2002). Implications of future climate change for stone deterioration. In S. Siegesmund, T. Weiss and A. Vollbrecht (eds.), *Natural Stone, Weathering Phenomena,*

Conservation Strategies and Case Studies. Geological Society of London Special Publication **205**, 407–418.

Viles, H. A. and Gorbushina, A. A. (2003). Soiling and microbial colonisation on urban roadside limestone: a three year study in Oxford, England. *Building and Environment*, **38**, 1217–1224.

Wang, K. (2006). Damaging effects of deicing chemicals on concrete materials. *Cement and Concrete Composites*, **28**, 173–188.

Webster, A. and May, E. (2006). Bioremediation of weathered building stone surfaces. *Trends in Biotechnology*, **24**, 255–260.

Wedekind, W. and Ruedrich, J. (2006). Salt-weathering, conservation techniques and strategies to protect the rock cut façades in Petra/Jordan. In R. Fort, M. Alvarez de Buergo, M. Gomez-Heras and C. Vazquez-Calvo (eds.), *Heritage, Weathering and Conservation*. London: Taylor & Francis, pp. 261–267.

Widhalm, C., Tschegg, E. and Eppensteiner, W. (1996). Anisotrophic thermal expansion causes deformation of marble claddings. *Journal of Performance of Constructed Facilities*, **10**, 5–10.

Widhalm, C., Tschegg, E. and Eppensteiner, W. (1997). Acoustic emission and anisotropic expansion when heating marble. *Journal of Performance of Constructed Facilities*, **11**, 35–40.

Wilson, J. F. (2003). Lightning-induced fracture of masonry and rock. *International Journal of Solids and Structures*, **40**, 5305–5318.

Wüst, R. A. J. and Schlüchter, C. (2000). The origin of soluble salts in rocks of the Thebes Mountains, Egypt: the damage potential to ancient Egyptian wall art. *Journal of Archaeological Science*, **27**, 1161–1172.

Zador, M. (1992). Experience with cleaning and consolidating stone facades in Hungary. In R. G. M. Webster (ed.), *Stone Cleaning and the Nature, Soiling and Decay Mechanisms of Stone*. London: Donhead, pp. 146–152.

13 Hazards associated with karst

Francisco Gutiérrez

13.1 Introduction: why are hazards associated with karst important?

Karst refers to terrains in which the geomorphology and hydrology, both at the surface and in the subsurface, are largely governed by dissolution of carbonate and/or evaporite rocks. Some distinctive characteristics of karst environments include: (1) the presence of enclosed depressions (dolines or sinkholes and poljes), swallow holes and large springs; (2) the prevalence of underground drainage through channels resulting from dissolutional enlargement of discontinuity planes. Groundwater flow in these interconnected conduits circulates much faster than in aquifers controlled by granular or fracture permeability (Ford and Williams, 2007). Three main types of karst settings can be differentiated: bare karst, covered or mantled karst and interstratal karst, depending on whether the soluble rocks are exposed at the surface, covered by unconsolidated deposits or overlain by non-karst rocks (caprocks), respectively.

Karst developed in evaporite rocks has distinctive features primarily due to the higher solubility and lower mechanical strength of the evaporites (Gutiérrez et al., 2008a). The equilibrium solubilities of gypsum ($CaSO_4 \cdot 2H_2O$) and halite (NaCl) in distilled water are 2.4 and 360 g/l, respectively. By comparison, the solubilities of calcite ($CaCO_3$) and dolomite ($MgCa[CO_3]_2$) minerals in regular meteoric waters are commonly lower than 0.1 g/l. The solubilities of these carbonate minerals are largely dependent on the pH of the water, which is generally controlled by the carbon dioxide partial pressure. Solutional bedrock retreat rates in limestone and gypsum permanently in contact with normal flowing meteoric water may reach values of the order of 0.1–1 mm/year and 1–10's mm/year, respectively (Dreybrodt, 2004). In favourable conditions

(turbulent freshwater flow), dissolution rates in halite may be as high as 1.2 cm/minute (Frumkin and Ford, 1995). Some of the peculiarities that differentiate the evaporite karsts from those developed in carbonate rocks include: (1) in evaporite terrains, dissolution and subsidence processes commonly occur at a much faster rate and subsidence mechanisms show a higher diversity. Dissolution may create cavities and substantially reduce the mechanical strength of the rocks at a human time scale (Gutiérrez et al., 2008a); (2) outcrops of evaporites are proportionally less abundant and interstratal karstification is much more important in evaporites; (3) interstratal dissolution of evaporites may cause ground subsidence at a regional scale producing extensive dissolution-collapse breccias and morpho-structures (depositional basins, grabens, monoclines, etc.) up to several hundred kilometres in length and several hundred metres in structural relief (e.g. Warren, 2006).

Karst processes and landforms pose multiple hazards that commonly require specific investigation and mitigation methods due to the singular and highly variable characteristics of karst environments. Unlike other geomorphic systems, in karst crucial keys to understanding and solving the problems are usually beneath the surface. Sinkholes are the most important hazard in karst. Other potentially hazardous processes and problems include: flooding of depressions and discharge areas; differential settlement due to compaction of soil over irregular rockhead; landslides; water losses in reservoirs; and flooding of tunnels and mines. In numerous regions of the world, karst hazards have a notorious economic impact, usually higher than expected, largely because of the dispersed and hidden character of the damage. Karst in carbonate and evaporite rocks occurs over ~20% of the Earth's ice-free continental area and around a quarter of the global population depends on karst water

Geomorphological Hazards and Disaster Prevention, eds. Irasema Alcántara-Ayala and Andrew S. Goudie. Published by Cambridge University Press. © Cambridge University Press 2010.

supply (Ford and Williams, 2007). Despite the fact that the scientific community has substantially improved the understanding of the karst systems and the ability to prevent and alleviate the hazards associated with them, the amount of damage can be expected to rise in the future due to several reasons: (1) the numbers of people and human structures exposed to karst processes will keep on increasing; (2) the impact of human activities that induce or favour hazardous processes will be progressively higher; (3) increased demand for water and energy has gradually augmented the use of karst regions for large hydrological projects (Milanovic, 2000, 2002); (4) technical and political decisions are frequently taken without an adequate understanding of the karst system.

13.2 Sinkhole hazard

13.2.1 Sinkhole types, processes and detrimental effects

Sinkholes or dolines are closed depressions with internal drainage that may display a wide range of morphologies (cylindrical, conical, bowl- or pan-shaped) and may reach hundreds of metres in diameter and depth (Williams, 2004). Several genetic classifications of sinkholes have recently been published (Williams, 2004; Beck, 2005a; Waltham *et al.*, 2005; Gutiérrez *et al.*, 2008b). These classifications distinguish two main groups of sinkholes. One of these groups corresponds to the *solution sinkholes*, generated by differential corrosional lowering of the ground surface where karst rocks are exposed at the surface or merely soil-mantled (bare karst). The development of these sinkholes is governed by focused centripetal flow towards higher permeability zones (e.g. dissolutional conduits) in the epikarst (Ford and Williams, 2007) (Figure 13.1). The other group includes different types of sinkholes resulting from both subsurface dissolution and downward gravitational movement (internal erosion and deformation) of the overlying material. Obviously, these sinkholes that cause ground to subside are the most important from a hazard and engineering perspective. The classification proposed by Gutiérrez *et al.* (2008b), applicable to both carbonate and evaporite karst terrains, describes the subsidence sinkholes using two terms: the first descriptor refers to the material affected by internal erosion and/or deformation processes (*cover, bedrock* or *caprock*), and the second descriptor indicates the main type of subsidence process (*collapse, suffosion* or *sagging*) (Figure 13.1). *Cover* refers to allogenic unconsolidated deposits or residual soil material, *bedrock* to karst rocks and *caprock* to non-karst rocks. *Collapse* is the brittle deformation of soil or rock material either by brecciation or

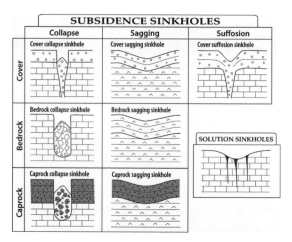

FIGURE 13.1. Genetic classification of sinkholes applicable to carbonate and evaporite karst terrains. (Modified from Gutiérrez *et al.*, 2008b.)

the development of well-defined failure planes, *suffosion* is the downward migration of cover deposits through conduits and its progressive settling, and *sagging* is the ductile flexure (bending) of sediments caused by the lack of basal support. In practice, more than one material type and several processes can be involved in the generation of sinkholes. These complex sinkholes are described using combinations of the proposed terms with the dominant material and/or process followed by the secondary one (e.g. cover and bedrock collapse sinkhole, bedrock sagging and collapse sinkhole).

The cover material may be affected by any of the three subsidence mechanisms. The progressive corrosional lowering of the rockhead may cause the gradual settlement of the overlying deposits by sagging, producing *cover sagging sinkholes*. An important applied aspect is that the generation of these sinkholes does not require the formation of cavities. These depressions are commonly shallow, have poorly defined edges and may reach several hundred metres across.

Cover deposits may migrate downward into fissures and conduits developed in the rockhead by action of a wide range of processes collectively designated as suffosion: downwashing of particles by percolating waters, cohesionless granular flows, viscous gravity flows (non-Newtonian), fall of particles, and sediment-laden water flows (Figure 13.2A). The downward transport of the cover material through pipes and fissures may produce two main types of sinkholes depending on the rheological behaviour of the mantling deposits. Where the cover behaves as a ductile or loose granular material, it may settle gradually as undermining by suffosion progresses,

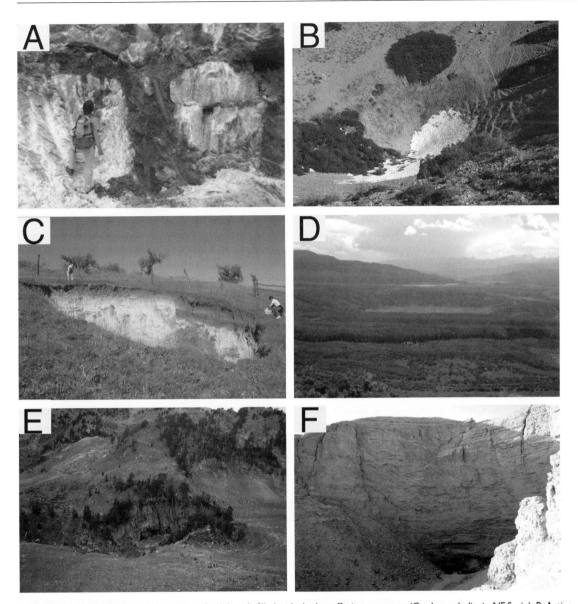

FIGURE 13.2. A: Dissolutional conduits in rock salt largely filled with clay by suffosion processes (Cardona salt diapir, NE Spain). B: Active suffosion sinkhole in the bottom of a large bedrock collapse sinkhole in limestone (Villar del Cobo, Iberian Range, Spain). C: Dankivsky cover collapse sinkhole formed on January 11, 1998 (western Ukraine). The sinkhole resulted from the collapse of Miocene clays and loess (cover material from a mechanical point of view) into a cave developed in Miocene gypsum. Most likely the sinkhole was controlled by the Quaternary fault exposed in the sinkhole walls. Note the faulted and buried soil in the downthrown block. (Photograph taken on 19 May 1999.) D: Shippes Bowl Sag, a caprock sagging sinkhole around 2 km long generated by the downward bending of Tertiary basalts due to the interstratal karstification of the underlying Carboniferous evaporites (Rocky Mountains of Colorado). (Courtesy of Bob Kirkham.) E: Forau de Aigualluts, a bedrock collapse sinkhole 80 m long developed in Paleozoic marbles at the bottom of the Esera glacier valley (Spanish Pyrenees). The water that flows into this sinkhole crosses the main topographic divide of the Pyrenees and reappears 4 km to the NE in France. This is probably the place in which the hydraulic connection between a ponor and a spring was demonstrated for the first time by Norbert Casteret in 1931 using fluorescein. (Courtesy of Pedro Lucha.) F: Dahal Hit, a caprock collapse sinkhole 120 m deep generated by the interstratal karstification of Jurassic evaporites (anhydrite and gypsum) of the Hit Formation, exposed in the lower part of the sinkhole wall (Interior Homocline of Central Saudi Arabia).

generating funnel- or bowl-shaped *cover suffosion sink-holes*, typically a few metres in diameter (Figure 13.2B). When the cover behaves in a brittle way, internal erosion of the mantling deposits above a dissolutional pipe results in an arched cavity whose upward migration by successive failures eventually leads to the formation of a *cover collapse sinkhole* (Figure 13.2C). If the stoping cavity reaches a more resistant material like a duricrust or an artificial pavement, it may expand laterally until the weight of the suspended roof exceeds its mechanical strength. The mechanics of the upward propagation of soil cavities has been analysed by Tharp (1995). These sinkholes, typically less than 10 m across, commonly appear in a sudden way and have scarped or overhanging sides at the time of formation. Mass wasting processes acting on their margins may cause their rapid enlargement and transform them into funnel- or bowl-shaped depressions. The cover collapse and cover suffosion sinkholes cause the vast majority of the sinkhole damage, since these are the sinkhole types with the higher probability of occurrence (Beck, 2005a; Waltham *et al.*, 2005).

Bedrocks and caprocks may be affected by ductile sagging or brittle collapse. The subsidence mechanism essentially depends on the mechanical strength of the rock mass, largely determined by the lithology and discontinuity planes, the size and geometry of the cavities (mainly the span), the thickness of the roof and the strata, and the hydraulic conditions (either vadose or phreatic). The sagging subsidence mechanism is particularly frequent in gypsum-bearing sequences, since gypsum has a lower Young's modulus than carbonate rocks and consequently a more ductile rheology. Interstratal differential dissolution may be accompanied by continuous sagging of the overlying rocks generating *bedrock* or *caprock sagging sinkholes*. This process results in the generation of basin structures with centripetal dips and vaguely edged depressions that may reach more than 1 km in length (Figure 13.2D). These basin structures may be affected by synthetic normal faults in the outer margins and reverse antithetic faults in the inner zone (bending-moment ring faults).

When the roofs of dissolutional cavities developed within the bedrock reach a critical span, they tend to propagate upwards by progressive collapse processes (stoping). The tension zone over the roof created by the deflection of the gravitational stresses around the cavity may determine the development of cupola-shaped failure planes and the generation of arched roofs. Waltham *et al.* (2005) present several graphs that relate the stability of rock cavity roofs (safety factor) to the cave width, roof or bed thickness, various rock mass ratings (RMR) and an imposed load.

The stoping of the cavity results in a breccia pipe (also called collapse chimney, breakdown column or geological organ) that, when rooted in thick evaporites, may reach several hundred metres in height (e.g. Johnson, 1997; Warren, 2006). The upward propagation of the cavity may cease if the cavity roof and the breakdown pile meet due to the increase in volume that the rock undergoes during breakdown (bulking effect). These highly permeable breccia pipes may act as zones of preferential groundwater flow and dissolution. Consequently, the initial clast-supported chaotic breccias (packbreccia) may turn into a matrix-supported mass of corroded blocks embedded in a karstic residue (floatbreccia) and finally into a massive karstic residue with some floaters (Loucks, 1999; Warren, 2006). The volume reduction of the collapse breccia by dissolution and compaction may also contribute to subsidence at the ground surface. Another option is the *en bloc* foundering of rock strata over large voids with almost no internal deformation. These collapse processes lead to the sudden formation of steep-walled *bedrock* or *caprock collapse sinkholes*, typically several tens of metres across (Figures 13.2E and 13.2F). These sinkholes commonly show a very low probability of occurrence (Beck, 2005a; Waltham *et al.*, 2005).

An additional widely used term with no genetic meaning is *buried sinkhole*. These may correspond to any sinkhole with no geomorphic expression overlain by natural or anthropogenic material. These sinkholes may pose subsidence problems due to differential compaction, typically more pronounced at the marginal sectors, or reactivation, especially when human activities involve changes in the natural hydrological regime or the application of loads (De Bruyn and Bell, 2001). The same considerations are applicable to soils underlain by pinnacled or irregular rockhead.

The settlement of the ground during the development of subsidence sinkholes may severely affect the integrity of any human structure including buildings, linear infrastructure, dams or nuclear power stations. Abundant information on sinkhole damage can be found in Waltham *et al.* (2005) and in the proceedings of the biannual sinkhole conferences edited by Barry Beck (e.g. Beck, 2005b). Furthermore, sinkholes that occur in a sudden way may cause the loss of human lives. For example, in the Far West Rand of South Africa, collapse sinkholes induced by dewatering of dolomite aquifers for gold mining have killed a total of 38 people (De Bruyn and Bell, 2001).

13.2.2 Controlling factors and impact of human activity

Two types of processes are involved in the generation of subsidence sinkholes: subsurface dissolution (hydrogeological component) and downward gravitational movement of the overlying material (mechanical component). In carbonate karst areas, where dissolution processes operate at relatively slow rates, the effects attributable to active corrosion alone over a human time scale are uncommon (Beck, 2005a). In contrast, dissolution may be very rapid in evaporite karst areas, specially those with turbulent freshwater flows and those with highly soluble salts. Subsidence processes, particularly collapse and suffosion, can be very rapid in any karst environment and may be related to the presence of relict voids generated in the past.

The main factors that control evaporite and carbonate karstification include: (1) composition of the karst rocks and that of the adjacent lithologies (lithology, mineralogy); (2) the structure (e.g. discontinuity planes) and texture of the karst rocks; (3) the amount of water flowing in contact with the evaporites and its physico-chemical properties (saturation index, pH, temperature); (4) hydrogeological and hydrodynamic conditions (gradient, phreatic or vadose, laminar or turbulent). Information on the role played by these factors on dissolution may be found in Dreybrodt (2004) and Ford and Williams (2007). The internal erosion and deformation processes are primarily controlled by: (1) thickness and mechanical properties of the rocks and deposits overlying the cavities and karstification zones; (2) geometry and size of the subsurface voids, primarily the span of the cavity roofs; (3) position and changes of the water table.

Natural and especially anthropogenic changes in the karst environment can activate or accelerate the processes involved in the generation of sinkholes, triggering or favouring their occurrence or reactivation. According to Waltham et al. (2005), human-induced sinkholes constitute the vast majority of new subsidence depressions. Table 13.1 presents the main natural or artificial changes that may induce the formation of sinkholes, together with their potential effects. This table is conceived as a checklist that may be used to elucidate the factors that may have contributed to the generation of sinkholes.

13.2.3 Identification and investigation of sinkholes and karst features

The application of sinkhole risk mitigation measures requires the recognition of the existing sinkholes (identification) and the delineation of the areas where new sinkholes are likely to occur (prediction). The construction of a karst inventory as complete and representative as possible is commonly the first step in the sinkhole hazard analysis. This inventory should include as much information as possible on the identified sinkholes (location, typology, chronology, morphometry, subsidence rate, relation with causal factors). A good knowledge of the geology and hydrogeology of the area, as well as of the human activities that may influence dissolution and subsidence processes, is crucial to understanding the subsidence phenomena. The identification of sinkholes is usually a difficult task since these geomorphic features are frequently masked by anthropogenic activities (filling, construction) or natural processes (aggradation, erosion). Additionally, shallow cavities and active subsidence structures that may lead to the generation of sinkholes in the near future may not have any topographic expression. For these reasons, it is essential to investigate as many sources of surface and subsurface information as possible.

Surface investigation

Aerial photographs, especially large-scale colour stereoscopic images, are an extremely useful tool for mapping sinkholes. Old aerial photographs may allow identification of sinkholes obliterated by human activity (e.g. Brinkmann et al., 2007; Galve et al., 2009b). The detailed interpretation of aerial photographs taken on different dates may allow us to: (1) constrain the chronology of recently formed sinkholes; (2) obtain minimum estimates of the probability of sinkhole occurrence (e.g. number of sinkholes per km^2 per year); (3) analyse the spatio-temporal distribution patterns of the subsidence phenomena. Airborne and satellite *multispectral and thermal images* may be used to distinguish surface terrain patterns and to extract variations in moisture, vegetation, colour and temperature that may be related to sinkholes and subsidence areas (e.g. Cooper, 1989).

Thorough *field surveys* are essential to locate sinkholes not identifiable on aerial photographs and to check depressions of uncertain origin detected by remote sensing techniques. Some features like anthropogenic fills with subcircular patterns, swampy areas or palustrine vegetation, may help in the detection of shallow subsidence depressions and filled sinkholes (Gutiérrez et al., 2007). Additionally, instability signs detected in the field, such as cracks, scarplets or pipes, may serve as premonitory indicators for anticipating the location of future sinkholes. Active subsidence in built-up areas may be studied through the production of *subsidence damage maps* using a damage ranking system (Cooper, 2008). These maps provide information on the spatial distribution of the subsidence, and may help to infer the main natural and anthropogenic factors that control the dissolution and subsidence processes (Gutiérrez and Cooper, 2002).

Detailed *topographic maps* may depict subsidence depressions not detectable by means of field surveys and aerial photographs (Kasting and Kasting, 2003). In some

TABLE 13.1. *Changes in the karst systems and their potential effects that may accelerate or trigger the development of sinkholes. Some illustrative selected references are indicated.*

Type of change	Effects	(1) Natural processes, (2) human activities
Increased water input to the ground (cover and bedrock)	Increases percolation accelerating suffosion Favours dissolution Increases the weight of sediments May reduce the mechanical strength of sediments	(1) Rainfall, floods (Hyatt and Jacobs, 1996), snow melting, permafrost thawing (2) Irrigation (Gutiérrez *et al.*, 2007), leakages from utilities (pipes, canals, ditches), impoundment of water (Milanovic, 2000), runoff concentration (urbanisation, soakaways) or diversion (White *et al.*, 1986), vegetation removal, drilling operations (Johnson, 1989), unsealed wells, injection of fluids, solution mining (Ege, 1984)
Water table decline	Increases the effective weight of the sediments (loss of buoyant support) Slow phreatic flow replaced by more rapid downward percolation favouring suffosion, especially when the water table is lowered below the rockhead May reduce the mechanical strength by desiccation and crystallisation of salts Suction effect	(1) Climate change, sea level decline, entrenchment of drainage network, tectonic uplift (2) Water abstraction (LaMoreaux and Newton, 1986) or de-watering for mining operations (Li and Zhou, 1999), decline of the water level in lakes (Yechieli *et al.*, 2006)
Impoundment of water	May create very high hydraulic gradients favouring dissolution and internal erosion processes Imposes a load	(1) Natural lakes (2) Reservoirs, lagoons (Milanovic, 2000; Romanov *et al.*, 2003)
Permafrost thawing	Favours dissolution Significant reduction in the strength of the sediments	(1) Climate change (2) Development, deforestation, water storage
Static loads	Favours the failure of cavity roofs and compaction processes	(1) Aggradation processes (2) Engineered structures, dumping, heavy vehicles (Waltham *et al.*, 2005)
Dynamic loads	Favours the failure of cavity roofs and may cause liquefaction-fluidisation processes involving a sharp reduction in the strength of soils	(1) Earthquakes, explosive volcanic eruptions (2) Artificial vibrations (blasting, explosions)
Thinning of the sediments over voids	Reduces the mechanical strength of cavity roofs May concentrate runoff and create a local base level for groundwater flows	(1) Erosion processes (2) Excavations
Underground excavations	Disturb groundwater flows May intercept phreatic conduits May weaken sediments over voids	(1) Biogenic pipes (2) Mining (Li and Zhou, 1999; Lucha *et al.*, 2008), tunnelling (Milanovic, 2000)
Waterproofing of reservoir bottoms	Trapped pressurised air in the vadose zone during rapid rises of the water table	(2) Waterproofing blankets in reservoirs (Milanovic, 2000)

areas, the contour lines and local names on old topographic maps may allow recognition of sinkholes obliterated by artificial fill or development (e.g. Galve *et al.*, 2009b). Several *geodetic techniques*, such as InSAR (Abelson *et al.*, 2003; Closson *et al.*, 2003; Castañeda *et al.*, 2009), GPS, photogrammetry, and high-resolution digital elevation models (DEMs) such as those produced by LIDAR (Light Detecting And Ranging) systems, may be applied to locate sinkholes and estimate subsidence rates with accuracies of the order of millimetres per year.

Oral and documentary information obtained from local residents and written documents (technical reports, newspapers) may substantially improve the sinkhole inventory, providing data on the characteristics and spatial and temporal distribution of undetected and filled sinkholes (Beck, 1991).

Palaeosinkholes and dissolution and subsidence features exposed in natural and artificial outcrops offer valuable information about sinkhole formation (spatial distribution, approximate sizes, subsidence mechanisms). Furthermore, since sinkholes commonly result from the reactivation of pre-existing cavities and subsidence structures, these features may be used as a predictive tool for identifying locations highly susceptible to subsidence (Gutiérrez *et al.*, 2008a). Guerrero *et al.* (2009), working in the salt-bearing evaporite karst of the Ebro Basin, in NE Spain, propose a sinkhole susceptibility zonation for a 24 km long stretch of the high-speed Madrid–Barcelona railway based on the type and distribution of the dissolution and subsidence features exposed in the adjacent cuttings.

Subsurface investigation

Speleological explorations may provide highly valuable information. The examination and mapping of underground cavities provides data on groundwater flow paths, the distribution of accessible voids and the location of the points where active deformation and internal erosion processes (stoping, suffosion and sagging) affecting the cavity ceilings may create new sinkholes in the near future. These unstable areas are revealed by the presence of collapse chimneys and sagging structures in the cavity ceilings, and debris cones in the cavity floors. Detailed maps of caves showing the distribution of breakdown chimneys and cones may be used to anticipate the location of future sinkholes (e.g. Klimchouk and Andrejchuk, 2005).

Geophysical exploration can be used to detect anomalies and changes in the physical properties of the ground that may correspond to cavities, subsidence structures (suffosion zones, breccia pipes, synclinal sags, downthrown blocks), irregular rockhead topography or buried sinkholes. In most cases, the characteristics of the anomalies need to be confirmed by intrusive methods such as probing, drilling or trenching. It is advisable to use a phased sequence of investigation using geophysics on sites prior to drilling and probing. A good option is to apply two or more geophysical methods and compare the results. The most widely used methods include: electrical resistivity, electromagnetic conductivity, ground penetrating radar (GPR), microgravimetry, seismic reflection and cross-hole tomography. Waltham

et al. (2005) review these geophysical methods and Gutiérrez *et al.* (2008a) present a summary of their main advantages and disadvantages.

Probing and drilling provide valuable information on the nature and geotechnical properties of the ground and allow the recognition of voids and sediments disturbed by subsidence processes including suffosion zones and breccia pipes. These may be identified in the core or detected in the borehole by the loss of penetration resistance or drilling fluids. However, these expensive and time-consuming techniques have other limitations: (1) Even deep and closely spaced boring programmes may miss cavities and subsidence structures. Waltham and Fookes (2003) indicate that 2,500 borings per hectare would be necessary for detecting a cavity 2.5 m in diameter with a 90% probability. According to Milanovic (2000), in Keban Dam, Turkey, 36 km of exploratory drilling and 11 km of adits were not enough to detect a cavern of over 600,000 m^3 in the dam site, which severely affected the watertightness of the reservoir (losses of 26 m^3/s). (2) Boreholes may not allow the identification of sagging subsidence structures. (3) The interpretations derived from borehole data may have a high degree of uncertainty due to the complex, sometimes chaotic, underlying geology in karst areas. Boreholes should be properly grouted, otherwise they can become the focus for dissolution and may lead to the occurrence of sinkholes.

Trenching, a commonly used technique in paleoseismological studies, involves the detailed study of the sediments and deformation structures exposed in the walls of trenches, in combination with the application of absolute dating techniques, mainly radiocarbon and luminescence (OSL and TL). This methodology may provide objective practical information on particular sinkholes in mantled karst settings (Gutiérrez *et al.*, 2009): (1) the nature of geophysical anomalies and topographic depressions that have an uncertain origin; (2) the precise limits of sinkholes and the ground affected by subsidence; (3) the subsidence mechanism and magnitude (cumulative displacement); (4) retro-deformation analysis of the deposits by means of the successive restoration of the sedimentary bodies may allow the interpretation of progressive or episodic subsidence; (5) numerical ages of selected stratigraphic units may be used to obtain mean subsidence rates and constrain the timing of the subsidence events. The inferred evolution of particular sinkholes from trenching may be used as an objective basis to forecast their future behaviour.

13.2.4 Temporal and spatial prediction

Once the pre-existing sinkholes and areas affected by subsidence have been identified and mapped, the next step in the hazard analysis is to predict the spatial and temporal distribution of future sinkholes. It would be desirable to know where sinkholes will occur in the future, when they will form, with what frequency, what size they will reach and their likely subsidence mechanism.

Temporal prediction

The temporal prediction of sinkholes has two facets; one is the anticipation of the precise moment or time interval when sinkholes will occur, and the other is the assessment of their frequency or probability of occurrence. At the present time, it is not possible to satisfactorily predict when an individual sinkhole will form at a specific site. Monitoring systems that provide continuous records of potential precursors such as subsurface microdeformations, variations in the water table, and subtle changes in elevation might help one to anticipate individual collapse sinkholes in the future. Another predictive strategy is the use of a good understanding of the temporal patterns of hydrological triggering factors, such as rainstorms, floods or major irrigation and water table decline periods. Correlation with these events may be used to forecast the periods that are susceptible to a higher frequency of sinkhole formation.

The sinkhole frequency, or probability of occurrence, can be regarded as the number of sinkhole events per year per unit area. Chronological information about sinkhole occurrences (either a precise age or an age range) is strictly necessary to be able to estimate temporal frequency values. The calculation of the probability of occurrence must be based on a sinkhole inventory, which should be as complete as possible covering a representative time period (Beck, 1991). The validity of the obtained frequency will depend on the completeness and quality of the available data. In most cases we obtain minimum or optimistic sinkhole frequency values since we are not able to record all the sinkhole events that occurred during the considered timespan. A minimum probability of 44 cover collapse sinkholes per km^2 per year has been calculated by Gutiérrez et al. (2007) in an intensely irrigated terrace of the Ebro River in NE Spain.

Spatial prediction

Several strategies have been applied to construct *sinkhole susceptibility maps*, which represent the relative probability for the formation of new sinkholes: (1) Quantification of the sinkhole clustering by means of nearest neighbour analysis (e.g. Zhou et al., 2003). This analysis may be applied to test whether new sinkholes tend to occur in the vicinity of the pre-existing ones and consequently if the spatial distribution of the latter has any predictive utility (Hyatt et al.,

1999). (2) The establishment by means of expert criteria of threshold values for a given set of contributing variables (e.g. Gao and Alexander, 2008). (3) The intuitive application of a scoring system to a group of conditioning factors (e.g. Buttrick and van Schalkwyck, 1998). The main drawbacks of these methodologies are that they are largely based on relatively subjective judgements and assumptions and that they do not allow quantitative assessment of the probability of occurrence of sinkholes and the predictive capability of the models.

A more objective approach has been proposed by Galve et al. (2009a) using two temporal populations of sinkholes, one for the analyses and the other for evaluating the models. Initially, multiple susceptibility models are generated analysing the statistical relationships between the analysed sinkhole population and different sets of conditioning factors. The predictive capability of the models is evaluated quantitatively and independently comparing the distribution of the susceptibility zones with that of the evaluation sinkhole population. If the evaluation indicates that one of the models is essentially valid, the best susceptibility model can be transformed into a hazard model considering the frequency of sinkholes of the evaluation sample occurring in each susceptibility class and their average size. This hazard map provides a minimum quantitative assessment of the spatio-temporal probability of sinkholes for each point of the territory (Figure 13.3). Both susceptibility and hazard maps may be used to identify the areas where the application of mitigation measures would have better cost/benefit ratios. The main advantages of this approach are that it provides a quantitative assessment of the sinkhole hazard and of the predictive capability of the model. The main drawback is that a large amount of information may be necessary to obtain satisfactory results.

13.2.5 Mitigation

The safest mitigation strategy is the avoidance of the subsidence features and areas most susceptible to sinkholes. Frequently, a setback distance is established around the sinkhole edges (Zhou and Beck, 2008). This preventive measure may be applied prohibiting or limiting development in the most hazardous areas through land use planning and regulations (Paukstys et al., 1999; Richardson, 2003). Preventive planning is commonly most effective when developed by local administrations (Paukstys et al., 1999).

When sinkhole-prone areas are occupied by people, vulnerable buildings or infrastructure, the risk should be mitigated by reducing the activity of the processes (hazard), the vulnerability of the human elements, or both. Since controlling the subsurface dissolution and subsidence

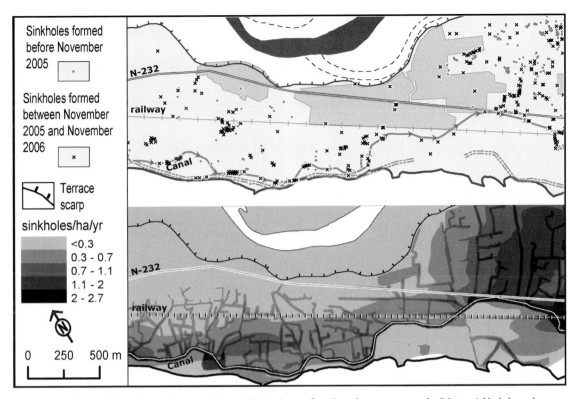

FIGURE 13.3. Temporal populations of cover collapse sinkholes (upper figure) used to construct and validate a sinkhole hazard map (lower figure) in a sector of the Ebro valley (NE Spain). See explanation in the text. (Modified from Galve *et al.*, 2008.)

processes involved in the generation of sinkholes may not be feasible, safe mitigation commonly requires careful planning and the application of subsidence-protected engineering designs. A critical design parameter is the maximum diameter of the sinkholes at the time of formation, as it determines the distance that has to be spanned to prevent the deformation of the engineered structure. Risk assessments and cost–benefit analyses may be used as the basis for estimating the cost effectiveness of different mitigation measures and selecting the most economically and socially advantageous one. Cooper and Calow (1998) present an excellent example of a cost–benefit analysis.

Some corrective measures aimed at diminishing the activity of the processes include: (1) preventing or controlling water withdrawal and the decline of the water table; (2) controlling irrigation; (3) lining of canals and ditches; (4) using flexible pipes with telescopic joints; (5) making the surface impermeable with geomembranes or geotextiles; (6) using efficient drainage systems and diverting surface runoff; (7) filling cavities in the soil or rock by grouting. Pressurised water or air may be used to wash out the fine-grained sediments in the rock voids and surface before grouting. The main disadvantage of grouting is that

it may block most of the flow paths, favouring back-flooding and focused internal erosion and dissolution (Zhou and Beck, 2008). (8) Improving the ground by compaction grouting to increase the strength and bearing capacity of the soils; (9) clogging swallow holes; (10) remediating sinkholes. Zhou and Beck (2008) propose several sinkhole remediation methods. The alternatives for the treatment of the sinkhole throat in shallow sinkholes (<10 m) include: (a) excavating the throat and plugging it with large blocks and concrete or grout and (b) excavating and filling the fractures by dental filling with grout. The throat of deep sinkholes may be treated by: (a) compaction grouting, (b) jet grouting or (c) cap grouting. The next step is to fill the sinkhole depression. Initially, the bottom and walls are lined with a geotextile filter fabric and a drainage structure is constructed if necessary. The sinkhole may be filled with compacted clay or granular material with layers forming a fining upward sequence following the inverted filter concept. The filled sinkhole is commonly capped by a layer of compacted clay or a rubber membrane.

Different types of engineering measures have been applied to protect structures from sinkhole development.

FIGURE 13.4. Wooden framework constructed next to several sinkholes to prevent catastrophic collapse in a track at Cardona salt diapir (NE Spain). Note the plastic membrane and clay blanket installed to prevent infiltration.

FIGURE 13.5. Zafarraya Polje flooded in 1996–7 (Betic Chain, southern Spain). (Courtesy of Professor Joaquín Rodríguez-Vidal.)

These include: (1) Special foundations for buildings including raft, slab, reinforced strip and ring-beam foundations. These are strong foundations that distribute the load of the structures over large areas. Beam extensions to these foundations, especially at the corners of the structures, can offer more protection and prevent a cantilever situation on the edges of structures. Skin friction and end-bearing piles are commonly used to transfer the structural load to the soil cover or solid bedrock, respectively (Cooper and Calow, 1998; Waltham *et al.*, 2005). Another option is jackable foundations that allow levelling of the structure. (2) Linear infrastructures including roads and railways can be reinforced by incorporating tensile geogrids in the sub-base and embankments. This technique prevents catastrophic collapse and acts as a warning mechanism (Cooper and Saunders 2002). Another option is structures acting as ground bridges, such as reinforced concrete slabs or wooden frameworks (Figure 13.4). An added degree of security could be gained by piling the slabs. (4) Sinkhole-resistant bridges can be built incorporating oversized foundation pads in the piers and a sacrificial pier design, so that the structure will withstand the loss of a pier (Cooper and Saunders 2002).

Other non-structural measures aimed at reducing the financial losses and harm to people include: (1) insurance policies to spread the cost generated by sinkholes among the people at risk; (2) the installation of monitoring and warning systems in problematic locations with highly vulnerable structures (inclinometers, extensometers, geodetic measurements, laser or light transmitters and receptors); (3) educational programmes oriented to alert the public and decision makers to the objective likelihood of sinkhole occurrence (Buskirk *et al.*, 1999); (4) the fencing and provision of warning signposts of sinkholes and sinkhole-prone areas.

13.3 Other hazards associated with karst

13.3.1 Flooding hazard

Flooding of closed depressions such as sinkholes or poljes may be caused by several processes including: (1) surface water inflow when this is greater than infiltration and the swallowing capacity of the absorption features (swallow holes or ponors); (2) water table rises above the ground surface. This situation may be induced artificially by the construction of surface or underground reservoirs and the submergence of springs (Milanovic, 2000, 2002); (3) back-flooding in the karst conduit network when the incoming groundwater flow exceeds the discharge capacity of karst channels. Flood hazard analyses of karst depressions taking into account the peculiar behaviour of the karst hydrological systems should be incorporated in land-use planning (Crawford, 1984). Digging out clogged sinkholes, constructing retention basins, diverting the drainage, and installing injection wells may be applied to mitigate the flooding hazard (Zhou, 2007).

Seasonal or occasional flooding caused by any of the processes indicated above constitutes an inherent characteristic of many poljes (López-Chicano, 2002; Milanovic, 2004) (Figure 13.5). Poljes are large closed depressions with flat and alluviated floors that typically have an elongated geometry parallel to the structural grain. Most or a significant part of the drainage in these basins is evacuated through ponors, whose intake capacity may reach more than 100 m^3/s (Milanovic, 2004). The ponors that temporarily function as springs are called estavelles. In the Dinaric karst, the natural hydrological regime of a significant number of poljes, including floods, has been modified through the construction of reservoirs, canals and tunnels (Milanovic, 2002, 2004).

FIGURE 13.6. Rock fall derived from a gypsum escarpment in which the rock mass strength has been significantly reduced by dissolutional enlargement of discontinuity planes (northern margin of the Ebro River valley in Zaragoza city area, Spain).

FIGURE 13.7. Montejaque Dam in the south of Spain. The dam, built for hydroelectric purposes between 1920 and 1923, has never retained water due to rapid infiltration and underground drainage through a previously known cave system. The small channel dissecting the reservoir fill deposit terminates in a swallow hole located at the foot of the dam.

Karst aquifers with a well-developed conduit network may concentrate large volumes of storm-derived water in springs, producing flash floods with typically short lag times and high specific peak discharges (Bonacci *et al.*, 2006). For example, in 1988 a flash flood largely related to spring discharge in Nimes city, France, killed nine people and caused damage with an estimated cost of 600 million euros. A monitoring and alert system that integrates the surface and underground hydrology has been installed in this city to anticipate flash flooding and mitigate the flood risk (Maréchal *et al.*, 2008).

13.3.2 Landslide hazard

Carbonate and gypsum rocks are lithologies that are particularly prone to the development of very to extremely rapid landslides with velocities higher than 5 m/s (planar rock slides, rock falls and rock avalanches). These are landslides that may cause the loss of human lives in addition to material damage; 5 m/s is roughly the running speed of a person (Hungr, 2007). Some of the factors and processes characteristic of karst rocks that may contribute to the formation of landslides include: (1) Stratification planes, especially in carbonates, may be very extensive and extremely planar favouring the development of large planar rock slides (Ford and Williams, 2007). In 1903, the Frank rock-slide avalanche, around 30 hm³ in volume, whose sliding surface was largely controlled by stratification planes, partially buried with limestone blocks the town of Frank in the Rocky Mountains of Alberta (Cruden and Krahn, 1978). In 1963, a 270 million m³ planar slide of karstified limestone fell catastrophically into the Vaiont reservoir (Italy) producing a flood that killed 2,000 people. (2) The widening of discontinuity planes by dissolution

may result in a significant reduction in the mechanical strength of rock slopes. This karstification-related weakening of the rock mass may be notably rapid in gypsum rock (Figure 13.6). In Azagra village, Spain, four rock-fall events from a gypsum escarpment killed a total of 106 people (Gutiérrez *et al.*, 2008c). (3) The rapid infiltration and circulation of water in the karst conduit network may induce high fluid pressures reducing the normal effective stresses and shear strength of potential failure planes. (4) The loss of basal support related to dissolution-induced subsidence and undermining due to spring sapping at the foot of slopes may contribute to slope instability. (5) Deep-seated dissolution of evaporites may reduce basal support (Tsui and Cruden, 1984) and oversteepen the slopes.

13.3.3 Reservoirs and dams: expect the unexpected

According to Milanovic (2000), the most frequent difficulties encountered in the construction of reservoirs and dams in carbonate karst areas include: (1) Water leakage through ponors, sinkholes and karst conduits. The unnatural high hydraulic gradients induced by the impoundment of water in reservoirs may cause the flushing out of sediments that block karst conduits and the rapid dissolutional enlargement of discontinuity planes reaching break-through dimensions (turbulent flow) and increasing significantly the hydraulic conductivity of the karst system at an engineering time scale (Romanov *et al.*, 2003). Water losses may reach rates over 20 m³/s, thereby compromising the operation of the hydraulic structure (Figure 13.7). (2) Development of human-induced sinkholes. In evaporite karst terrains, rapid dissolution in the abutments or foundations of a dam may cause its collapse and the development of catastrophic floods (e.g. Johnson, 2008). The flood caused by the failure of St Francis Dam, California, in 1928, killed more than 400 people. The collapse of this dam was attributed to the prompt dissolution of gypsum veins cutting the conglomerates that formed one of the abutments (James, 1992).

Two complementary strategies can be applied to reduce water losses from reservoirs: preventing water infiltration and sealing the underground karst conduits (Milanovic, 2000). The options to make the surface impermeable include: isolating ponors with annular concrete dams or dykes; closing estavelles with concrete plugs equipped with one-way valves; lining overburden with PVC foils equipped with aeration systems; blanketing karstified limestone by shotcrete; and surface compaction. The alternative methods to plug karst conduits include: grouting with cement mortar, polyurethane foam or asphalt; filling large voids with granular material that can be subsequently grouted; construction of plugs in karst channels with concrete or reinforced concrete; and excavating a deep trench in the dam foundation and filling it with impervious material (cut-off or diaphragm screens).

13.3.4 Tunnels and mines

The excavation of tunnels and mine galleries in karst rocks may be adversely affected by various problems and may induce hazardous processes such as the formation of sinkholes (Milanovic, 2000; Marinos, 2001). The main difficulties related to these underground excavations arise from the presence of cavities and cavernous rock, especially when the operations are developed below the groundwater level. Tunnels and mines may be flooded and violent inrushes of water under pressure from phreatic conduits (at rates as high as several hundreds of litres per second) may endanger human lives and the excavation equipment (e.g. Li and Zhou, 1999). Large caverns encountered during excavation works may need to be filled, grouted, bridged or bypassed, or may even force the relocation of tunnels. Furthermore, poor karst rock mass quality related to the presence of faults, cavities and karstified discontinuity planes may favour the development of collapse processes. Several measures may be applied to avoid or remediate the inflow of water in the excavations: (1) cut tunnels and galleries with a gentle inclination to allow them to drain gravitationally; (2) water pumping from the excavation; (3) grouting the rock mass around the tunnel or excavation. Modern practice is to drill a 360° array of grouting holes forward and then excavate and seal a section of tunnel

FIGURE 13.8. Collapse sinkhole around 110 m across formed on 16 July 2008 at the site of a brine well in Eddy County, New Mexico. The sinkhole results from the upward stoping of a large cavity generated in the Permian Salado Formation by injecting freshwater and recovering a brine for use as oil drilling fluid. The top of the salt formation is 120 m deep. A seismograph located 13 km from the well recorded ground motion attributable to premonitory subsurface collapse before the abrupt occurrence of the depression. (Image and personal communication from Lewis Land.)

inside this grout curtain (Ford and Williams, 2007); (4) dewatering the mine or tunnel zone creating a large depression cone by pumping. Dewatering may trigger the development of a huge number of catastrophic sinkholes at the surface (e.g. Li and Zhou, 1999; De Bruyn and Bell, 2001).

The problems may be particularly severe when freshwater flows into salt mines coming from an adjacent or overlying aquifer (Kappel, 1999; Andrejchouk, 2002) or from a surface water body (Autin, 2002; Lucha et al., 2008). The inflow of water may be favoured by breakdown processes induced by the mined voids. The aggressive water may cause massive dissolution of salt and uncontrollable ground subsidence, leading to the abandonment of the mine. In the Cardona salt diapir, NE Spain, the interception of a phreatic conduit by a salt mine gallery, whose final goal was the disposal of hazardous wastes, caused the inflow of freshwater from an nearby river producing massive dissolution, widespread subsidence, and the abandonment of the mine (Lucha et al., 2008).

Solution mining, applied to highly soluble salts, involves the injection of freshwater at depth and the recovery of a brine, with the consequent creation of voids that may reach more than 100 m in width and several hundred metres in height (Warren, 2006). In a large number of cases, these cavities have propagated upward through several hundred metres of overlying strata in a few decades, eventually leading to the sudden occurrence of large sinkholes (Ege, 1984; Johnson, 1997) (Figure 13.8). In the city of Tuzla, Bosnia and Herzegovina, solution mining has caused subsidence at rates as high as 40 cm/year. Here, 2,000 buildings have been demolished and 1,500 people relocated (Mancini *et al.*, 2009). The experience suggests that the disposal of hazardous wastes in salt is very probably not a safe practice.

13.4 Conclusions

Karst environments, which occur over 20% of the Earth's ice-free continental area, pose specific hazards largely determined by their distinctive geomorphological, hydrological and geotechnical characteristics (i.e. soluble rocks, enclosed depressions, rapid underground flow through conduit networks). The main hazards and problems associated with karst include: (1) sinkholes; (2) differential settlement due to compaction of soil over irregular rockhead or in buried sinkholes; (3) flooding of depressions due to runoff concentration, water table rise or back-flooding in the conduit network; (4) flash floods from major springs; (5) flooding of tunnels and mines, including violent inrushes of water from phreatic conduits; (6) water losses in reservoirs; and (7) landslides.

Sinkhole activity is commonly the most relevant hazard in karst areas. The main subsidence mechanisms involved in the development of sinkholes are: (1) collapse of soil and rock cavity roofs; (2) suffosion, which is the downward migration of unconsolidated deposits through voids and the consequent gradual settlement of the ground; (3) gradual sagging due to interstratal karstification or differential lowering of the rockhead. Collapse sinkholes that form in a sudden way are particularly dangerous since they may cause the loss of human lives. Sinkhole hazard is commonly higher in evaporite areas than in carbonate terrains due to the lower solubility and mechanical strength of evaporite rocks. Generally, a great number of the damaging sinkholes are induced by various types of anthropogenic activities. Three main tasks should be performed for an effective sinkhole risk management: (1) identification and characterisation of the existing sinkholes and karst features. A karst inventory as complete as possible should be constructed using as many sources of surface and subsurface data as possible; (2) prediction of the spatio-temporal distribution of future sinkholes, their subsidence mechanism and size. The predictive capability of sinkhole susceptibility maps, regardless of the approach used for their elaboration, should be assessed quantitatively and independently. Susceptibility zonations with an acceptable reliability can be transformed into hazard maps using independent temporal sinkhole populations; (3) selection of mitigation measures of either a preventive or corrective nature. Generally, it is not possible to control the subsurface dissolution and subsidence processes involved in the development of sinkholes (hazard reduction). Consequently, safe development requires the application of subsidence protection engineering designs (vulnerability reduction). Given the frequently poor understanding of karst systems and the inadequate perception of the hazards they pose, it is also important to develop educational programmes aimed at the public, technical community and decision makers.

Acknowledgements

The original version of this chapter was improved thanks to the thorough reviews carried out by Professor Andrew Goudie, Dr Peter Milanovic and Dr Yoseph Yechieli.

Jorge Pedro Galve is thanked for drafting Figure 13.1. This work has been co-financed by the Spanish Education and Science Ministry and the FEDER (project CGL2007–60766), as well as by the Aragón Government (project PM008/2007).

References

Abelson, M., Baer, G., Shtivelman, V. *et al.* (2003). Collapse-sinkholes and radar interferometry reveal neotectonics concealed within the Dead Sea basin. *Geophysical Research Letters*, **30**(10), 1545.

Andrejchouk, V. (2002). Collapse above the world's largest potash mine (Ural, Russia). *International Journal of Speleology*, **31**, 137–158.

Autin, W. J. (2002). Landscape evolution of the Five Islands of south Louisiana: scientific policy and salt dome utilization and management. *Geomorphology*, **47**, 227–244.

Beck, B. F. (1991). On calculating the risk of sinkhole collapse. In E. H. Kastning and K. M. Kastning (eds.), *Proceedings of the Appalachian Karst Symposium. National Speleological Society*. Radford, Virginia, pp. 231–236.

Beck, B. F. (2005a). Soil piping and sinkhole failures. In W. B. White. (ed.), *Encyclopedia of Caves*. New York: Elsevier, pp. 523–528.

Beck, B. F. (ed.) (2005b). *Sinkholes and the Engineering and Environmental Impacts of Karst*. ASCE Geotechnical Special Publication 144, 677 p.

Bonacci, O., Ljubenkov, I. and Roje-Bonacci, T. (2006). Karst flash floods: an example from the Dinaric karst (Croatia). *Natural Hazards and Earth System Science*, **6**, 195–203.

Brinkmann, R., Wilson, K., Elko, N. *et al.* (2007). Sinkhole distribution based on pre-development mapping in urbanized Pinellas County, Florida, USA. In M. Parise and J. Gunn (eds.), *Natural and Anthropogenic Hazards in Karst Areas: Recognition, Analysis and Mitigation*. Geological Society of London Special Publication 279, pp. 5–11.

Buskirk, E. D., Pavelk, M. D. and Strasz, R. (1999). Education about and management of sinkholes in karst areas: initial efforts in Lebanon. In B. F. Beck. (ed.), *Proceedings of the Seventh Multidisciplinary Conference on Sinkholes and the*

Engineering and Environmental Impacts of Karst. Rotterdam: A. A. Balkema, pp. 263–266.

Buttrick, D. and van Schalkwyk, A. (1998). Hazard and risk assessment for sinkhole formation on dolomite land in South Africa. *Environmental Geology*, **36**, 170–178.

Closson, D., Abu Karaki, N., Hussein, M. J., Al-Fugha, H. and Ozer, A. (2003). Space-borne radar interferometric mapping of precursory deformations of a dyke collapse, Dead Sea area, Jordan. *International Journal of Remote Sensing*, **24**, 843–849.

Cooper, A. H. (1989). Airborne multispectral scanning of subsidence caused by Permian gypsum dissolution at Ripon, North Yorkshire. *Quarterly Journal of Engineering Geology*, **22**, 219–229.

Cooper, A. H. (2008). The classification, recording, databasing and use of information about building damage caused by subsidence and landslides. *Quarterly Journal of Engineering Geology and Hydrogeology*, **41**, 409–424.

Cooper, A. H. and Calow, R. C. (1998). *Avoiding Gypsum Geohazards: Guidance for Planning and Construction*. British Geological Survey Technical Report WC/98/5, 57 pp. http://www.bgs.ac.uk/dfid-kargeoscience/database/reports/colour/WC98005_COL.pdf

Cooper, A. H. and Saunders, J. M. (2002). Road and bridge construction across gypsum karst in England. *Engineering Geology*, **65**, 217–223.

Crawford, N. C. (1984). Sinkhole flooding associated with urban development upon karst terrain: Bowling Green, Kentucky. In B. F. Beck. (ed.), *Proceedings of the First Multidisciplinary Conference on Sinkholes*. Rotterdam: A. A. Balkema, pp. 283–292.

Cruden, D. M. and Krahn, J. (1978). Frank rockslide, Alberta, Canada. In B. Voight. (ed.), *Rockslides and Avalanches*, vol. 1. Amsterdam: Elsevier, pp. 97–112.

De Bruyn, I. A. and Bell, F. G. (2001). The occurrence of sinkholes and subsidence depressions in the Far West Rand and Gauteng Province, South Africa, and their engineering implications. *Environmental Engineering Geoscience*, **7**, 281–295.

Dreybrodt, W. (2004). Dissolution: evaporite and carbonate rocks. In J. Gunn. (ed.), *Encyclopedia of Caves and Karst Science*. New York: Fitzroy Dearborn, pp. 295–300.

Ege, J. R. (1984). Mechanisms of surface subsidence resulting from solution extraction of salt. *Geological Society of America. Reviews in Engineering Geology*, **6**, 203–221.

Ford, D. and Williams, P. (2007). *Karst Hydrogeology and Geomorphology*. Chichester: Wiley.

Frumkin, A. and Ford, D. C. (1995). Rapid entrenchment of stream profiles in the salt caves of Mount Sedom, Israel. *Earth Surface Processes and Landforms*, **20**, 139–152.

Galve, J., Gutiérrez, F., Lucha, P. *et al.* (2009a). Probabilistic sinkhole modelling for hazard assessment. *Earth Surface Processes and Landforms*, **34**, 437–452.

Galve, J. P., Gutiérrez, F., Lucha, P. *et al.* (2009b). Sinkholes in the salt-bearing evaporite karst of the Ebro River valley

upstream of Zaragoza city (NE Spain): geomorphological mapping and analysis as a basis for risk management. *Geomorphology*, **108**, 145–158.

Gao, Y. and Alexander, E. C. (2008). Sinkhole hazard assessment in Minnesota using a decision tree model. *Environmental Geology*, **54**, 945–956.

Guerrero, J., Gutiérrez, F., Bonachea, J. and Lucha, P. (2009). A sinkhole susceptibility zonation based on paleokarst analysis along a stretch of the Madrid-Barcelona high-speed railway built over gypsum- and salt-bearing evaporites (NE Spain). *Engineering Geology*, **102**, 62–73.

Gutiérrez, F. and Cooper, A. H. (2002). Evaporite dissolution subsidence in the historical city of Calatayud, Spain: damage appraisal and prevention. *Natural Hazards*, **25**, 259–288.

Gutiérrez, F., Galve, J. P., Guerrero, J. *et al.* (2007). The origin, typology, spatial distribution, and detrimental effects of the sinkholes developed in the alluvial evaporite karst of the Ebro River valley downstream Zaragoza city (NE Spain). *Earth Surface Processes and Landforms*, **32**, 912–928. doi 10.1007/s00254-008-1590-8.

Gutiérrez, F., Cooper, A. H. and Johnson, K. S. (2008a). Identification, prediction and mitigation of sinkhole hazards in evaporite karst areas. *Environmental Geology*, **53**, 1007–1022.

Gutiérrez, F., Guerrero, J. and Lucha, P. (2008b). A genetic classification of sinkholes illustrated from evaporite paleokarst exposures in Spain. *Environmental Geology*, **53**, 993–1006.

Gutiérrez, F., Calaforra, J. M., Cardona, F. *et al.* (2008c). Geological and environmental implications of evaporite karst in Spain. *Environmental Geology*, **53**, 951–965.

Gutiérrez, F., Galve, J. P., Lucha, P. *et al.* (2009). Investigation of a large collapse sinkhole affecting a multi-storey building by means of geophysics and the trenching technique (Zaragoza city, NE Spain). *Environmental Geology*, in press.

Hungr, O. (2007). Dynamics of rapid landslides. In K. Sassa, H. Fukuoka, F. Wang and G. Wang (eds.), *Progress in Landslide Science*. Berlin: Springer, pp. 47–57.

Hyatt, J. and Jacobs, P. (1996). Distribution and morphology of sinkholes triggered by flooding following Tropical Storm Alberto at Albany, Georgia, USA. *Geomorphology*, **17**, 305–316.

Hyatt, J., Wilkes, H. and Jacobs, P. (1999). Spatial relationship between new and old sinkholes in covered karst, Albany, Georgia, USA. In B. F. Beck, A. J. Pettit and J. G. Herring (eds.), *Proceedings of the Seventh Multidisciplinary Conference on Sinkholes and the Engineering and Environmental Impacts of Karst*. Rotterdam: A. A. Balkema, pp. 37–44.

James, A. N. (1992). *Soluble Materials in Civil Engineering*. New York: Ellis Horwood.

Johnson, K. S. (1989). Development of the Wink Sink in Texas, U.S.A., due to salt dissolution and collapse. *Environmental Geology and Water Science*, **14**, 81–92.

Johnson, K. S. (1997). Evaporite karst in the United States. *Carbonates and Evaporites*, **12**, 2–14.

Johnson, K. S. (2008). Gypsum-karst problems in constructing dams in the USA. *Environmental Geology*, **53**, 945–950.

Kappel, W. M. (1999). The Retsof salt mine collapse. In D. Galloway, D. R. Jones and S. E. Ingebritsen (eds.), *Land Subsidence in the United States*. U.S. Geological Survey Circular 1182, pp. 111–120.

Kasting, K. M. and Kasting, E. H. (2003). Site characterization of sinkholes based on resolution of mapping. In B. F. Beck (ed.), *Sinkholes and the Engineering and Environmental Impacts of Karst*. Reston:ASCE, pp. 72–81.

Klimchouk, A. B. and Andrejchuk, V. (2005). Karst breakdown mechanisms from observations in the gypsum caves of the Western Ukraine: implications for subsidence hazard assessment. *Environmental Geology*, **48**, 336–359.

LaMoreaux, P. E. and Newton, J. G. (1986). Catastrophic subsidence: an environmental hazard, Shelby County, Alabama. *Environmental Geology and Water Sciences*, **8**, 25–40.

Li, G. and Zhou, W. (1999). Sinkhole in karst mining areas in China and some methods of prevention. *Engineering Geology*, **52**, 45–50.

López-Chicano, M., Calvache, M. L., Martín-Rosales, W. and Gisbert, J. (2002). Conditioning factors in flooding of karstic poljes, the case of the Zafarraya polje (South Spain). *Catena*, **49**, 331–352.

Loucks, R. G. (1999). Paleocave carbonate reservoirs: origins, burial-depth modifications, spatial complexity and reservoir implications. *AAPG Bulletin*, **83**, 1795–1834.

Lucha, P., Cardona, F., Gutiérrez, F. and Guerrero, J. (2008). Natural and human-induced dissolution and subsidence processes in the salt outcrop of the Cardona Diapir (NE Spain). *Environmental Geology*, **53**, 1023–1035.

Mancini, F., Stecchi, F., Zanni, M. and Gabbianelli, G. (2009). Monitoring ground subsidence induced by salt mining in the city of Tuzla (Bosnia and Herzegovina). *Environmental Geology*, **58**, 381–389.

Maréchal, J. C., Ladouche, B. and Dörfliger, N. (2008). Karst flash floods in a Mediterranean karst, the example of Fontaine de Nimes. *Engineering Geology*, **99**, 138–146.

Marinos, P. G. (2001). Tunnelling and mining in karstic terrain: an engineering challange. In B. F. Beck and J. G. Herring (eds.), *Proceeding of the Eighth Multidisciplinary Conference on Sinkholes and the Engineering and Environmental Impacts of Karst*. Rotterdam: Balkema, pp. 3–16.

Milanovic, P. T. (2000). *Geological Engineering in Karst*. Belgrade: Zebra.

Milanovic, P. (2002). The environmental impacts of human activities and engineering constructions in karst regions. *Episodes*, **25**, 13–21.

Milanovic, P. (2004). Dinaride polje. In J. Gunn (ed.), *Encyclopedia of Caves and Karst Science*. New York: Fitzroy Dearborn, pp. 291–293.

Paukstys, B., Cooper, A. H. and Arustiene, J. (1999). Planning for gypsum geohazard in Lithuania and England. *Engineering Geology*, **52**, 93–103.

Richardson, J. J. (2003). Local land use regulation of karst in the United States. In B. F. Beck (ed.), *Sinkholes and the Engineering and Environmental Impacts of Karst*. ASCE Special Publication 112, pp. 492–501.

Romanov, D., Gabrovsek, F. and Dreybrodt, W. (2003). Dam sites in soluble rocks: a model of increasing leakage by dissolutional widening of fractures beneath a dam. *Engineering Geology*, **70**, 17–35.

Tharp, T. M. (1995). Mechanics of upward propagation of cover-collapse sinkholes. *Engineering Geology*, **52**, 23–33.

Tsui, P. C. and Cruden, D. M. (1984). Deformation associated with gypsum karst in the Salt River Escarpment, northeastern Alberta. *Canadian Journal of Earth Sciences*, **21**, 949–959.

Waltham, A. C. and Fookes, P. G. (2003). Engineering classification of karst ground conditions. *Quarterly Journal of Engineering Geology and Hydrogeology*, **36**, 101–118.

Waltham, T., Bell, F. and Culshaw, M. (2005). *Sinkholes and Subsidence*. Chichester: Springer-Praxis.

Warren, J. K. (2006). *Evaporites: Sediments, Resources and Hydrocarbons*. Berlin: Springer.

White, E. L., Aron, G. and White, W. B. (1986). The influence of urbanization on sinkhole development in central Pennsylvania. *Environmental Geology and Water Sciences*, **8**, 91–97.

Williams, P. (2004). Dolines. In *Encyclopedia of Caves and Karst Science*, ed. J. Gunn. New York: Fitzroy Dearborn, pp. 304–310.

Yechieli, Y., Abelson, M., Bein, A., Crouvi, O. and Shtivelman, V. (2006). Sinkholes "swarms" along the Dead Sea coast: reflection of disturbance of lake and adjacent groundwater systems. *Geological Society of America Bulletin*, **118**, 1075–1087.

Zhou, W. (2007). Drainage and flooding in karst terranes. *Environmental Geology*, **51**, 963–973.

Zhou, W. and Beck, B. F. (2008). Management and mitigation of sinkholes on karst lands: an overview of practical applications. *Environmental Geology*, **55**, 837–851.

Zhou, W., Beck, B. F. and Adams, A. (2003). Application of matrix analysis in delineating sinkhole risk areas along highway (I-70 near Frederick Maryland). *Environmental Geology*, **44**, 834–842.

14 Soil erosion

Andrew S. Goudie and John Boardman

14.1 Introduction: the nature of the problem

Erosion is a natural phenomenon, but one that has been accelerated by human activities. The scale of accelerated soil erosion that has been achieved by humans has been summarized by Myers (1988, p. 6):

Since the development of agriculture some 12,000 years ago, soil erosion is said by some to have ruined 4.3 million km² of agricultural lands, or an area equivalent to rather more than one-third of today's crop-lands … the amount of agricultural land now being lost through soil erosion, in conjunction with other forms of degradation, can already be put at a minimum of 200,000 km² per year.

Accelerated soil erosion is indeed a serious aspect of environmental change and there has been a long history of research (see e.g. Marsh, 1864; Sauer, 1938; Bennett, 1939; Jacks and Whyte, 1939; Morgan, 2005). Although many techniques have been developed to mitigate the problem it appears to be intractable. As Carter (1977, p. 409) reported for the USA:

Although nearly $15 billion has been spent on soil conservation since the mid-1930s, the erosion of croplands by wind and water … remains one of the biggest, most pervasive environmental problems the nation faces. The problem's surprising persistence apparently can be attributed at least in part to the fact that, in the calculation of many farmers, the hope of maximizing short-term crop yields and profits has taken precedence over the longer term advantages of conserving the soil. For even where the loss of topsoil has begun to reduce the land's natural fertility and productivity, the effect is often masked by the positive response to heavy application of fertilizer and pesticides, which keep crop yields relatively high.

Pimentel (1976) estimated that in the USA soil erosion on agricultural land operates at an average rate of about 30 tonnes per hectare per year, which is approximately eight times quicker than topsoil is formed. He calculated that water runoff delivers around 4 billion tonnes of soil each year to the rivers of the 48 contiguous states, and that three-quarters of this comes from agricultural land. More recently, Pimentel et al. (1995) have argued that about 90 per cent of US cropland is losing soil above the sustainable rate, that about 54 per cent of US pasture land is overgrazed and subject to high rates of erosion, and that erosion costs about $44 billion each year. However, as Trimble and Crosson (2000) and Boardman (1998) point out, determination of general rates of soil erosion is fraught with uncertainties.

14.2 Forms of erosion

The principal agents of erosion are water, wind and gravity giving rise to mass movements. Also, we are now more aware of the importance of cultivation as an erosional agent where the repeated action of the plough moves soil down slope (tillage erosion). In some environments all of these major forms of erosion will interact.

Water is the most important agent. Some erosional forms are well known such as gullies and rills. Ephemeral gullies are now recognized as important on arable land (Poesen et al., 2006). Badlands or dense permanent gully networks have been described from many parts of the world, e.g. the Mediterranean, USA, Canada, Iceland, South Africa. Piping is important on some collapsible or dispersive soils and may be a major factor in the development of gullies (Faulkner, 2006). The significance of sheetflow or interrill flow has probably been exaggerated: Govers and Poesen (1988) show that it is responsible for about 20 per cent of total erosion by water though it plays an important role in the removal of agricultural chemicals from fields and therefore in river pollution (e.g. Harrod, 1994; Evans, 2006). Definitions and discussion of forms of erosion are given in

Geomorphological Hazards and Disaster Prevention, eds. Irasema Alcántara-Ayala and Andrew S. Goudie. Published by Cambridge University Press. © Cambridge University Press 2010.

ISSS (1996). In steeply sloping areas, mass movements may be an important form of erosion (e.g. as a result of Cyclone Bola, March 1988, on formerly forested rangeland in New Zealand).

14.3   Rates of erosion: natural and anthropogenic

There are huge difficulties in estimating erosion rates in pre-human times, but in a recent analysis McLennan (1993) has estimated that the pre-human suspended sediment discharge from the continents was about 12.6×10^{15} grams per year, which is about 60 per cent of the present figure.

In Colorado, USA, Carrara and Carroll (1979), using dendrogeomorphological techniques, found that rates over the last 100 years have been about 1.8 mm per year, whereas in the previous 300 years rates were between 0.2 and 0.5 mm per year, indicating an acceleration of about sixfold. This has been attributed to the introduction of large numbers of cattle to the area.

Another way of obtaining long-term rates of soil erosion is to look at rates of sedimentation on continental shelves. This method was employed by Milliman *et al.* (1987) to evaluate sediment removal down the Yellow River in China during the Holocene. They found that, because of accelerated erosion, rates of sediment accumulation on the shelf over the last 2,300 years have been ten times higher than those for the rest of the Holocene.

Another example of using long-term sedimentation rates to infer long-term rates of erosion is provided by Hughes *et al.*'s (1991) study of the Kuk Swamp in Papua New Guinea. They identify low rates of erosion until 9000 BP, when, with the onset of the first phase of forest clearance, erosion rates increased from 0.15 cm per thousand years to about 1.2 cm per thousand years. Rates remained relatively stable until the last few decades when, following European contact, the extension of anthropogenic grasslands, subsistence gardens and coffee plantations has produced a rate that is very markedly higher: 34 cm per thousand years.

A further good long-term study of the response of rates of erosion to land use changes is that undertaken on the North Island of New Zealand by Page and Trustrum (1997). During the last 2,000 years of human settlement their catchment has undergone a change from indigenous forest to fern/scrub following Polynesian settlement (*c.* 560 BP) and then a change to pasture following European settlement (AD 1878). Sedimentation rates under European pastoral land use are between five and six times the rates that occurred under fern/scrub and between eight and seventeen times the rate under indigenous forest. In a broadly comparable study in another part of New Zealand, Sheffield *et al.* (1995) looked

at rates of infilling of an estuary. In pre-Polynesian times rates of sedimentation were 0.1 mm per year, during Polynesian times the rate climbed to 0.3 mm per year, while since European land clearance in the 1880s the rate has shot up to 11 mm per year.

Various attempts have also been made to establish rates of accelerated erosion on the plainlands of Russia (Sidorchuk and Golosov, 2003). It has been calculated that during the period 1696–1796 a total of $19.5 \times 10^{9}\,\mathrm{m}^{3}$ of soil was mobilized by sheet and rill erosion; for 1796–1887 it was $36.7 \times 10^{9}\,\mathrm{m}^{3}$; and for 1887–1980 it was $42.5 \times 10^{9}\,\mathrm{m}^{3}$. This increasing trend was due to an increase in the area under cultivation and the assimilation of land more prone to erosion.

The use of rates of sedimentation as an analogue for erosion is fraught with difficulty. Even at small spatial scales it is clear that soil losses from agricultural fields are texture dependent (Evans, 1990). With an increase in catchment size a greater proportion of sediment is stored within the catchment: storage will depend on slope, natural features such as flood plains, and cultural features such as hedges and walls. Amounts stored will vary through time and release from storage likewise (Trimble, 1983). Attempts to quantify the relationship between erosion and sediment yield to a given point in the catchment, in terms of a sediment delivery ratio (SDR) (Walling, 1983), are in practice unreliable.

14.4   Assessment of current erosion

Methods of assessing current erosion both quantitative and qualitative are reviewed in Boardman (2007) and are summarized here:

1. *Experimental plots.* Small experimental plots (e.g. 22 × 2 m) have been widely used both to answer research questions regarding process and to estimate rates of erosion. The latter procedure is questionable on grounds of their representativeness of wider areas of the landscape, and therefore extrapolation of plot results is not to be recommended (Boardman, 1998).
2. *Sediment yield of rivers.* While of great interest in itself (e.g. Milliman and Meade, 1983), such data cannot be equated with erosion on hillslopes or fields – problems of storage and SDR have already been referred to. Many erosion studies have used sediment trapped in reservoirs in order to reconstruct erosional histories.
3. *Field measurement of erosional forms.* Measurements of cross-sectional area and length of rills and gullies can be combined to give volumetric estimates of soil lost (m^{3}/ha) and converted to t/ha by means of sampling for bulk density. In badland environments erosion pins are

used to estimate rates of erosion and deposition, e.g. Keay-Bright and Boardman (2009). Simple methods of erosion assessment, such as plant/tree root exposure, are reviewed by Stocking and Murnaghan (2001).

4. *Remote sensing*. This approach has been widely used in assessments of erosion. Scale, data availability and ground truthing issues have always to be addressed. Outstanding surveys date from as early as Talbot's (1947) to the recent Landsat-based survey of Iceland

accumu-
rates of
used is
are also
diments,
2007).
cuments.
h use of
vium and
ded soil
different
ncluding
erosional
Slovakia

erlooked
issues as
own use
sment of
d the oft-
s of com-
a regional
ll (2001)

om com-
ed meas-

urement and remote sensing. The uncritical use of models resulting in unrealistic predictions can be avoided by comparison of results with those from field-based surveys and observations (Boardman, 1998; Evans and Brazier, 2005).

14.5 Consequences of erosion

Accelerated erosion has a number of consequences, some of which are on site and some of which are off site. Soil erosion has serious implications for soil productivity albeit over the medium and long term. Fertile soils may be covered down slope by less fertile eroded material. A reduction in soil thickness reduces available water capacity and the depth through which root development can occur. The water-holding properties of the soil may be lessened as a result of the preferential removal of organic material and fine sediment. Hardpans, duricrusts or bedrock may become exposed at the surface, and provide a barrier to root penetration. Furthermore, splash erosion may cause soil compaction and crusting, both of which are unfavourable to germination and seedling establishment. Erosion also removes nutrients preferentially from the soil. Extreme erosion may lead to wholesale removal of both seeds and fertilizer and necessitate replanting.

Off site, erosion causes sedimentation in reservoirs, shortening their lives and reducing their capacity (Verstraeten *et al.*, 2006). It may also cause siltation that reduces the vitality of coral reefs. In Western Europe, flooding of property by runoff from agricultural fields is of widespread concern (Boardman *et al.*, 2006). Soil erosion by creating high turbidity levels in streams can greatly reduce the quality of water supplies. Costs of providing clean supplies of drinking water are considerable (Forster *et al.*, 1987; Evans, 1996). Eroded material may also contain nitrates and phosphates that can cause accelerated eutrophication of water bodies. Stocking (1984) reviews these problems. Recent European legislation, the Water Framework Directive, requires states to ensure that watercourses reach 'good ecological status' in a series of timesteps. Since much sediment and chemical pollution is from agricultural land, this will require soil conservation and runoff prevention measures by national governments.

Erosion also produces gully systems that can encroach on agricultural land and on engineering structures. For example, *donga* is a term used in southern Africa to describe a gully or badland area caused by severe erosion. They are widespread in colluvium and in deeply weathered bedrock in areas where the mean annual precipitation lies between *c.* 600 and 800 mm. Where the materials in which they are developed have high ESP (exchangeable sodium percentage) contents, they may have highly fluted 'organ pipe' sides (Watson *et al.*, 1984). Repeated oscillations have taken place in colluvium deposition and palaeosol formation on the one hand, and incision on the other. Causes of incision may include climatic change, and land cover changes brought about by human activities, the latter including the spread of pastoralism and deforestation for iron smelting. Piping is probably an important process in their development (Rienks *et al.*, 2000).

Catastrophic misuse of natural resources including the soil are detailed by Diamond (2005) and implicated in the collapse of various societies. On the other hand, he shows that wise use of resources leads to stable and successful civilizations.

A debate concerning the relative importance of on-site and off-site damage has developed. A tendency to over-estimate erosion rates based on model output suggested that the lifespan of soils was rather limited (e.g. less than 100 years: Morgan (1987)). More recently the emphasis has shifted toward the social, political and economic costs of off-site impacts of erosion (Boardman and Poesen, 2006).

14.6 Causation: soil erosion associated with deforestation and agriculture

Forest removal is a land cover change that impacts upon rates of erosion, and was identified by Marsh (1864) as being perhaps the most potent way in which humans transform the landscape. Forests protect the underlying soil from the direct effects of rainfall, generating what is generally an environment in which erosion rates tend to be low. The canopy shortens the fall of raindrops, decreasing their velocity and thus reducing kinetic energy and splash erosion. Possibly more important than the canopy in reducing erosion rates in forests is the presence of humus in forest soils (Trimble, 1988). This both absorbs the impact of raindrops and has an extremely high permeability. Thus forest soils have high infiltration capacities and so may generate less erosive runoff. Another reason that forest soils have an ability to transmit large quantities of water through their fabrics is that they have many macro-pores produced by roots and their rich soil fauna. Forest soils are also well aggregated, making them resistant to both wetting and water drop impact. This superior degree of aggradation is a result of the presence of considerable organic material, which is an important cementing agent in the formation of large water-stable aggregates. Deep-rooted trees help to stabilize steep slopes by increasing the total shear strength of the soils.

In some cases the erosion produced by deforestation will be in the form of widespread surface stripping. In other cases the erosion will occur as more spectacular forms of mass movement, such as mudflows, landslides and debris avalanches. Substantial effects may be created by clear-cutting and by the construction of logging roads. Different methods of cutting and removal have been shown to have different impacts on erosion. It is indeed probable that a large proportion of the erosion associated with forestry operations is caused by road construction, and care needs to be exercised to minimize these effects (Anderson and Spencer, 1991). The digging of drainage ditches in upland pastures and peat moors to permit tree-planting in central Wales has also been found to cause accelerated erosion (Clarke and McCulloch, 1979).

Water-induced soil erosion is widespread in the UK, in spite of the relatively low erosivity of the rainfall. Other factors are important and many of these are common to other intensively farmed arable landscapes in western Europe. The severity of erosion varies also with topography and soil type, and in terms of off-site damage the proximity of watercourses and property is important. The following farming practices contribute to this developing problem:

1. Ploughing up of steep slopes that were formerly under grass, in order to increase the area of arable cultivation.
2. Use of larger and heavier agricultural machinery, which has a tendency to increase soil compaction, and reduce the size and interconnectivity of soil pores, thereby impeding infiltration.
3. Removal of hedgerows and the associated increase in field size. Larger fields cause an increase in slope length with a concomitant increase in erosion risk.
4. Declining levels of organic matter resulting from intensive cultivation and reliance on chemical fertilizers, which in turn lead to reduced aggregate stability.
5. Availability of more powerful machinery that permits cultivation in the direction of maximum slope rather than along the contour. Rills often develop along tractor and implement wheelings and along drill lines.
6. Use of powered harrows in seedbed preparation and the rolling of fields after drilling.
7. Widespread introduction of autumn-sown cereals to replace spring-sown cereals. Because of their longer growing season, winter cereals produce greater yields and are therefore more profitable. The change means that seedbeds are exposed with little vegetation cover throughout the period of winter rainfall.

Since the 1980s when large areas of winter cereals seemed to be the main problem, the growth in importance of high-value crops such as maize and potatoes and of outdoor pigs on vulnerable soils has led to erosion and runoff with resulting off-site damage especially to watercourses. While extreme precipitation events (e.g. over 100 mm in 24 hours) may lead to serious erosion, other farming-practice factors are always involved (e.g. Boardman et al., 1996; Boardman, 2001).

14.7 Soil erosion produced by fire

Many fires are started by humans and because they remove vegetation and expose the ground they tend to increase rates of soil erosion. Burnt forests often have rates a whole order of magnitude higher than those of protected areas. Combining the effects of increased flow rate and sediment yield, it was found that, after fire, the total sediment load could be increased as much as 1,000 times (Pereira, 1973). The causes of the marked erosion associated with chaparral burning in the western USA are particularly interesting. There is normally a distinctive 'non-wettable' layer in the soils supporting chaparral. This layer, composed of soil particles coated by hydrophobic substances leached from the shrubs or their litter, is normally associated with the upper part of the soil profile (Mooney and Parsons, 1973), and builds up through time in the unburned chaparral. The high temperatures that accompany chaparral fires cause these hydrophobic substances to be distilled so that they condense on lower soil layers. This process results in a shallow layer of wettable soil overlying a non-wettable layer. Such a condition, especially on steep slopes, can result in severe surface erosion (DeBano, 2000; Shakesby et al., 2000; Letey, 2001).

14.8 Soil erosion associated with construction and urbanization

Urbanization can create significant changes in erosion rates. The highest rates of erosion are produced in the construction phase, when there is a large amount of exposed ground and much disturbance produced by vehicle movements and excavations. Wolman and Schick (1967) showed that the equivalent of many decades of natural or even agricultural erosion may take place during a single year in areas cleared for construction. In Maryland they found that sediment yields during construction reached 55,000 tonnes per square kilometre per year, while in the same area rates under forest were around 80–200 tonnes per square kilometre per year and those under farming 400 tonnes per square kilometre per year. In Virginia, USA, Vice et al. (1969) noted equally high rates of erosion during construction and reported that they were 10 times those from agricultural land, 200 times those from grassland and 2,000 times those from forest in the same area. However, construction eventually ceases, roads are surfaced, and gardens and lawns are cultivated. The rates of erosion then fall dramatically.

14.9 Humans or nature?

It is extremely difficult to disentangle the relative importance of human and natural causation in causing erosion. For example, in the Mediterranean valleys there have been controversies surrounding the age and causes of alternating phases of aggradation and erosion. Vita-Finzi (1969) suggested that at some stage during historical times many of the streams in the Mediterranean area, which had hitherto been engaged primarily in downcutting, began to build up their beds. Renewed downcutting, still seemingly in operation today, has since incised the channels into the alluvial fill. He proposed that the reversal of the downcutting trend

in the Middle Ages was both ubiquitous and confined in time, and that some universal and time-specific agency was required to explain it. He believed that vegetation removal by humans was not a medieval innovation and that some other mechanism was required: precipitation change during the climatic fluctuation known as the Little Ice Age (AD 1550–1850). This was disputed by Butzer (1974). He reported plenty of post-Classical and pre-1500 alluviation (which could not therefore be ascribed to the Little Ice Age), and he doubted whether Vita-Finzi's dating was precise enough to warrant a 1550–1850 date. Instead, he suggested that humans from as early as the middle of the first millennium BC were responsible for multiple phases of erosion from slopes and accelerated sedimentation in valley bottoms. Similarly, van Andel et al. (1990) detected an intermittent and complex record of cut and fill during the late Holocene in Greece, associated with landscape destabilization by local economic and political conditions. This is a view shared in the context of the Algarve in Portugal by Chester and James (1991). In a recent review of historical erosion around the Mediterranean basin, Montgomery (2007) concludes that the main causes of soil erosion are human mismanagement in the form of movement of arable farming onto steep slopes, failure of terrace systems, and overgrazing, rather than climate change. Similarly, in southern Britain, Bell (1982) shows conclusively that colluviation of valleys was initiated and was occurring at all periods from the Neolithic to Post-Medieval and that this reflects changes in land use not climate. However, the climate versus land use debate is ongoing and complicated by the fact that climate, and in particular extreme events, are effective in 'prepared landscapes' where vegetation cover is absent or limited. A balanced review of the debate is provided by Bell and Walker (2005).

A further location with spectacular gullies, locally called *lavaka*, is Madagascar. Here too there have been debates about cultural versus natural causation (Wells and Andriamihaja, 1993). Proponents of cultural causes have argued that since humans arrived on the island in the last two thousand years, there has been excessive cattle grazing, removal of forest for charcoal and for slash-and-burn cultivation, devastating winter (dry season) burning of grasslands, and erosion along tracks and trails. However, the situation is more complex than that and the *lavaka* are polygenetic. Tectonics and natural climatic fluctuations may have been at least as important, and given the climatic and soil types of the island many *lavaka* are a natural part of the landscape's evolution. Some of them also clearly predate primary (i.e. uncut) rain forest.

Another example that demonstrates the problem of disentangling the human from the natural causes of erosion is provided by the eroding peat bogs of highland Britain (Bragg and Tallis, 2001). These are gashed by erosion scars, and many rivers draining them are discoloured by the presence of eroded peat particles (Labadz *et al.*, 1991). Some of the observed peat erosion may be an essentially natural process, for the high water content and low cohesion of undrained peat masses make them inherently unstable. Moreover, the instability must normally become more pronounced as peat continues to accumulate, leading to bog slides and bursts around the margins of expanded peat blankets. Tallis (1985) believes that there have been two main phases of erosion in the Pennines. The first, initiated 1,000–1,200 years ago, may have been caused by natural instability of the type outlined above. However, there has been a second stage of erosion, initiated 200–300 years ago, in which miscellaneous human activities appear to have been important, such as heavy sheep grazing, regular burning, peat cutting, the digging of boundary ditches, the incision of packhorse tracks, military manoeuvres during the First World War, footpath erosion (Wishart and Warburton, 2001) and severe air pollution (Tallis, 1965), the last causing the loss of a very important peat-forming moss, *Sphagnum*.

Overgrazing of the British uplands by sheep has been particularly severe since the Second World War driven by a system of headage payments – a subsidy based on the number of animals (Evans, 1997). Erosion of both peat and mineral soils accompanied ecological damage and reservoir siltation. A more recent move to financial support linked to sustainable numbers of sheep has seen significant improvements in the landscape.

In the western USA many broad valleys and plains became rapidly and deeply incised with gullies (*arroyos*) between 1865 and 1915, with the 1880s being especially important (Cooke and Reeves, 1976). There has been a long history of debate as to the causes of this incision (Elliott *et al.*, 1999; Gonzalez, 2001) and an increasing appreciation of the scale, frequency and impact of climatic changes in the Holocene (McFadden and McAuliffe, 1997). For example, Waters and Haynes (2001) have argued that arroyos first appeared in the American southwest after *c.* 8,000 years ago, and that a dramatic increase in cutting and filling episodes occurred after *c.* 4,000 years ago. They believe that this could be related to a change in the frequency and strength of El Niño events. Elliott *et al.* (1999) recognize various Holocene phases of channel incision at 700–1,200, 1,700–2,300 and 6,500–7,400 years ago.

It is possible that human actions (e.g. timber-felling, over-grazing, cutting grass for hay in valley bottoms, compaction along well-travelled routes, channelling of runoff from trails and railways, disruption of valley-bottom sods

by animals' feet, and the invasion of grasslands by scrub) caused the entrenchment, and the apparent coincidence of white settlement and arroyo development in the late nineteenth century tended to give support to this viewpoint. On the other hand, the Holocene history of the fills shows that there have been repeated phases of aggradation and incision and that some of these took place before human actions could have been a significant factor (see, for example, Leopold, 1951; Hereford, 1986; Balling and Wells, 1990; Mann and Meltzer, 2007).

There is however a pattern in the impact of European-style farming systems on semi-arid environments that caused significant changes to the ecology and led to increased runoff and erosion. The impacts were due to several factors such as selective grazing and either localized or more general excessive numbers of sheep or cattle (overgrazing). The pattern is seen in Colorado, USA (Womack and Schumm 1977), in New South Wales, Australia (Fanning 1999), in the Karoo, South Africa (Boardman and Foster 2008), and in the Caldenal, Argentina (Fernandez and Gill, 2009).

14.10 Soil erosion by wind

So far in this chapter, the emphasis has been on soil erosion by water. However, wind erosion of soils is also an important phenomenon (see also, Chapter 16). Neff *et al.* (2008) used analyses of lake cores in the San Juan Mountains of southwestern Colorado, USA, to show that dust levels increased by 500% above the late Holocene average following the increased western settlement and livestock grazing during the nineteenth and early twentieth centuries.

Possibly the most famous case of soil erosion by wind was the Dust Bowl of the 1930s in the USA. This was caused by (1) a series of hot, dry years that depleted the vegetation cover and made the soils dry enough to be susceptible to wind erosion and (2) by years of over-grazing and poor farming techniques. However, perhaps the prime cause was the rapid expansion of wheat cultivation in the Great Plains during and after the First World War. Over large areas, the tough sod that exasperated the earlier homesteaders had been busted and had given way to friable soils of high erosion potential (Worster, 1979).

Dust storms are still a serious problem in parts of the USA. Thus, for example, in the San Joaquin Valley area of California in 1977 a dust storm caused extensive damage and erosion over an area of about 2,000 square kilometres. While the combination of drought and a very high wind provided the predisposing natural conditions for the stripping to occur, over-grazing and the general lack of wind-breaks in the agricultural land played a more significant role. In addition, broad areas of land had recently been stripped of vegetation, levelled or ploughed up prior to planting. Other quantitatively less important factors included stripping of vegetation for urban expansion, extensive denudation of land in the vicinity of oilfields, and local denudation of land by vehicular recreation (Wilshire *et al.*, 1981). Elsewhere in California dust yield has been considerably increased by mining operations in dry lake beds (Wilshire, 1980) and by disturbance of playas (Gill, 1996).

A comparable acceleration of dust storm activity occurred in the former Soviet Union. After the 'Virgin Lands' programme of agricultural expansion in the 1950s, dust storm frequencies in the southern Omsk region increased on average by a factor of 2.5 and locally by factors of 5 to 6. Data on trends elsewhere are evaluated by Goudie and Middleton (1992). A good review of wind erosion of agricultural land is provided by Warren (2002).

14.11 Global hotspots of erosion

In general it is quite clear that the major areas of intense erosion are associated with both human and natural factors. The map produced by Milliman and Meade (1983: Figure 1) of sediment yield of rivers to the seas, while having the disadvantages noted above as a record of erosion, makes it clear that major soil loss is associated with areas of high rainfall that are usually coincident with areas of high topography. Steep slopes in active tectonic belts (the Himalayas, the East Indies, New Zealand and the Andes) exacerbate the problem. Many of these regions also have population pressure and intensively farmed slopes. Vulnerable, unstable soils as in the Loess Plateau of China are an additional factor.

There is no reliable map of global erosion and data deficiencies in many areas make the construction of such a map difficult. Avoiding the use of GLASOD and relying as far as possible on published observational evidence, Boardman (2006) suggested the following candidates as global erosion hotspots:

Loess Plateau, the Yangtze basin and the southern hilly
    country of China
Ethiopia
Swaziland and Lesotho
Andes
South and East Asia
Mediterranean basin
Iceland
Madagascar
Himalayas

West African Sahel
Caribbean and Central America

14.12 Soil conservation: water erosion

Soil conservation has a long history: terrace systems were being used in the Mediterranean basin thousands of years ago. In the modern era, techniques to prevent erosion are described by Bennett (1939) and few have been added since. The exception has been the growth of no-till or minimum-till methods in many parts of the world. But erosion continues and the principle reason is that techniques to prevent erosion are only instigated or maintained if the social, political and economic will is in place. Terrace systems decay and are themselves eroded if they are not constantly maintained. In the developed world, with its technological ability to farm unfavourable landscapes, to redesign and to provide water, there are many examples of the unwise adoption of crops and practices that lead inevitably to erosion. The changes are driven by agricultural subsidies or powerful market forces; thus we have had almond and olive cultivation in the Mediterranean, maize in western Europe, over-grazing by sheep in Iceland and widespread collectivization in eastern Europe, all leading to erosion. Powerful economic forces have in these cases led directly to ecological damage including erosion, and soil conservation has been neglected (Boardman *et al.*, 2003).

Many techniques are available to conserve soil (Hudson, 1987). Some are of some antiquity, and traditional techniques have both a wide range of types and many virtues (see Critchley *et al.*, 1994; Reij *et al.*, 1996). The following are some of the main ways in which soil cover may be conserved:

1. Revegetation:
   (a) Deliberate planting;
   (b) Suppression of fire, grazing, etc., to allow regeneration.
2. Measures to stop stream bank erosion.
3. Measures to stop gully enlargement:
   (a) Planting of trailing plants, etc.;
   (b) Weirs, dams, gabions, etc.
4. Crop management:
   (a) Maintaining cover at critical times of year;
   (b) Rotation;
   (c) Cover crops.
5. Slope runoff control:
   (a) Terracing;
   (b) Deep tillage and application of humus;
   (c) Transverse hillside ditches to interrupt runoff;

(d) Contour ploughing;

(e) Preservation of vegetation strips (to limit field width).

6. Prevention of erosion from point sources such as roads, feedlots:

(a) Intelligent geomorphic location;

(b) Channelling of drainage water to non-susceptible areas;

(c) Covering of banks, cuttings, etc. with vegetation, geotextiles, etc.

Some attempts at soil conservation have been particularly successful. For example, in Wisconsin, Trimble and Lund (1982) showed that in the Coon Creek Basin, erosion rates declined fourfold between the 1930s and the 1970s. One of the main reasons for this was the progressive adoption of contour-strip ploughing.

14.13 Soil conservation: wind erosion

Various attempts have been made to control wind erosion and the occurrence of dust storms (Bennett, 1939; Middleton, 1990; Riksen et al., 2003; Sterk, 2003; Nordstrom and Hotta, 2004). Techniques are frequently classified into three categories: (1) crop management practices (2) mechanical tillage operations, and (3) vegetative barriers. All of these methods aim to decrease wind speed at the soil surface by increasing surface roughness and/or increasing the threshold velocity that is required to initiate particle movement by wind (Morgan, 2005).

Agronomic measures for controlling erosion use living vegetation or the residues from harvested crops to protect the soil. When a vegetative cover is sufficiently high and dense to prevent the wind stress on adjacent exposed land exceeding the threshold for particle movement, then the soil will not erode. Roots also help to prevent erosion through their contribution to the mechanical strength of the soil. Maintaining a sufficient vegetative cover is the 'cardinal rule' for controlling wind erosion (Skidmore, 1986). The wise management of crop residues is widely used in dryland agriculture in many parts of the world. In the Sahel of West Africa millet mulches of around 2 tonnes per hectare have proved to be highly effective (Bielders et al., 2001). In an experiment to determine the loss of topsoil prevented by millet mulch in Niger, Michels et al. (1995) found a relative difference in surface elevation of 33 mm after just one year between bare millet plots and those spread with $2,000 \, \text{kg ha}^{-1}$ of mulch, as a result of wind erosion and sediment deposition. Dung, which is widely used in subsistence agriculture because of its fertilizing properties, also provides effective protection against particle creep and

saltation – initiators of suspension – even at a very low level of cover (de Rouw and Rajot, 2004). Pebble and gravel mulches have been used by farmers in northwest China for more than 300 years, to dampen down soil erosion and to trap dust carried by the wind (Li et al., 2001). The accumulation of dust may supply valuable additional nutrients to gravel-mulched fields (Li and Liu, 2003). In some countries, sandy soils can be stabilized by the addition of clay to the soil. This process is often called marling, and it reduces erosion risk by increasing aggregate stability.

Soil management techniques focus on ways of preparing the soil to promote good vegetative growth and to improve soil structure. Applying organic matter can decrease soil erodibility as well as enhance its fertility, but most soil management methods are concerned with the effects of various forms of conservation tillage on erosion rates, soil conditions and crop yields (see, for example, Merrill et al., 1999) and the results show the success of the system to be highly soil specific and also to depend on how well weeds, pests and diseases are controlled (Morgan, 2005). Significant differences in dust production from field experiments in semi-arid northeast Spain were detected by López et al. (1998) when conventional tillage operations (mouldboard ploughing) were compared to reduced tillage (chisel ploughing).

Mechanical approaches to wind erosion control manipulate the surface topography to control wind flow. Such techniques include the creation of barriers such as fences and windbreaks (known as shelterbelts when composed of living plants) and altering surface topography such as by ploughing furrows. Barriers to wind flow aid erosion control by decreasing surface shear stress in their lee and by acting as a trap to moving particles, although barriers also create turbulence in their lee which can reduce their effective protection. The most efficient barrier is semipermeable because, although its velocity reduction is less than for an impermeable fence, the amounts of eddies and turbulence in its lee are reduced (Cooke et al. 1982). In the same way windbreaks and shelterbelts should be designed to optimize the interaction between height, density, porosity, shape and width of the plant barrier (Cornelis and Gabriels, 2005).

14.14 Conclusions

Erosion by water and wind reduces soil quality, and causes off-site sedimentation and changes in land cover, and deterioration in water and air quality. Ground incision in the form of miscellaneous types of gully, including dongas and arroyos, can create loss of land and erosion of engineering structures. However, there are many causes of accelerated erosion, and it is often difficult to disentangle the relative importance of natural processes and human actions. Nonetheless, there are various studies of long-term erosion rates that suggest that anthropogenic actions have caused dramatic increases in erosion rates. However, there are also various techniques that have been developed to try to reduce the erosion hazard caused by the action of water and wind.

References

Andel, T. H. van, Zangger, E. and Demitrack, A. (1990). Land use and soil erosion in prehistoric and historical Greece. *Journal of Field Archaeology*, **17**, 379–396.

Anderson, J. M. and Spencer, T. (1991). Carbon, nutrient and water balances of tropical rain forest ecosystems subject to disturbance: management implications and research proposals. *MAB Digest 7*. Paris: UNESCO.

Arnalds, O., Thorarinsdottir, E. F., Metusalemsson, S. *et al.* (2001). *Soil Erosion in Iceland*. Soil Conservation Service/Agricultural Research Institute, Iceland.

Balling, R. C. and Wells, S. G. (1990). Historical rainfall patterns and arroyo activity within the Zuni river drainage basin, New Mexico. *Annals of the Association of American Geographers*, **80**, 603–617.

Bell M. (1982). The effects of land-use and climate on valley sedimentation. In A. F. Harding (ed.), *Climatic Change in Later Prehistory*. Edinburgh: Edinburgh University Press, pp. 127–142.

Bell, M. and Walker, M. J. C. (2005). People, climate and erosion. In M. Bell and M. J. C. Walker (eds.), *Late Quaternary Environmental Change: Physical and Human Perspectives*, 2nd edition. Harlow, UK: Pearson, pp. 226–243.

Bennett, H. H. (1939). *Soil Conservation*. New York: McGraw-Hill.

Bielders, C. L., Lamers, J. P. A. and Michels, K. (2001). Wind erosion control technologies in the West African Sahel: the effectiveness of windbreaks, mulching and soil tillage, and the perspective of farmers. *Annals of Arid Zone*, **40**, 369–394.

Boardman, J. (1998). An average soil erosion rate for Europe: myth or reality? *Journal of Soil and Water Conservation*, **53**, 46–50.

Boardman, J. (2001). Storms, floods and soil erosion on the South Downs, East Sussex, autumn and winter 2000–01. *Geography*, **84**, 346–355.

Boardman, J. (2006). Soil erosion science: reflections on the limitations of current approaches. *Catena*, **68**, 73–86.

Boardman, J. (2007). Soil erosion: the challenge of assessing variation through space and time. In A. S. Goudie and J. Kalvoda (eds.), *Geomorphological Variations*. Nakladatelsti P3K, Prague, pp. 205–220.

Boardman, J. and Foster, I. D. L. (2008). Badland and gully erosion in the Karoo, South Africa. *Journal of Soil and Water Conservation*, **63**, 121–125.

Boardman, J. and Poesen, J. (2006). Soil erosion in Europe: major processes, causes and consequences. In J. Boardman and J. Poesen (eds.), *Soil Erosion in Europe*. Chichester: Wiley, pp. 480–487.

Boardman, J., Burt, T., Evans, R., Slattery, M. C. and Shuttleworth, H. (1996). Soil erosion and flooding as a result of a summer thunderstorm in Oxfordshire and Berkshire, May 1993. *Applied Geography*, **16**, 21–34.

Boardman, J., Poesen, J. and Evans, R. (2003). Socio-economic factors in soil erosion and conservation. *Environmental Science and Policy*, **6**, 1–6.

Boardman, J., Verstraeten, G. and Bielders, C. (2006). Muddy floods. In J. Boardman and J. Poesen (eds.), *Soil Erosion in Europe*. Chichester: Wiley, pp. 743–755.

Bork, H-R. (1989). Soil erosion during the past millennium in Central Europe and its significance within the geodynamics of the Holocene. *Catena*, **15**, 121–131.

Bragg, O. M. and Tallis, J. H. (2001). The sensitivity of peat-covered upland landscapes. *Catena*, **42**, 345–60.

Butzer, K. W. (1974). Accelerated soil erosion: a problem of man-land relationships. In I. R. Manners and M. W. Mikesell (eds.), *Perspectives on Environments*. Washington, D.C.: Association of American Geographers, pp. 57–78.

Carrara, P. E. and Carroll, T. R. (1979). The determination of erosion rates from exposed tree roots in the Piceance Basin, Colorado. *Earth Surface Processes*, **4**, 407–417.

Carter, L. J. (1977). Soil erosion: the problem persists despite the billions spent on it. *Science*, **196**, 409–411.

Chester, D. K. and James, P. A. (1991). Holocene alluviation in the Algarve, southern Portugal: the case for an anthropogenic cause. *Journal of Archaeological Science*, **18**, 73–87.

Clarke, R. T. and McCulloch, J. S. G. (1979). The effect of land use on the hydrology of small upland catchments. In G. E. Hollis (ed.), *Man's Impact on the Hydrological Cycle in the United Kingdom*. Norwich: Geo Abstracts, pp. 71–8.

Cooke, R. U. and Reeves, R. W. (1976). *Arroyos and Environmental Change in the American South-west*. Oxford: Clarendon Press.

Cooke, R. U., Brunsden, D., Doornkamp, J. C. and Jones, D. K. C. (1982). *Urban Geomorphology in Drylands*. Oxford: Oxford University Press.

Cornelis, W. M. and Gabriels, D. (2005). Optimal windbreak design for wind-erosion control. *Journal of Arid Environments*, **61**, 315–332.

Critchley, W. R. S., Reij, C. and Willcocks, T. J. (1994). Indigenous soil and water conservation: a review of the state of knowledge and prospects for building on traditions. *Land Degradation and Rehabilitation*, **5**, 293–314.

De Rouw, A. and Rajot, J-L. (2004) Soil organic matter, surface crusting and erosion in Sahelian farming systems based on manuring or fallowing. *Agriculture, Ecosystems & Environment*, **104**, 263–276

DeBano, L. F. (2000). The role of fire and soil heating on water repellency in wildland environments: a review. *Journal of Hydrology*, **231/2**, 195–206.

Diamond J. (2005). *Collapse: How Societies Choose to Fail or Survive*. London: Allen Lane.

Elliott, J. G., Gillis, A. C. and Aby, S. B. (1999). Evolution of arroyos: incised channels of the southwestern United States. In S. E. Darby and A. Simon (eds.), *Incised River Channels*. Chichester: Wiley, pp. 153–185.

Evans, R. (1990). Water erosion in British farmers' fields: some causes, impacts, predictions. *Progress in Physical Geography*, **14**, 199–219.

Evans, R. (1996). *Soil Erosion and its Impacts in England and Wales*. London: Friends of the Earth.

Evans, R. (1997). Soil erosion in the UK initiated by grazing animals: a need for a national survey. *Applied Geography*, **17**, 127–141.

Evans, R. (2006). Land use, sediment delivery and sediment yield in England and Wales. In P. N. Owens and A. J. Collins (eds.), *Soil Erosion and Sediment Redistribution in River Catchments*. Wallingford, UK: CABI, pp. 70–84.

Evans, R. and Brazier, R. (2005). Evaluation of modelled spatially distributed predictions of soil erosion by water versus field-based assessments. *Environmental Science and Policy*, **8**, 493–501.

Fanning, P. (1999). Recent landscape history in arid western New South Wales, Australia: a model for regional change. *Geomorphology*, **29**, 191–209.

Faulkner, H. (2006). Piping hazard on collapsible and dispersive soils in Europe. In J. Boardman and J. Poesen (eds.), *Soil Erosion in Europe*. Chichester: Wiley, pp. 537–562.

Fernández, O. A. and Gil, M. E. (2009). The challenge of rangeland degradation in a temperate semiarid region of Argentina: the Caldenal. *Land Degradation and Development*, **20**, 431–440.

Forster, D. L., Bardos, C. P. and Southgate, D. D. (1987). Soil erosion and water treatment costs. *Journal of Soil and Water Conservation*, **42**, 349–352.

Foster, I. D. L., Boardman, J. and Keay-Bright, J. (2007). Sediment tracing and environmental history for two small catchments, Karoo uplands, South Africa. *Geomorphology*, **90**, 126–143.

Gill, T. E. (1996). Eolian sediments generated by anthropogenic disturbance of playas: human impacts on the geomorphic system and geomorphic impacts on the human system. *Geomorphology*, **17**, 207–28.

Gonzalez, M. A. (2001). Recent formation of arroyos in the Little Missouri Badlands of southwestern Dakota. *Geomorphology*, **38**, 63–84.

Goudie, A. S. and Middleton, N. J. (1992). The changing frequency of dust storms through time. *Climate Change*, **20**, 197–225.

Govers, G. and Poesen, J. (1988). Assessment of the interrill and rill contributions to total soil loss from an upland field plot. *Geomorphology*, **1**, 343–354.

Harrod, T. R. (1994). Runoff, soil erosion and pesticide pollution in Cornwall. In R. J. Rickson (ed.), *Conserving Soil Resources: European Perspectives*. Wallingford, UK: CABI, pp. 105–115.

Hereford, R. (1986). Modern alluvial history of the Paria River drainage basin, southern Utah. *Quaternary Research*, **25**, 293–311.

Hoffman, T. and Ashwell, A. (2001). *Nature Divided: Land Degradation in South Africa*. Cape Town: University of Cape Town Press.

Hudson, N. (1987). Soil and water conservation in semi-arid areas. *FAO Soils Bulletin*, **55**.

Hughes, R. J., Sullivan, M. E. and Yok, D. (1991). Human-induced erosion in a highlands catchment in Papua New Guinea: the prehistoric and contemporary records. *Zeitschrift für Geomorphologie, Supplement-band*, **83**, 227–239.

ISSS (1996). *Terminology for Soil Erosion and Conservation*. Wageningen: International Society for Soil Science.

Jacks, G. V. and Whyte, R. O. (1939). *The Rape of the Earth: A World Survey of Soil Erosion*. London: Faber and Faber.

Keay-Bright, J. and Boardman, J. (2009). Evidence from field based studies of rates of erosion on degraded land in the central Karoo, South Africa. *Geomorphology*, **103**, 455–465.

Labadz, J. C., Burt, T. P. and Potter, A. W. L. (1991). Sediment yield and delivery in the blanket peat moorlands of the southern Pennines. *Earth Surface Processes and Landforms*, **16**, 255–71.

Leopold, L. B. (1951). Rainfall frequency: an aspect of climatic variation. *Transactions of the American Geophysics Union*, **32**, 347–357.

Letey, J. (2001). Causes and consequences of fire-induced soil water repellency. *Hydrological Processes*, **15**, 2867–2875.

Li, X-Y. and Liu, L-Y. (2003). Effect of gravel mulch on aeolian dust accumulation in the semiarid region of northwest China. *Soil and Tillage Research*, **70**, 73–81.

Li, X-Y., Liu, L-Y. and Gong, J-D. (2001). Influence of pebble mulch on soil erosion by wind and trapping capacity for windblown sediment. *Soil and Tillage Research*, **59**, 137–142.

López, M. V., Sabre, M., Gracia, R., Arrúe, J. L. and Gomes, L. (1998). Tillage effects on soil surface conditions and dust emission by wind erosion in semiarid Aragón (NE Spain). *Soil and Tillage Research*, **45**, 91–105.

Mann, D. H. and Meltzer, D. J. (2007). Millennial-scale dynamics of valley fills over the past 12000 ^{14}C yr in northeastern New Mexico. *Bulletin of the Geological Society of America*, **119**, 1433–1448.

Marsh, G. P. (1864). *Man and Nature*. New York: Scribner.

McFadden, L. D. and McAuliffe, J. R. (1997). Lithologically influenced geomorphic responses to Holocene climatic changes in the southern Colorado Plateau, Arizona: a soil-geomorphic and ecologic perspective. *Geomorphology*, **19**, 303–332.

McLennan, S. M. (1993). Weathering and global denudation. *Journal of Geology*, **101**, 295–303.

Merrill, S. D., Black, A. L., Fryrear, D. W. *et al.* (1999). Soil wind erosion hazard of spring wheat-fallow as affected by long-term climate and tillage. *Journal of the Soil Science Society of America*, **63**, 1768–1777.

Michels, K., Sivakumar, M. V. K. and Allison, B. E. (1995). Wind erosion control using crop residue: I. Effects on soil flux and soil properties. *Field Crops Research*, **40**, 101–110.

Middleton, N. J. (1990). Wind erosion and dust storm prevention. In A. S. Goudie (ed.), *Desert Reclamation*. Chichester: Wiley, pp. 87–108.

Milliman, J. D. and Meade, R. H. (1983). Worldwide delivery of river sediment to the oceans. *Journal of Geology*, **91**, 1–21.

Milliman, J. D., Qin, Y. S., Ren, M. E. and Yoshiki, S. (1987). Man's influence on erosion and transport of sediment by Asian rivers: the Yellow River (Huanghe) example. *Journal of Geology*, **95**, 751–762.

Montgomery, D. R. (2007). *Dirt: The Erosion of Civilisations*. Berkeley: University of California Press.

Mooney, H. A. and Parsons, D. J. (1973). Structure and function of the California Chaparral: an example from San Dimas. *Ecological Studies*, **7**, 83–112.

Morgan, R. P. C. (1987). Sensitivity of European soils to ultimate physical degradation. In H. Barth and P. l'Hermite (eds.), *Scientific Basis for Soil Protection in the European Community*. Amsterdam: Elsevier, pp. 147–157.

Morgan, R. P. C. (2005). *Soil Erosion and Conservation*, 3rd edition. Oxford: Blackwell.

Myers, N. (1988). *Natural Resource Systems and Human Exploitation Systems: Physiobiotic and Ecological Linkages*. World Bank Policy Planning and Research Staff, Environment Department Working Paper, 12.

Neff, J. C., Ballantyne, A. P., Farmer, G. L. *et al.* (2008). Increasing eolian dust deposition in the western United States linked to human activity. *Nature Geoscience*, **1**, 189–195.

Nordstrom, K. F. and Hotta, S. (2004). Wind erosion from cropland in the USA: a review of problems, solutions and prospects. *Geoderma*, **121**, 157–167.

Oldeman, L. R., Hakkeling, R. T. A. and Sombroek, W. G. (1991). *World Map of the Status of Human-induced Soil Degradation: An Explanatory Note*. Wageningen: International Soil Reference and Information Centre; Nairobi, United Nations Environment Programme.

Page, M. J. and Trustrum, N. A. (1997). A late Holocene lake sediment record of the erosion response to land use change in a steepland catchment, New Zealand. *Zeitschrift für Geomorphologie*, **41**, 369–392.

Pereira, H. C. (1973). *Land Use and Water Resources in Temperate and Tropical Climates*. Cambridge: Cambridge University Press.

Pimentel, D. (1976). Land degradation: effects on food and energy resources. *Science*, **194**, 149–155.

Pimentel, D., Harvey, C., Resosuddarmo, P. *et al.* (1995). Environmental and economic costs of soil erosion and conservation benefits. *Science*, **267**, 1117–1122.

Poesen, J., Vanwalleghem, T., de Vente, J. *et al.* (2006). Gully erosion in Europe. In J. Boardman and J. Poesen (eds.), *Soil Erosion in Europe*. Chichester: Wiley, pp. 515–536.

Reij, C., Scoones, I. and Toulmin, C. (eds.) (1996). *Sustaining the Soil: Indigenous Soil and Water Conservation in Africa*. London: Earthscan.

Rienks, S. M., Botha, G. A. and Hughes, J. C. (2000). Some physical and chemical properties of sediments exposed in a gully (donga) in northern Kwazulu-Natal, South Africa, and their relationship to the erodibility of the colluvial layers. *Catena*, **39**, 11–31.

Riksen, M., Spaan, W., Arrué, J. L. and López, M. V. (2003). What to do about wind erosion. In A. Warren (ed.), *Wind Erosion on Agricultural Land in Europe*. Luxembourg: European Commission, pp. 39–52.

Sauer, C. O. (1938). Destructive exploitation in modern colonial expansion. In *Comptes Rendus du Congrès International de Geographie, Amsterdam*, Vol. II, Section IIIC, pp. 494–499.

Shakesby, R. A., Doerr, S. H. and Walsh, R. P. D. (2000). The erosional impact of soil hydrophobicity: current problems and future research directions. *Journal of Hydrology*, **231/2**, 178–191.

Sheffield, A. T., Healy, T. R. and McGlone, M. S. (1995). Infilling rates of a steepland catchment estuary, Whangamata, New Zealand. *Journal of Coastal Research*, **11**, 1294–1308.

Sidorchuk, A. Y. and Golosov, V. N. (2003). Erosion and sedimentation on the Russian Plain, II: The history of erosion and sedimentation during the period of intensive agriculture. *Hydrological Processes*, **17**, 3347–3358.

Skidmore, E. L. (1986). Wind erosion control. *Climatic Change*, **9**, 209–218.

Stankoviansky, M. (2003). Historical evolution of permanent gullies in the Myjava Hill Land, Slovakia. *Catena*, **51**, 223–239.

Sterk, G. (2003). Causes, consequences and control of wind erosion in Sahelian Africa: a review. *Land Degradation and Development*, **14**, 95–108.

Stocking, M. (1984). *Erosion and Soil Productivity: A Review*. FAO Soil Conservation Programme Land and Water Development Division Consultants' Working Paper, 1.

Stocking, M. and Murnaghan, N. (2001). *Handbook for the Field Assessment of Land Degradation*. London: Earthscan.

Talbot, W. J. (1947). *Swartland and Sandvelt: A Survey of Land Utilisation and Soil Erosion in the Western Lowland of the Cape Province*. Cape Town: Oxford University Press.

Tallis, J. H. (1965). Studies on southern Pennine peats, IV: Evidence of recent erosion. *Journal of Ecology*, **53**, 509–20.

Tallis, J. H. (1985). Erosion of blanket peat in the southern Pennines: new light on an old problem. In R. H. Johnson (ed.), *The Geomorphology of North-west England*. Manchester: Manchester University Press, pp. 313–336.

Trimble, S. W. (1983). A sediment budget for Coon Creek basin in the Driftless Area, Wisconsin, 1853–1977. *American Journal of Science*, **283**, 454–474.

Trimble, S. W. (1988). The impact of organisms on overall erosion rates within catchments in temperate regions. In H. A. Viles (ed.), *Biogeomorphology*. Oxford: Basil Blackwell, pp. 83–142.

Trimble, S. W. and Crosson, S. (2000). US soil erosion rates: myth and reality. *Science*, **289**, 248–250.

Trimble, S. W. and Lund, S. W. (1982). *Soil Conservation and the Reduction of Erosion and Sedimentation in the Coon Creek Basin, Wisconsin*. US Geological Survey Professional Paper, 1234.

Verstraeten, G., Bazzoffi, P., Lajczak, A. *et al.* (2006). Reservoir and pond sedimentation in Europe. In J. Boardman and J. Poesen (eds.), *Soil Erosion in Europe*. Chichester: Wiley, pp. 759–774.

Vice, R. B., Guy, H. P. and Ferguson, G. E. (1969). *Sediment Movement in an Area of Suburban Highway Construction, Scott Run Basin, Fairfax, County, Virginia, 1961–64*. US Geological Survey Water Supply Paper, 1591-E.

Vita-Finzi, C. (1969). *The Mediterranean Valleys*. Cambridge: Cambridge University Press.

Walling, D. E. (1983). The sediment delivery problem. *Journal of Hydrology*, **65**, 209–237.

Warren, A. (ed.) (2002). *Wind Erosion on Agricultural Land in Europe*. Brussels: European Commission.

Waters, M. R. and Haynes, C. V. (2001). Late Quaternary arroyo formation and climate change in the American southwest. *Geology*, **29**, 399–402.

Watson, A., Price-Williams, D. and Goudie, A. S. (1984). The palaeoenvironmental interpretation of colluvial sediments and palaeosols of the Late Pleistocene hypothermal in southern Africa. *Palaeogeography, Palaeoclimatology, Palaeoecology*, **5**, 225–249.

Wells, N. A. and Andriamihaja, B. (1993). The initiation and growth of gullies in Madagascar: are humans to blame? *Geomorphology*, **8**, 1–46.

Wilshire, H. G. (1980). Human causes of accelerated wind erosion in California's deserts. In D. R. Coates and J. D. Vitek (eds.), *Geomorphic Thresholds*. Stroudsburg: Dowden, Hutchinson & Ross, pp. 415–433.

Wilshire, H. G., Nakata, J. K. and Hallet, B. (1981). Field observations of the December 1977 wind storm, San Joaquin Valley, California. In T. L. Péwé (ed.), *Desert Dust: Origin, Characteristics and Effects on Man*. Boulder, Colorado: Geological Society of America, pp. 233–251.

Wishart, D. and Warburton, J. (2001). An assessment of blanket mire degradation and peatland gully development in the Cheviot Hills, Northumberland. *Scottish Geographical Magazine*, **117**, 185–206.

Wolman, M. G. and Schick, A. P. (1967). Effects of construction on fluvial sediment, urban and suburban areas of Maryland. *Water Resources Research*, **3**, 451–464.

Womack, W. R. and Schumm, S. A. (1977). Terraces of Douglas Creek, Northwestern Colorado: an example of episodic erosion. *Geology*, **5**, 72–76.

Worster D. (1979). *Dust Bowl: The Southern Plains in the 1930s*. New York: Oxford University Press.

15 Desertification and land degradation in arid and semi-arid regions

YANG Xiaoping

15.1 Introduction

Land degradation occurring in particular climate zones, i.e. arid, semi-arid, and dry sub-humid areas, is defined as desertification by the United Nations Convention to Combat Desertification (UNCCD, 1999). As desertification is widely viewed as one of the leading environmental issues facing the world today, many individual researchers and international organizations have tried to clarify and to update the definition of this term. Among various conceptions, the ones from the United Nations Authority have been most often used. The earlier definition adopted by the United Nations Environment Programme (UNEP, 1990) was practically the same as the one given by Conacher and Conacher (2000), limiting the cause of degradation to human activities. At the UN Conference on Environment and Development in Rio de Janeiro in 1992, both climatic variations and human activities were accepted as factors triggering desertification (Williams and Balling, 1995).

Independent of the exact causes of desertification in each individual case, desertification is a distinct geomorphological hazard due to its impacts on landforms and on geomorphological processes. Among various geomorphological hazards, desertification has been given probably the greatest attention by the United Nations. In 1994 the United Nations Convention to Combat Desertification (UNCCD) was adopted. In 1996 the agreement legally entered into force following its 50th ratification. The year 2006, the 10th anniversary since the UNCCD came into effect, was declared by the United Nations as the International Year of Deserts and Desertification. However, some scientists suggested that the term 'desertification' was a buzzword and should be replaced by land degradation or landscape degradation (e.g., Seuffert, 2001).

In general, areas with a mean annual precipitation below 200 mm are defined as arid, and those with a mean annual precipitation between 200 and 500 mm as semi-arid. As the aridity, i.e. water deficit in air and soil, is closely related to the temperature as well, the bioclimatic aridity index P/PET is used to differentiate the variations of drylands (FAO/UNESCO, 1971–1979). Here P refers to annual precipitation and PET, the potential evapotranspiration that is the quantity of water lost by evaporation directly from the soil and by transpiration of plant cover. The value of P is derived directly from records of weather stations, while PET is calculated using various models and equations (FAO/UNESCO, 1971–1979). The quantitative classification is as follows:

Hyperarid: $P/PET < 0.03$,
Arid: $0.03 < P/PET < 0.20$,
Semi-arid: $0.20 < P/PET < 0.50$.

Arid (including hyperarid) and semi-arid regions, commonly described as drylands, occupy roughly one-third of the Earth's land surface. They are found from the vicinity of the equator to more than $50°$ N and S latitude. There are two major concentrations of drylands. One, directly related to the subtropical high-pressure cells, is centered on the Tropics of Cancer and Capricorn and extends $10°$ to $15°$ poleward and equatorward from there. The other occupies continental interiors, particularly in the Northern Hemisphere (Figure 15.1). Hyperarid and arid climate zones correspond to desert, and semi-arid zones to steppe.

Agriculture and animal grazing have been established in arid regions all around the world wherever river or well water is available. Rains in semi-arid regions are usually sufficient to maintain a vegetation cover that can feed the roaming herbivores that graze on them.

Geomorphological Hazards and Disaster Prevention, eds. Irasema Alcántara-Ayala and Andrew S. Goudie. Published by Cambridge University Press. © Cambridge University Press 2010.

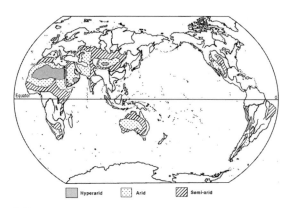

FIGURE 15.1. Global distribution of hyperarid, arid and semi-arid regions.

FIGURE 15.2. Road in the desert of Chaidamu Basin of western China, endangered by sand encroachment.

15.2 Regional-scale hazards and causes

15.2.1 Sand encroachment

To many people sand dunes are one of nature's most beautiful landforms. The movement of sand, however, is a constant threat to those who live in arid and semi-arid regions. There is little doubt that sand seas or dune fields expanded or shrank naturally due to climate changes during various geological eras (Zhu *et al.*, 1980; Williams *et al.*, 1998; Goudie, 2002; Yang *et al.*, 2004; Brook *et al.*, 2007 Unkel *et al.*, 2007). But the modern expansion of sand seas, or the process of sand encroachment, may bury railways, roads (Figure 15.2), agricultural and grazing land, as well as human settlements, causing serious environmental disasters on a regional scale. Sand encroachment can also be caused by the reactivation of fixed dunes and sand sheet under human impacts (Yang *et al.*, 2007; Figure 15.3).

FIGURE 15.3. Reactivated dunes due to overgrazing in the Hunshandake Sandy Land of eastern Inner Mongolia.

Sand encroachment takes places by a combined process of sand transportation and accumulation. It occurs particularly in areas where sand seas exist. On Earth the proportions of sand dunes in various deserts vary considerably. Active and stable sand dunes occupy 45% of deserts in China (Zhu *et al.*, 1980). By contrast, aeolian sand covers less than 1% of the arid zone in the Americas (Lancaster, 1995). Active dunes cover between 15% and 30% of arid regions in the Sahara, Arabian Peninsula, Australia, and Southern Africa (Goudie, 2002). Some of the sand seas are enormous, such as the Arabian sand seas covering nearly 800,000 km² and the Taklamakan occupying 330,000 km² in western China. A huge amount of loose sand is deposited in these sand seas, offering great potential for sand encroachment.

The causes of sand encroachment have their explanations in physics. It was found that changes in wind direction and dune collisions could destabilize the dunes and generate surface waves on the barchans. Such surface waves can produce a series of new, small barchans by breaking the horns of large dunes, because these waves propagate at a higher speed than the dunes themselves. The formation of these new dunes provides a mechanism for sand loss that prevents sand seas from merging into a single megadune (Elbelrhiti *et al.*, 2005).

One should also bear in mind that sand encroachment occurs not only in the form of dune movement but also in the form of new accumulation of aeolian sand. Sand will accumulate around obstacles in the ground where sand flow velocities are reduced by the presence of a topographical barrier such as a hill or an inselberg or even a building. On the windward side of a barrier the air current will be diverted over or around it. As a result, a comparatively calm space exists in front of the barrier and in that location sand will accumulate. To the lee of small topographical obstacles sand may also accumulate when sand is blown over and around them. As time goes on, the sand can accumulate to a considerabe amount.

Here sand encroachment includes both dune movement and dune advancement, as differentiated by Embabi (2004). Dune movement is referred to as the displacement of barchans irrespective of the changes that might occur in size or shape, while dune advancement is limited to indicate the extension of linear dunes, which might occur at their heads (Embabi, 2004). Several studies have contributed to understanding the dune movement in the Western Desert of Egypt. In the period between 1907 and 1908, dunes moved 15 m year^{-1} on average in the Kharga Oasis in the Western Desert (Beadnell, 1910). Later, Embabi (2004) undertook detailed observations of the movement of 25 dunes in the same oasis, and found out that dune migration was between 20 and 100 m year^{-1} with an average of 48 m in the early 1970s. The higher speed is thought to be related to the selection of dunes for measurements. The dunes observed by Embabi (2004) are much smaller than those studied by Beadnell (1910). It was reported that the advancement of linear dunes is generally slower than the movement of barchans. But there are observations showing higher speeds (30 m year^{-1}) of linear dunes as well (Embabi, 2004).

Field observations in the southern margin of the Taklamakan Desert confirmed that taller dunes move faster than lower ones. The annual mean distance of the movement was 9.8–12.5 m for 2 m high dunes but 1.4–2.4 m for 12 m high dunes during the early 1960s (Zhu *et al.*, 1981; Figure 15.4). Wind is a strong transportation agent in desert environments because soil, vegetation, or moisture do not

seal the surface. The strength of the wind, however, differs considerably from desert to desert. The drift potentials produced by Fryberger (1979) offer a quantitative indicator describing effectiveness of winds for sand transportation. The deserts of the world are divided into high-energy, intermediate-energy and low-energy wind environments. Globally, Taklamakan belongs to a low-energy wind environment (Fryberger, 1979).

Even capital cities, like one called Tongwan Town, can be destroyed by sand encroachment. Tongwan Town was the capital of a small kingdom named Helianxia occupying the Erduosi Plateau of northern China in the fifth century. At that time, there was no desert but beautiful grassland around Tongwan Town. In the ninth century the town was facing the problem of sand encroachment, but still acted as an important center in the region. In AD 994 the town was totally covered by aeolian sand according to historical literature (Hou, 1973). Only the tall defense towers are visible nowadays.

Another extremely dramatic case of sand encroachment is the abandonment of the ancient cities in the interior of the Taklamakan Desert of western China, where several relatively large settlements have been buried under aeolian sand for centuries. For example, archaeologists were able to excavate an old settlement, called Yuansha, in the central part of the Taklamakan. This settlement is marked by a 3–4 m high, *c*. 1 km long city wall, and was an important city 2,000 years ago (Sino-French Expedition Team, 1997; Yang *et al.*, 2006b). It was destroyed by shifting sand. The road in the southern Taklamakan has had to be rebuilt in southward locations several times during the last 200 years as sand was transported southwards by northerly winds (Yang, 1991).

The direction of the sand movement is mainly dependent on the wind direction near the land surface. Among all geodynamic processes, wind is the only agent that can transport sand in all directions, i.e. horizontally, downwards, and upwards. Therefore it is hard to set efficient barriers to stop sand encroachment. Plants and mechanical barriers, however, can considerably reduce the intensity of sand movement as sand is transported mainly within 20 cm of the surface.

Methods for mitigating the problems of dune movement and sand accumulation are provided by Watson (1990).

15.2.2 Deflation

Deflation refers to the process by which the wind removes clay, silt, and sand particles from the land surface. The three primary conditions that are needed for wind to become an effective geomorphological agent are spare vegetation

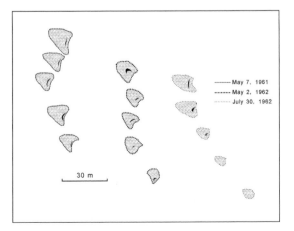

FIGURE 15.4. Rapid movement of small dunes (height < 2 m) under the dominant northwestern winds on the southwestern margins of the Taklamakan Desert in western China. (Modified from Zhu *et al.*, 1981.)

cover, the presence of dry, loose materials at the surface, and a wind velocity that is high enough to pick up and move those surface materials. These three conditions are most widespread in arid and semi-arid regions. Experiments in wind tunnels show that wind cannot move the sand grains if the water content in the sand is higher than 1% (He and Shen, 1988). Observations in the African Kalahari suggest that a vegetation cover of over 14% can significantly prevent the movement of sand (Wiggs et al., 1995). It has been estimated that the degraded land caused by wind erosion amounts to 5.05×10^6 km^2, accounting for 46.4% of the global degraded land (UNEP and ISRIC, 1990).

Sediments are transported by winds in three forms, i.e. suspension, saltation, and creep. In erosion, the ability of winds is enhanced by the presence of hard particles in the wind. As these wind-blown sediments hit a mass of weaker material, they erode the surface. Such a process is called abrasion. Due to the high percentage of quartz in the desert sands, quartz sand is a very effective and common abrasive agent in aeolian processes. The impact of natural sandblast is limited to a range close to the ground, because sand grains are mostly transported by saltation and creep and are rarely lifted higher than 1 m above the surface.

Deflation is generally hazardous as it causes the loss of surface soil that is crucial to plant communities. The nutrients and organic matter, particularly in the topsoil, will be lost on site. It is not difficult to assume that deflation, particularly abrasion, can cause considerable damage to crops, too. For example, in the San Joaquin valley area of California, a dust storm caused extensive damage in 1977. It was estimated that more than 25 million tonnes of soil were stripped away from an area of about 2,000 km^2 of grazing land within a 24-hour period (Wilshire et al., 1981; Goudie, 2002).

Sandy loess is widely distributed in the margins of the deserts of northern China. In such a zone deflation often removes the loess particles and causes an increase in the grain size of soils. As the coarsening processes continue, aeolian sand will appear in these soils. The coarsening of topsoil has formed sand sheets and dunes in many places along the Great Wall in northwestern China (Zhu and Chen, 1994).

A very infamous case of deflation was the Dust Bowl of the 1930s in the USA. It was partly caused by a series of hot, dry years which depleted the vegetation cover and made the soils dry. The effects of this drought were exacerbated by years of intensive human activities, i.e. overgrazing and rapid expansion of wheat cultivation in the Great Plains (Goudie, 2002).

Although the documentation of wind erosion and its disastrous consequences in China dates back 2,000 years

(Shi et al., 2004), precise and quantitative estimates of deflation are still difficult tasks. Mathematical interpolations, based on the measurements made with erosion pins, and experiments in wind tunnels, can often over- or under-estimate the reality of wind erosion over a large area. For example, the annual rate of wind erosion in the Keerqin Sandy Land of eastern Inner Mongolia is estimated to be 80 tonnes ha^{-1} by Xu et al. (1993), and 174–349 tonnes ha^{-1} by Zhao et al. (1989). Both of these two studies were based on erosion pins.

Various landforms are closely related to deflation processes. If the sediment at the land surface is made up only of fine particles, the erosion of these particles by the wind can lower the land surface substantially. Thick deposits of aeolian loess are evidence that wind erosion and deposition have played a major role in modifying the Earth's surface. A blowout is a depression on a land surface formed by deflation. The enormous Qattara Depression in northwestern Egypt, about 45,000 km^2 in area, 145 m below sea level at the deepest point, was created partly by deflation (Embabi, 2004). Some of the lakes in the Badain Jaran Desert of northwestern China were formed when the bottom of such blowouts reached the groundwater table (Yang and Williams, 2003). Yardangs are landforms that also owe their formation largely to wind erosion. They look like prows of upturned ships and occur often in old lake beds. The initiation of parabolic dunes is also closely related to deflation. Parabolic dunes are those with U- or V-shapes opening towards the wind. Their formation is summarized in four successive stages: (a) blowouts occur in the sandy areas where vegetation is destroyed; (b) sand is swept out from the windward side and accumulates on the leeside slope of the blowouts; (c) deflation of the windward slope and deposition on the leeside take place concurrently; (d) a U- or V-shape develops by migration of their bodies in the wind direction and by thinning out of their wings (Mainguet, 1999).

Not only wind erosion but also water erosion can be intensified in drylands by desertification. Desertification was conceptualized as an issue in need of global political attention following the severe drought and associated famine in the Sahel region of northern Africa in the late 1960s and the early 1970s (Thomas and Middleton, 1994). Within this zone millions of people live on subsistence farming and livestock grazing. In addition to the erosion of fine materials by wind, the hardening of the land surface, the increased erosion by water, and reactivation of old dunes were the main geomorphological changes caused by desertification in the Sahel region (Mensching, 1990).

Wind erosion causes hazardous effects not only in arid and semi-arid regions but also in humid environments.

Studies have shown that wind erosion has also caused heavy crop losses and additional inputs in the case of resowing in western Europe (Riksen and De Graaff, 2001).

Methods for controlling wind erosion and dust storm generation are discussed in Goudie and Middleton (2006, chapter 8).

15.2.3 Salinization and alkalinization

Salinization, or the accumulation of salts in the soil, is described as the most acute aspect of the degradation of drylands, the ultimate stage that is difficult to reverse (Mainguet, 1999). Although the problem occurs in all types of climate, it is particularly serious in arid and semi-arid regions, because the potential evaporation rate of water from the soil exceeds the amount of water arriving as rainfall, allowing salts to accumulate near the surface as the soil dries. The global extent of salinization in irrigated lands is estimated to be quite alarming. Kayasseh and Schenk (1989) reported that about 25×10^6 ha of the entire 92×10^6 ha of irrigated land on Earth are affected by salinization. Another study suggested that 30–50% of the irrigated soils or 30–46×10^6 ha are facing this problem (Barrow, 1991).

A soil is seen as salinized when its content of soluble salts is above 1–2% in the top 20 cm (WMO, 1983). Generally speaking, there are two kinds of salts affecting soils. The first is neutral salts, mainly chloride and sulfate of sodium, magnesium or calcium, inhibiting the growth of plants. The second is mainly the alkali salt sodium carbonate, causing the clay particles in soils to disperse and resulting in a deterioration of soil structure. When the cations of the salts are essentially sodium the effect is referred to as alkalinization or sodication. Sodic soils have a low permeability to water and air and a high pH, consequently reducing nutrient availability to plants.

The origins of salts causing salinization can be salty dust, dissolved in rainwater and/or by-products of chemical weathering. Under hyperarid conditions and at low temperatures, evaporites of calcium and magnesium chlorides, nitrates, and other soluble salts can accumulate in soils. Salts sometimes may not accumulate in uplands due to deep water tables, or insufficient weathering to release them from rocks. But in lowland settings where the groundwater table is shallow, as in stream valleys, playas, and the bottoms of interior basins, the most mobile salts will move to the highest part of the profile. The high groundwater table is often associated with geomorphological settings, but can also be a consequence of intensive irrigation. The water balance between rainfall, stream flow, groundwater, and evaporation and transpiration exists in the natural state

of a catchment basin. When large quantities of extra water are applied in the basin for irrigation, extra water seeps into the soil and causes an increase in the water table. Salts accumulate in the root zone over time due to lack of drainage. Owing to this process, white salt patches are common in many irrigated fields all over the world. Frequently people apply too much water to the irrigated fields. Water in excess of the needs of plants percolates deeper than the roots of crops, dissolving salts in the deeper layers of soil profiles and bringing these salts to the top by waterlogging. During this process soils above a shallow water table become salinized.

Changes of land use can cause salinization also. In Australia the clearing of deep-rooted vegetation and its replacement with annual agricultural species is the principal cause of dryland salinity. Agricultural species use much less water than native trees and shrubs, so the evapotranspiration rate of the landscape is lowered and the hydrological balances are altered. Over time the water table rises and salts formerly stored deep in the soils dissolve. Finally these salts percolate through saline subsurface sediments to horizontal layers, producing salty, waterlogged soils.

Methods for reclaiming salt affected lands are discussed by Rhoades (1990).

15.2.4 Disappearance of wetlands

Rivers and lakes, or wetlands, in arid and semi-arid regions are particularly valued because they store water, provide a habitat for many organisms, and cycle carbon, nitrogen, and other elements through different Earth systems. Climatic changes and human activities, however, may easily cause severe shrinkage or even disappearance of rivers and lakes in such regions. For instance, human activities have caused abrupt changes of the river courses and lakes in the lower reaches of the Black River in western China in the past four decades. The catchment of the Black River is characterized by an arid climate. In its lower reaches the river is divided into two separate courses which have their mouths in different endorheic basins. Historical literature shows that the locations of the endorheic lakes in the lower reaches have changed in the past 2,000 years, but at least one of the lakes was filled with water. Since 1973 there have been years when all these lakes have been completely dry, as have been the river courses. The earlier changes of lake locations were caused by sedimentation processes associated with river dynamics in drylands. But the complete disappearance of these lakes was directly triggered by human activities. To develop agriculture in the catchment, a large number of reservoirs have been constructed. Consequently, river flow ceased in

FIGURE 15.5. Deflation occurring in the dried shore of the endorheic lake in the lower reaches of the Black River. Plant roots mark the thickness of soil lost due to wind erosion.

the lower reaches for most of the year and the endorheic lakes downstream dried out completely in the 1980s, causing severe degradation of the floodplain ecosystem in its lower reaches (Yang *et al.*, 2006a). The dried floodplain and lake basins became a key source of dust storms in northern China (Figure 15.5). To rehabilitate the environment, in 2000 the Chinese Government implemented a compulsory measure to let some water flow downwards. Nowadays there is some water in the formerly dried lakes, but it will take a long time to see a full recovery of the ecosystem.

Another case of the dramatic degradation of wetlands is the desiccation of the Aral Sea, formerly the fourth largest inland water body on Earth. The Aral Sea is the endorheic lake of two major Central Asian rivers, the Amudarya and the Syrdarya. Between the 1950s and 2005, the Aral Sea lost approximately 90% of its volume and 75% of its area (Zavialov, 2005). Intensive expansion of irrigation has been attributed as the cause of the degradation (Mensching, 1990; Mainguet, 1999). Archaeological and geomorphological investigations show that the lake level of the Aral Sea has fluctuated between *c.* 73 m and *c.* 30 m (at present) above sea level since the late Pleistocene. The fluctuations of the lake level in the Aral Sea, except for the period after the 1960s, are assumed to have been driven by climate changes, although humans have been active on the shorelines since the Palaeolithic era (Boroffka *et al.*, 2006).

It should be mentioned that the lengths of rivers and extents of lakes in arid and semi-arid regions have changed considerably in response to global climate changes. According to the historical literature, the former Lop Nuer Lake, located in the eastern Tarim Basin in western China, decreased from a large lake around 2,000 years ago to a much smaller one around 300 years ago. Archaeological evidence shows that irrigated agriculture was widely practised in the interior of the Tarim Basin 2,000 years ago. Therefore, it was suggested that a possible climatic change towards drier conditions at *c.* AD 500 caused the decrease of lake level in the Lop Nuer region and the abandonment of the ancient oasis in the the Tarim Basin. There is no doubt that the desiccation of the Lop Nuer in recent decades has been directly caused by human activities (Yang *et al.*, 2006b).

15.3 Global-scale hazards

15.3.1 Assessment of the extent and severity

The UNCCD (1999) reported that 70% of global drylands are affected by desertification that threatens some one billion people. There is no doubt that there are severe cases of desertification in all continents but data about its global extent and severity need new assessment and modifications. Only five attempts to assess land degradation at a global scale have been made by individual researchers or international organizations since 1977 (Safriel, 2007). Dregne (2002) admitted that the first assessment, done by himself in 1977, was based on few data and has only historical significance. Safriel (2007) commented that none of the five assessments can be considered a reliable source of information. The UN's estimate is also questionable because it is mainly derived from data of varying authenticity and consistency (Thomas and Middleton, 1994). For example, the data from the State Forestry Administration of China, where the national office of UNCCD is based, show that about 2.64 million km^2 or 27.46% of China's total land area should be affected by desertification (State Forestry Administration, 2005). Another report concludes that a total area of 861.6×10^3 km^2 has been desertified in China. Within this frame, an area of 664×10^3 km^2 is distributed in arid, semi-arid, and dry sub-humid regions and accounts for 6.92% of the total Chinese land territories (Study Group, 1998). During the UN Year of Deserts and Desertification in 2006, some geographers at the International Geographical Union (IGU) Congress in Brisbane pointed out that the UN should not confuse deserts with desertification. It was emphasized that deserts have biodiverse ecosystems and rich human cultures and histories (Conacher and Gisladottir, 2006). Indeed, this aspect was clearly identified as the UN International Year of Deserts and Desertification was declared.

Uncertainties concerning the extent of land degradation arise partly from the lack of quantified assessments at regional scales. The disparities between various estimates

FIGURE 16.3. Sand fences constructed in an attempt to control the movement of transverse (barchanoid) dunes near Walvis Bay, Namibia.

(3) Reduction of the sand supply by surface treatments (e.g. water spraying, chemical stabilizers, mulches), fences, and vegetation strips. Among the plants that can be used are marram grass (*Ammophila arenaria*), *Tamarix* spp., *Eucalyptus* spp., *Hippophae rhamnoides*, *Prosopis juliflora*, and *Acacia cyanophylla* (Pye and Tsoar, 1990, pp. 303–306).

(4) Deflection of the moving sand by fences, barriers and tree belts.

With regard to the control of moving dunes, the main techniques that are available are:

(1) Removal of the dunes by mechanical excavation and transportation to a new location.

(2) The dissipation of a mobile dune by disrupting its aerodynamic profile (a process termed dissipation) by means of reshaping, trenching, or surface treatment.

(3) Dune immobilization by surface strips, fences, etc. (Figure 16.3).

Frequently these techniques are not particularly successful and very often the best solution is to site and design engineering structures to allow free movement of sand across them. Alternatively, by mapping different dune types and knowing their direction and rate of movement, structures should be located out of harm's way. Avoidance may be better than defence.

Various studies of the comparative effectiveness of different stabilization techniques have been undertaken in recent years. For example, Zhang *et al.* (2004) found that the best means of stabilizing moving dunes in Inner Mongolia, China, were wheat-straw checkerboards and the planting of *Artemisia halodendron*. This finding was confirmed by a study in the Kerqin Sandy Land of northern China (Li *et al.*, 2009). Along a major highway in the

Taklamakan desert, checkerboards, reed fences, and nylon nets were found to be effective (Dong *et al.*, 2004). In north west Nigeria, Raji *et al.* (2004) found that shelterbelts were the most effective technique, and were superior to mechanical fencing. Some success has also been claimed for chemical stabilizers (Han *et al.*, 2007), and also for geotextiles (Escalente and Pimentel, 2008). Some devices are prohibitively expensive (e.g. chemical fixers) (Dong *et al.*, 2004), while others, such as checkerboards, are not.

16.5 Conclusions

An array of techniques has been employed to try and estimate the speed with which mobile dunes move, and there are now plenty of data on this with respect to barchans. Rates of dune movement depend on a variety of factors of which dune type and dune size are important. However, land cover changes, changes in wind energy conditions, and changes in moisture levels mean that rates vary considerably in time and space. Numerous techniques have been developed to try to reduce problems posed by dune mobility, though they have met mixed success, and can sometimes be either costly or unsightly. Finally, it is likely that in a warmer world dune activity may be greater. This is a topic addressed briefly in Chapter 20.

References

Anthonsen, K. L., Clemmensen, L. B. and Jensen, J. H. (1996). Evolution of a dune from crescentic to parabolic form in response to short-term climatic changes: Råbjerg Mile, Skagen Odde, Denmark. *Geomorphology*, **17**, 63–77.

Bailey, S. D. and Bristow, C. S. (2004). Migration of parabolic dunes at Aberffraw, Anglesey, north Wales. *Geomorphology*, **59**, 165–174.

Barbosa, L. M. and Dominguez, M. L. (2004). Coastal dune fields at the Sao Francisco River Strandplain, northeastern Brazil: morphology and environmental controls. *Earth Surface Processes and Landforms*, **29**, 443–456.

Bristow, C. S. and Lancaster, N. (2004). Movement of a small slipfaceless dome dune in the Namib Sand Sea, Namibia. *Geomorphology*, **59**, 189–196.

Bristow, C. S., Lancaster, N. and Duller, G. A. T. (2005). Combining ground penetrating radar surveys and optical dating to determine dune migration in Namibia. *Journal of the Geological Society*, **162**, 315–321.

Bristow, C. S., Duller, G. A. T. and Lancaster, N. (2007). Age and dynamics of linear dunes in the Namib Desert. *Geology*, **35**, 555–558.

Bullard, J. E., Thomas, D. S. G., Livingstone, I. and Wiggs, G. F. S. (1996). Wind energy variations in the southwestern Kalahari Desert and implications for linear dunefield activity. *Earth Surface Processes and Landforms*, **21**, 263–278.

Cooke, R. U., Warren, A. and Goudie, A. S. (1993). *Desert Geomorphology*. London: UCL Press.

Dong, Z., Chen, G., He, X., Han, Z. and Wang, X. (2004). Controlling blown sand along the highway crossing the Taklimakan desert. *Journal of Arid Environments*, **57**, 329–344.

Embabi, N. S. (1986/7). Dune movement in the Kharga and Dakhla oases depressions, the Western Desert, Egypt. *Bulletin de la Société de Géographie d'Egypte*, **59–60**, 35–70.

Escalante, S. A. and Pimentel, A. S. (2008). Coastal dune stabilization using geotextile tubes at Las Colorados. *Geosynthetics*, **26**, 16–24.

Forman, S. L., Sagintayev, Z., Sultan, M. *et al.* (2008). The twentieth-century migration of parabolic dunes and wetland formation at Cape Cod National Sea Shore, Massachusetts, USA: landscape response to a legacy of environmental disturbance. *The Holocene*, **18**, 765–774.

Han, Z., Wang, T., Sun, Q., Dong, Z. and Wang, X. (2003). Sand harm in Taklimakan Desert Highway and sand control. *Journal of Geographical Sciences*, **13**, 45–53.

Han Z., Wang, T., Dong Z., Hu, Y. and Yao, Z. (2007). Chemical stabilization of mobile dunefields along a highway in the Taklimakan desert of China. *Journal of Arid Environments*, **68**, 260–270.

Hanson, P. R., Joeckle, R. M., Young, A. R. and Horn, J. (2009). Late Holocene dune activity in the eastern Platte River Valley, Nebraska. *Geomorphology*, **103**, 555–561.

Hesp, P. A. and Thom, B. G. (1990). Geomorphology and evolution of active transgressive dunefields. In K. F. Nordstrom, N. P. Psuty and R. W. G. Carter (eds.), *Coastal Dunes: Form and Process*. Chichester: John Wiley and Sons, pp. 253–288.

Hugenholtz, C. H., Wolfe, S. A., Walker, I. J. and Moorman, B. J. (2009). Spatial and temporal patterns of aeolian sediment transport on an inland parabolic dune, Bigstick Sand Hills, Saskatchewan, Canada. *Geomorphology*, **105**, 158–170.

Jimenez, J. A., Maia, L. P., Serra, J. and Morais, J. (1998). Aeolian dune migration along the Ceará coast, north-eastern Brazil. *Sedimentology*, **46**, 689–701.

Kittredge, J. H. (1948). *Forest Influences*. New York: McGraw Hill.

Lancaster, N. (1989). *The Namib Sand Sea*. Rotterdam: Balkema.

Li, Y., Cui, J., Zhang, T., Okuro, T. and Drake, S. (2009). Effectiveness of sand-fixing measures on desert land restoration in Kerqin Sandy Land, northern China. *Ecological Engineering*, **35**, 118–127.

Livingstone, I. (1989). Monitoring change on a Namib linear dune. *Earth Surface Processes and Landforms*, **14**, 317–332.

Maia, L. P., Freire, G. S. S. and Lacerda, L. D. (2005). Accelerated dune migration and aeolian transport during El Niño events along the NE Brazilian coast. *Journal of Coastal Research*, **21**, 1121–1126.

Marín, L., Forman, S. L., Valdez, A. and Bunch, F. (2005). Twentieth century dune migration at the Great Sand Dunes National Park and Preserve, Colorado, relation to drought variability. *Geomorphology*, **70**, 163–183.

Marsh, G. P. (1864). *Man and Nature*. New York: Scribner.

Mitasova, H., Overton, M. and Harmon, R. S. (2005). Geospatial analysis of a coastal sand dune field evolution: Jockey's Ridge, North Carolina. *Geomorphology*, **72**, 204–221.

Pye, K. and Tsoar, H. (1990). *Aeolian Sand and Sand Dunes*. London: Unwin Hyman.

Raji, B. A., Utovbisere, E. O. and Momodu, A. B. (2004). Impact of sand dune stabilization structures on soil and yield of millet in the semi-arid region of NW Nigeria. *Environmental Monitoring and Assessment*, **99**, 181–196.

Stokes, S. and Bray, H. (2004). Reconciling lateral and vertical; dune migration rates and drift potential estimates: some examples from the Rub Al Khali. *Geophysical Research Abstracts*, **6**, 04012.

Stokes, S., Goudie, A. S., Ballard, J. *et al.* (1999). Accurate dune displacement and morphometric data using kinematic GPS. *Zeitschrift für Geomorphologie Supplementband*, **116**, 195–214.

Thomas, D. S. G. (1992). Desert dune activity: concepts and significance. *Journal of Arid Environments*, **22**, 31–38.

Watson, A. (1990). The control of blowing sand and mobile desert dunes. In A. S. Goudie (ed.), *Techniques for Desert Reclamation*. Chichester: Wiley, pp. 35–85.

Wiles, G. C., McAllister, R. P., Davi, N. K. and Jacoby, G. C. (2003). Eolian response to Little Ice Age climate change, Tana Dunes, Chugach Mountains., Alaska, U.S.A. *Arctic, Antarctic and Alpine Research*, **35**, 67–73.

Yao, Z. Y., Wang, T., Han, Z. W., Zhang, W. M. and Zhao, A. G. (2007). Migration of sand dunes on the northern Alxa Plateau, Inner Mongolia, China. *Journal of Arid Environments*, **70**, 80–93.

Zhang, T-H., Zhao, H-L., Li, S-G. *et al.* (2004). A comparison of different measures for stabilizing moving sand dunes in the Horqin Sandy Land of Inner Mongolia, China. *Journal of Arid Environments*, **58**, 203–214.

Part II
Processes and applications of geomorphology to risk assessment and management

17 GIS for the assessment of risk from geomorphological hazards

Cees J. van Westen

17.1 Introduction

As demonstrated by well-known disaster statistics (e.g. EM-DAT, 2008), the world is confronted with a rapidly increasing impact of disasters, many of those derived from geomorphological extreme events. The possible impact of hazardous events can no longer be ignored and there is an urgent need to include the concepts of disaster risk management into spatial planning, environmental impact assessment (EIA) and strategic environmental assessment (SEA). Tools need to be developed to evaluate the possible losses given different scenarios of development and climate change, which may lead to different patterns of risk than those that are prevalent today (IPCC, 2007). The Hyogo framework of action 2005–15 of the United Nations International Strategy for Disaster Reduction (UN-ISDR) indicates a number of priority actions for disaster risk reduction in the coming decade. One of these is to ensure that disaster risk reduction is a national and a local priority with a strong institutional basis. Another one is to identify, assess, and monitor disaster risks and enhance early warning. Disaster risk management requires the assessment of risk, which is a multi-disciplinary endeavor embedded in a good culture of risk governance involving all stakeholders (UN-ISDR, 2005).

The evaluation of the expected losses due to hazardous events requires a spatial analysis, as all components of a risk assessment differ in space and time. Therefore hazard, vulnerability, and risk assessments can only be carried out effectively when use is made of tools that handle spatial information, such as geographic information systems (GIS).

17.1.1 GIS and risk assessment

A geographic information system is defined traditionally as "a system for the input, management, analysis, and output of spatial data and information" (Aronof, 1993). Traditionally a GIS was seen as a software tool (e.g. ArcGIS, MAPINFO, GRASS, ILWIS). However, there are many more computer-based tools for handling spatial information, such as tools for collecting and storing locational data (GPS), local spatial and non-spatial information (MobileGIS, PGIS), remote sensing data (image processing), altitude data (photogrammetry, LiDAR, InSAR software), attribute data (RDMS) and statistical data (e.g. SPSS, R). Also tools are included for the modeling of spatial processes (e.g. Matlab, hazard models), spatial multi-criteria evaluation (e.g. ILWIS SMCE), decision support (SDSS), and visualization of spatial data (cartographic software, Web-GIS, Google Earth, Microsoft Bing). GIS software tools vary widely with respect to the underlying theory on spatial information, spatial data representation, types of data analysis components included, interfaces, user friendliness, and costs.

As far as GIS-related material related to multi-hazard risk assessment is concerned, the HAZUS methodology developed in the USA can be considered the standard. This comprehensive loss estimation software, which runs under ArcGIS, is a very good tool for carrying out loss estimations for earthquakes, flooding, and windstorms (FEMA, 2008). It allows the estimation of physical damage (e.g. to buildings, schools, critical facilities, and infrastructure), economic loss (e.g. lost jobs, business interruptions, repair and reconstruction costs) and social impacts (e.g. shelter requirements, displaced households). However the use of HAZUS is restricted to the USA because of data constraints, and standards used in the software are mostly only valid for the USA.

This chapter intends to provide an overview on the use of spatial information for the assessment of risk due to geomorphological hazards. It is illustrated with materials from

Geomorphological Hazards and Disaster Prevention, eds. Irasema Alcántara-Ayala and Andrew S. Goudie. Published by Cambridge University Press. © Cambridge University Press 2010.

a case study that has been developed by the United Nations University – ITC School on Disaster Geo-Information Management. This case study, called RiskCity, provides practical hands-on experience on the procedures for collecting and analyzing spatial information for multi-hazard risk assessment in an urban context.

17.1.2 Spatial risk assessment approach

UN-ISDR (2004) defines risk as "the probability of harmful consequences, or expected losses (deaths, injuries, property, livelihoods, economic activity disrupted or environment damaged) resulting from interactions between natural or human-induced hazards and vulnerable conditions." Risk can be presented conceptually with the following basic equation:

$$Risk = Hazard * Vulnerability$$
$$* Amount\ of\ element\text{-}at\text{-}risk \qquad (17.1)$$

The equation given above is not only a conceptual one, but can also be actually calculated with spatial data in a GIS to quantify risk from geomorphological hazards. The way in which the amounts of elements-at-risk are characterized (e.g. as number of buildings, number of people, economic value) also defines the way in which the risk is presented. Table 17.1 gives a more in-depth explanation of the various components involved. In order to calculate the specific risk (see Table 17.1) Equation (17.1) can be modified in the following way:

$$R_s = P_{(T:Hs)} * P_{(L:Hs)} * V_{(Es|Hs)} * A_{Es} \qquad (17.2)$$

in which:

$P_{(T:Hs)}$ is the temporal (e.g. annual) probability of occurrence of a specific hazard scenario (H_s) with a given return period in an area;

$P_{(L:Hs)}$ is the locational or spatial probability of occurrence of a specific hazard scenario with a given return period in an area impacting the elements-at-risk;

$V_{(Es|Hs)}$ is the physical vulnerability, specified as the degree of damage to a specific element-at-risk E_s given the local intensity caused due to the occurrence of hazard scenario H_s;

A_{Es} is the quantification of the specific type of element at risk evaluated (e.g. number of buildings).

In order to be able to evaluate these components, we need to have spatial information as all components of Equation (17.2) vary spatially, as well as temporally. Equation (17.2) also contains a term ($P_{(T:Hs)}$) indicating the

spatial probability of occurrence and impact. Although for most types of hazards this can be considered 1, for others, such as landslides, the location of future events cannot be identified exactly, because the areal unit used in assessing hazard is not always identical to the area specifically impacted by the hazard. The intensity of the hazard varies from place to place (e.g. flood depth or landslide volume), and the location of the elements-at-risk varies.

For a number of different hazard scenarios the consequences are plotted against the temporal probability of occurrence of the hazard events in a graph. Through these points a curve is fitted, the so-called risk curve, and the area below the curve presents the total risk. In a multi-hazard risk assessment this procedure is carried out for all individual hazard types, and care should be taken to evaluate also interrelations between hazards (e.g. domino effects, such as a landslide damming a river and causing a flood). The approach indicated in Table 17.1 is related to the estimation of physical vulnerability, and its use in quantitative risk assessment. There are other approaches that can be used to assess risk, leading to more qualitative results, involving aspects such as coping capacity, resilience, exposure, etc. (Birkmann, 2006). Figure 17.1 gives a schematic overview of the process of multi-hazard risk assessment, based on the case study of RiskCity.

The RiskCity training package on urban multi-hazard risk assessment is based on a dataset from the city of Tegucigalpa in Honduras. Tegucigalpa suffered severe damage from landslides and flooding during Hurricane Mitch in October 1998 (Harp *et al.*, 2002; Mastin and Olsen, 2002).

17.2 Spatial data requirements for risk assessment

As can be concluded from the above, a large amount of spatial and attribute data is required to carry out a comprehensive multi-hazard risk assessment. The amount and detail of these data depend on a number of interrelated factors, such as the objectives of the study, the stakeholders involved, the scale of the study, the size of the study area, the types of hazards, the methods used, the availability of resources (money and manpower), and the amount of available data. Table 17.2 gives an overview of the dataset used for the case study of RiskCity. In this example the analysis was carried out for a local authority, as the basis for a cost/benefit evaluation of risk reduction options, and as input in land use planning and disaster preparedness planning. The work was carried out at a large scale (1:10,000) in an urban area, subjected to several types of hazards (here we limit ourselves to

TABLE 17.1. *List of terms and definitions used in the GIS-based risk assessment presented in this chapter (based on IUGS, 1997; UN-ISDR, 2004)*

	Definition	Equations & explanation
Natural hazard (H)	A potentially damaging physical event, phenomenon or human activity that may cause the loss of life or injury, property damage, social and economic disruption or environmental degradation. This event has a probability of occurrence within a specified period of time and within a given area, and has a given intensity.	$P_{(T:Hs)}$ is the temporal (e.g. annual) probability of occurrence of a specific hazard scenario (H_s) with a given return period in an area; $P_{(L:Hs)}$ is the locational or spatial probability of occurrence of a specific hazard scenario with a given return period in an area impacting the elements-at-risk.
Elements-at-risk (E)	Population, properties, economic activities, including public services, or any other defined values exposed to hazards in a given area. Also referred to as "assets".	E_s is a specific type of elements-at-risk (e.g. masonry buildings of two floors).
Vulnerability (V)	The conditions determined by physical, social, economic and environmental factors or processes that increase the susceptibility of a community to the impact of hazards. Can be subdivided in physical, social, economic, and environmental vulnerability.	$V_{(Es|Hs)}$ is the physical vulnerability, specified as the degree of damage to E_s given the local intensity caused due to the occurrence of hazard scenario H_s. It is expressed on a scale from 0 (no damage) to 1 (total loss).
Amount of elements-at-risk (A_E)	Quantification of the elements-at-risk either in numbers (of buildings, people, etc.), in monetary value (replacement costs etc.), area or perception (importance of elements-at-risk).	A_{Es} is the quantification of the specific type of element at risk evaluated (e.g. number of buildings).
Consequence (C)	The expected losses (of which the quantification type is determined by A_E) in a given area as a result of a given hazard scenario.	C_s is the "specific consequence", or expected losses of the specific hazard scenario, which is the multiplication of $V_s * A_{Es}$.
Specific risk (R_s)	The expected losses in a given area and period of time (e.g. annual) for a specific set of elements-at-risk as a consequence of a specific hazard scenario with a specific return period.	$R_s = H_s * V_s * A_{Es}$ $R_s = H_s * C_s$ $R_s = P_{(T:Hs)} * P_{(L:Hs)} * V_{(Es|Hs)} * A_{Es}$
Total risk (R_T)	The probability of harmful consequences, or expected losses (deaths, injuries, property, livelihoods, economic activity disrupted or environment damaged) resulting from interactions between natural or human-induced hazards and vulnerable conditions in a given area and time period. It is calculated by first analyzing all specific risks. It is the integration of all specific consequences over all probabilities.	$R_T \approx \sum (R_T) = \sum (H_s * V_s * A_{Es})$. Or better: $R_T = \int (V_s * A_{Es})$ • for all hazard types; • for all return periods; • for all types of elements-at-risk. It is normally obtained by plotting consequences against probabilities, and constructing a risk curve. The area below the curve is the total risk.

flooding and landslides). As can be seen from Table 17.2, the dataset is subdivided into a number of thematic types: remote sensing data, altitude data, elements-at-risk data, and hazard data.

17.2.1 Remote sensing data

Remote sensing is a general term encompassing an extensive range of tools and products for Earth observation that are either spaceborne, airborne or terrestrial in origin and that measure electromagnetic energy reflected by the Earth's surface either using the radiation of the Sun (passive sensors) or using active emitted energy (active sensors such as radar or LiDAR). Remote sensing data are characterized by their spectral resolution (which part of the electromagnetic spectrum is covered), spatial resolution (the minimum surface element for which information can be collected), and temporal resolution (minimum time after which the

Spatial data collection

Remote sensing data	Elements-at-risk database: • Buildings • Population	Hazard database: • Inventory • Factors
Altitude data		

Hazard assessment

Flood modeling: • HEC-RAS, SOBEK • PGIS	Landslide hazard: • Statistical/heuristic • Physically based models

Vulnerability assessment

Flooding: • Stage damage curve • Multi-factor curve	Landslides: • Qualitative • $V = 1$

Qualitative risk assessment

• Spatial MC evaluation

• Risk matrix

Quantitative risk assessment

$$R_S = P_T * P_L * V * A_{E_S}$$

Multi-hazard risk assessment

Flooding 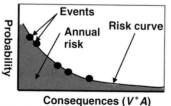 Landslides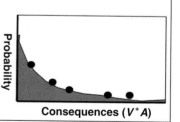

Risk evaluation

Flooding risk: • Risk reduction options • Cost/benefit analysis	Landslide risk: • Risk reduction options • Cost/benefit analysis

FIGURE 17.1. Schematic overview of the multi-hazard risk assessment for the RiskCity case study.

TABLE 17.2. *Overview of data layers used for multi-hazard risk assessment (with emphasis on landslides and floods) for the RiskCity case study*

Name	Type	Meaning
Image data		
Airphoto 1977 ortho	Raster	Orthorectified airphoto from 1977, with a georeference direct linear and resampled to the common georeference of the area.
Airphoto 1977 stereo	Stereopair	Stereopair generated from the Airphoto 1977 ortho and the LiDAR DEM. It can be visualized using a screen stereoscope or using anaglyphs.
Airphoto 1998 ortho	Raster	Orthorectified airphoto from 1998 taken just after the landslide and flood disaster.
Airphoto 1998 stereo	Stereopair	Stereopair generated from the Airphoto 1998 and the LiDAR DEM.
Image 2001 stereo	Raster image	Stereopair generated from a high-resolution colour image derived from an IKONOS image.
Image 2006 stereo	Raster image	Stereopair generated from a high-resolution image downloaded from Google Earth, which was georeferenced and resampled in order to use it for the stereo image interpretations.
ASTER 2005	Image data	Multi-spectral information from the 15 bands of ASTER with a spatial resolution of 15 meters, which are used for land use classification.
Elevation data		
LiDAR DEM	Raster map	A digital surface model derived from a laser scanning flight. The original data points have been interpolated into a 1-meter resolution raster map.
Contours	Segment map	Contour lines with 2.5-meter contour interval digitized from a series of 1:2,000 scale topographical maps.
TopoDEM	Raster map	Digital terrain model showing the elevation of the terrain made by interpolating contour lines into a raster map.
Slope	Raster map	Slope angle map derived from the TopoDEM.
Aspect		Aspect map derived from the TopoDEM.
Elements-at-risk		
Wards	Polygon map	A polygon map representing the administrative units within the city. In the accompanying table information is given on the number of buildings and number of people. Also on a number of characteristics such as age distribution, ethnic distribution, etc.
Mapping units	Polygon map and table	This map represents the mapping units used for elements-at-risk mapping as polygons. Each of the mapping units has a unique identifier, so that in the accompanying table information can be stored for each unit. The units may be individual large buildings or plots with a specific land use, although they are mostly grouping a number of buildings. In the accompanying table information is given on the number of buildings and number of people
Building map	Raster map	Building footprint map of the city prior to the 1998 Mitch event. The map still shows the buildings that were destroyed by landslides and flooding during the Mitch event.
Roads	Segment map	A segment map of the streets, roads, and paths, made by digitizing from topographic maps.
Hazard data		
Landslide ID	Raster map	Landslides in the study area, with an attribute table containing information on the landslides. Attributes are for landslide type, activity, return period of triggering event (50, 100, 200, 300, and 400 years), depth, area, volume, component (scarp or accumulation).
Flood year	Polygon map	Flood extent maps for five different return periods (5, 10, 25, 50, and 100), obtained through modeling with HEC-RAs hydrological software.
Geology	Polygon map	Lithological map digitized from existing paper maps.
Land use	Raster map	Land use map derived from image classification. From this map the roughness coefficients for flood modeling were derived.
Rivers	Segment map	A segment map of the drainage network in the city, digitized from topographic maps.

same part of the Earth can be observed again). These three characteristics form the basis for defining which remote sensing data can be used for which purpose for hazard and risk assessment (CEOS, 2001).

Remote sensing data form the basis for many of the input data layers required for risk assessment. Since remote sensing provides synoptic data over large areas, it is the ideal basis for generating base maps. High-resolution imagery (e.g. Ikonos, Quickbird, Cartosat, PRISM, GeoEye) is a good basis for mapping of hazard-related aspects (e.g. geomorphological mapping, landslide inventory mapping) as well as for the inventory of elements-at-risk (e.g. buildings, roads, agricultural fields, etc.), using visual interpretation. Tools such as Google Earth and Microsoft Bing can be used to download high-resolution images, or display terrain and building information in three dimensions and interpret them directly.

When high-resolution satellite imagery is also capable of providing high-quality digital elevation models (e.g. Cartosat-1, ALOS PRISM), or when this information is obtained from LiDAR flights, it is possible to automatically classify elements-at-risk, such as individual buildings or hazard features (such as landslides, erosion areas, etc.) with object-oriented classification using both spectral and altitude information.

Remote sensing data are very appropriate in hazard assessment both for obtaining inventories of past occurrences of hazardous events, as well as for the mapping of factors that can be used to model susceptibility and hazard. For inventory mapping it is crucial to have sufficient temporal coverage, in order to generate maps of events with different return periods. Imagery should be used that is taken during, or immediately after, a major event. This may be problematic for flood events, as the peak flood may not last very long, and the area is nearly always cloud covered, which makes it difficult to use optical data. If the study area is large and flood events have a duration of several days to weeks, the use of radar images for flood mapping is very suitable.

In the case of RiskCity several types of remote sensing data were used. Aerial photographs for several periods, including the period of the major disaster event in 1998, and two sets of satellite data from 2001 and 2006 were the basis for landslide mapping. High-resolution satellite data were used for mapping elements-at-risk, and medium resolution Aster data for generating a land use map of the area. Figure 17.2 gives an illustration of some of the remote sensing data used in the case study of RiskCity.

17.2.2 Altitude data

Altitude data play an important role in risk assessment for several reasons. A basic dataset is required for correcting remote sensing images, generating ortho-images, stereo-scopic images, and 3-D views. Altitude is a controlling factor for most geomorphological hazards, which are gravity controlled. Altitude information can also be used to monitor hazardous events over time. This is particularly so for mass movement and subsidence, which produce measurable changes in terrain elevation that can be monitored with a range of tools, such as GPS, differential radar interferometry, multi-temporal LiDAR, or photogrammetry. Altitude information can be used to characterize elements-at-risk in relation to the hazard intensity, in order to derive and apply vulnerability curves (see Table 17.2 and Figure 17.2).

17.2.3 Elements-at-risk data

Elements-at-risk data can be obtained at different levels of detail. In the RiskCity case study this is done at the urban level, where information needs to be as detailed as possible, preferably at the individual building level, or at a slightly more aggregated level of mapping units or building blocks with homogeneous land use type. In the RiskCity case study two different situations with respect to the availability of input data were simulated: a situation where the database should be constructed from scratch and a situation in which already detailed spatial and attribute information is available. The first situation (a data-scarce environment) assumed that apart from a high-resolution satellite image and basic altitude information, there were no other data on buildings and population available. In that case the generation of the elements-at-risk database was based on the on-screen digitizing of the basic units for risk assessment. For each mapping unit the main urban land use was interpreted based on image characteristics, using a legend expressing the different building types and temporal population density patterns. Also, for each land use type, characteristics of building types, average building size, and number of persons per unit area were recorded. The number of buildings per mapping unit and land use type was sampled and, based on these, average floorspace, number of floors and population numbers were estimated. Different population scenarios were considered using a daytime, nighttime, and commuter-time scenario, based on the method presented in HAZUS (FEMA, 2008).

The second approach for generating an elements-at-risk database considered the availability of detailed information

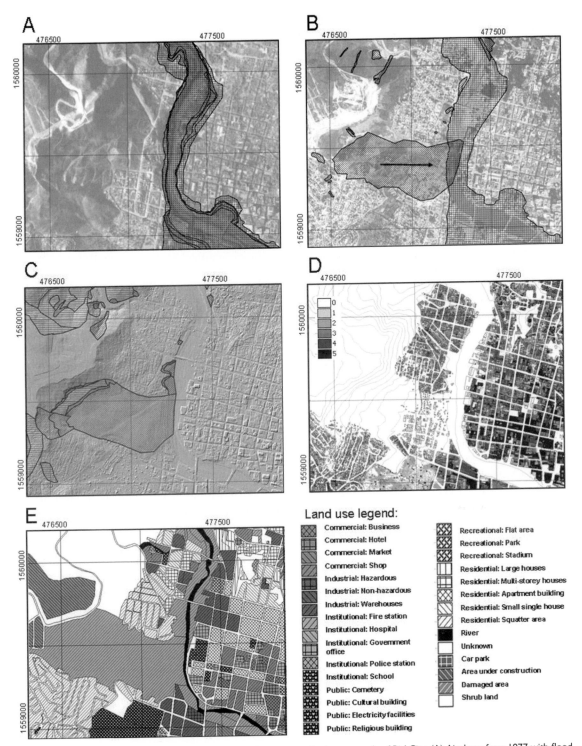

FIGURE 17.2. Different types of spatial information for risk assessment in the case study of RiskCity. (A) Airphoto from 1977 with flood scenarios of different return periods. (B) Post disaster airphoto of 1998 with flood and landslides. (C) Hillshading of LiDAR DSM with landslide inventory. (D) Building height, in number of storeys, from LiDAR DSM and building footprint map. (E) Mapping units, representing zones of more or less homogeneous urban land use and building types, with land use classification.

in the form of a building footprint map, a detailed digital surface model (DSM) derived from LiDAR, and a census database with population information. The building footprint map was used to calculate the building size, and the number of buildings per mapping unit. The LiDAR DSM was used to calculate the building height, which was combined with the footprint area to obtain the total floorspace per mapping unit. Standard values for the number of persons per floorspace area were used to distribute the population numbers available from the census data over the individual buildings and mapping units, again for three different temporal scenarios.

17.2.4 Hazard data

In the RiskCity case study landslide information was obtained from multi-temporal airphoto and satellite image interpretation, fieldwork, and existing maps and reports. One of the main sources of information was the airphotos taken after the major event of 1998, which had an estimated 100-year return period. For this period it was possible to generate a so-called "event-based inventory map". For other landslide-triggering events less detailed information was available, and expert judgment was used to estimate the age of (predominantly old) landslides. A landslide inventory database was made with attribute information regarding the type of landslide, activity, depth, volume, and return period of the triggering event. Information on past flood events was obtained from historical archives. The case study also simulated the results of a participatory mapping exercise where information on the location, timing, and intensity of flood and landslide events was obtained by interviewing the local population. This information was used in validating the results of the flood modeling presented in Section 17.3.1, and for the social vulnerability assessment (see Section 17.4).

The other dataset for hazard assessment consisted of the factor maps used as input data for the modeling of landslide and flood hazard. These maps consist of a geological map, a geotechnical map (e.g. soil depth and soil types with associated geotechnical and hydrological parameters), land use maps from different periods (e.g. derived from image classification) and maps derived from a DEM (slope gradient, slope direction, slope length, etc.).

17.3 Hazard assessment

There are many different techniques that can be used within a GIS for the assessment of susceptibility and hazard maps, such as inventory mapping, data integration (using heuristic or statistical methods) and modeling approaches. In the

RiskCity case study a combination of techniques was selected for the modeling of flood and landslide hazard.

17.3.1 Flood hazard assessment

Historic flood events were reconstructed using participatory GIS mapping, historic studies, and image interpretation of the 1998 airphotos. A frequency–magnitude analysis was made based on available discharge measurements for several stations. Future flood scenarios were modeled using two different models: HEC-RAS and SOBEK. HEC-RAS is a one-dimensional flood model (USACE, 2008) which requires as input a detailed characterization of the cross sections along the main river, generated using the LiDAR data (Mastin and Olsen, 2002), and discharge information. The results of the HEC-RAS modeling were polygon maps with the flood extent for five different return periods. The other model used was SOBEK, a dynamic hydraulic model that has a combination of one- and two-dimensional (1-D and 2-D) models. The 2-D flood model in SOBEK was designed to simulate the flow out of the channel through complex topography (WL/Delft Hydraulics, 2008). The input for this model consists of a detailed DSM, in which the buildings are included, surface roughness, derived from the urban land use map, and discharge information (Alkema, 2007). The output of this model consists of a series of hourly maps of flood depth and flow velocity. The individual flood water depth and flow velocity maps can be used to derive a number of useful indicator maps for flood hazard zonation, such as maximum flood depth and flow velocity, maximum flood impact, minimum time to flooding, flood duration, etc.

17.3.2 Landslide hazard assessment

Landslide hazard assessment was carried out in the RiskCity case study using two different approaches: a combined statistical/heuristic analysis and physical modeling (Figure 17.3). In the statistical analysis the weights of evidence method was used to evaluate the importance of several potential causal factor maps, such as lithology, soil depth, land use, slope gradient, etc. Scarps of active landslides selected from the landslide inventory map were overlain in GIS with the factor maps. For each class of these maps a weight value was calculated with the use of a script in GIS. Based on the weights a selection was made of the most relevant factor classes, which were regrouped into more meaningful factor maps. Eventually, after several iterations, a final susceptibility score map was generated, which was tested with a test dataset of the landslide population, generating success rate curves. These were also

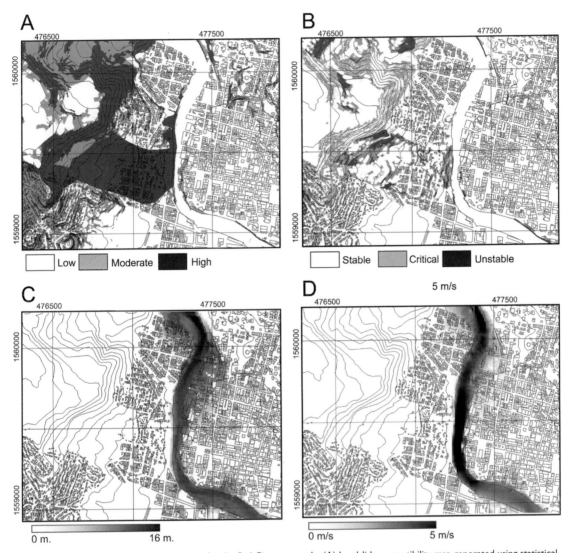

FIGURE 17.3. Main results of the hazard maps for the RiskCity case study. (A) Landslide susceptibility map generated using statistical analysis. (B) Landslide susceptibility map generated using physically based modeling. (C) Maximum flood depth map generated using SOBEK modeling. (D) Maximum flow velocity map generated using SOBEK modeling.

used to classify the susceptibility map into several classes of susceptibility, with predefined levels of event probability (chance that when a landslide happens it will happen in the particular susceptibility class). The susceptibility map was converted to a hazard map using three probabilities, which were multiplied, under the assumption that they are independent: event probability, spatial probability, and temporal probability. For each return period of triggering events, these three individual probabilities were determined using the landslide inventory database.

The second approach for landslide hazard mapping consisted of physically based modeling of slope stability using the infinite slope equation, which can be performed for each pixel individually. First the possible groundwater levels

were simulated using an open source GIS for dynamic modeling, called PCRaster (PCRaster, 2008). Inputs for the analysis were a soil depth map, a map with soil parameters, and the DEM, from which the slope gradient, flow accumulation, and contributing areas for each pixel were derived. Daily rainfall data were also used as input in the simplified version of the STARWARS model (van Beek *et al.*, 2004) which calculated the ratio of groundwater depth against soil depth for each time step. This was used as the input in the infinite slope calculation, resulting in an average safety factor for each pixel per day. The calculation also allowed one to obtain the failure probability, using the PROBSTAB model (van Beek *et al.*, 2004) and the probability of the triggering rainfall events.

17.4 Vulnerability and risk assessment

After generating hazard maps and an elements-at-risk database, the next step in the risk assessment was to combine them and define the level of vulnerability. Vulnerability is a concept that evolved from the social sciences and there is still no common understanding of this concept, illustrated by the fact that there are over 25 different definitions in the literature. This makes it difficult to measure or quantify, as it is also multi-dimensional, scale dependent and dynamic (Birkmann, 2006). As indicated in Table 17.1 and in Equation (17.1) the emphasis in many GIS-based risk assessment studies is on physical vulnerability assessment. However, as mentioned before, it is also important to include other types of vulnerability in the risk assessment, although this cannot be done using the procedure indicated in Table 17.1. For this a spatial multi-criteria evaluation is most suitable.

17.4.1 Spatial multi-criteria evaluation

Spatial multi-criteria evaluation is a technique that assists stakeholders in decision-making with respect to a particular goal (in this case a qualitative risk assessment). It is an ideal tool for transparent group decision-making, using spatial criteria, which are combined and weighted with respect to the overall goal. For implementing the analysis in the RiskCity case study, the SMCE module of ILWIS was used (ITC, 2001). The input is a set of maps that are the spatial representation of the criteria, which are grouped, standardized, and weighted in a criteria tree. The theoretical background for the multi-criteria evaluation is based on the analytical hierarchical process (AHP) developed by Saaty (1980).

In the analysis a number of steps were followed. First the problem was structured into a main goal (qualitative risk assessment) and a number of sub-goals. The main sub-goals identified were *social vulnerability*, *population vulnerability*, *physical vulnerability*, and *capacity*. An over-view of the criteria used for each sub-goal is presented in Figure 17.4. For each of these sub-goals a number of criteria were defined, which measure their performance. Once these were defined, a criteria tree was created, which represented the hierarchy of the main goal, sub-goals, and criteria. For each of the criteria a link was made with the relevant spatial and attribute information. In the RiskCity case study the vulnerability and capacity criteria are linked to three different spatial levels: mapping units, wards, and districts within the city. As the criteria were in different formats (nominal, ordinal, interval, etc.) they were normalized to a range of 0–1. The criteria classes were weighted against each other, then the criteria belonging to the same sub-goal and eventually also the sub-goals themselves were weighted, using either pair-wise comparison, or rank ordering methods. Once the standardization and weighting was done, a composite index map was calculated for each sub-goal, and eventually the qualitative risk map was produced, and classified into a number of classes.

17.4.2 Using fragility and vulnerability curves

If the aim of the risk assessment is to evaluate the consequences in quantitative terms, it is important to have a measure of the relationship between hazard intensity and expected damage for the various elements-at-risk.

Flood vulnerability can be evaluated using so-called stage–damage curves, which relate the degree of damage to a group of elements-at-risk with the same characteristics (e.g. wooden houses of one floor) with the intensity of the hazard (e.g. flood depth). An overview of the various methods for flood vulnerability assessment is presented in Alkema (2007). Most of the stage damage curves for flooding are still based on a single hazard indicator (e.g. flood depth), whereas it is the combination of several (e.g. flood depth, duration, flow velocity) that will actually determine the degree of damage. In the RiskCity case study, four different GIS-based approaches were followed for estimating the vulnerability of buildings to flooding, ranging from simple to more complex. The first approach used simply a vulnerability value of 1 in combination with the flood extent maps for different return periods obtained from HEC-RAS modeling (see Section 17.3.1), resulting in a quantification of the number of buildings affected by flooding for each scenario. The second approach used simple stage–damage relationships in the form of a table, which was linked to the urban land use, providing different values per land use type and flood return period. The third method used the output maps derived from the SOBEK modeling, and compared the depth of flooding to the height of each building, derived from the LiDAR survey, in order to determine what percentage of the building was affected. The fourth approach used a multi-dimensional stage–damage curve, with critical thresholds of flood depth and flow velocity for the safety of people, vehicles, and buildings in the urban area.

Landslide vulnerability studies have received considerable attention in recent years; however, apart from exceptions such as the curves developed for the Hong Kong area, there are still very few useable vulnerability curves for landslides available (Crozier and Glade, 2005).

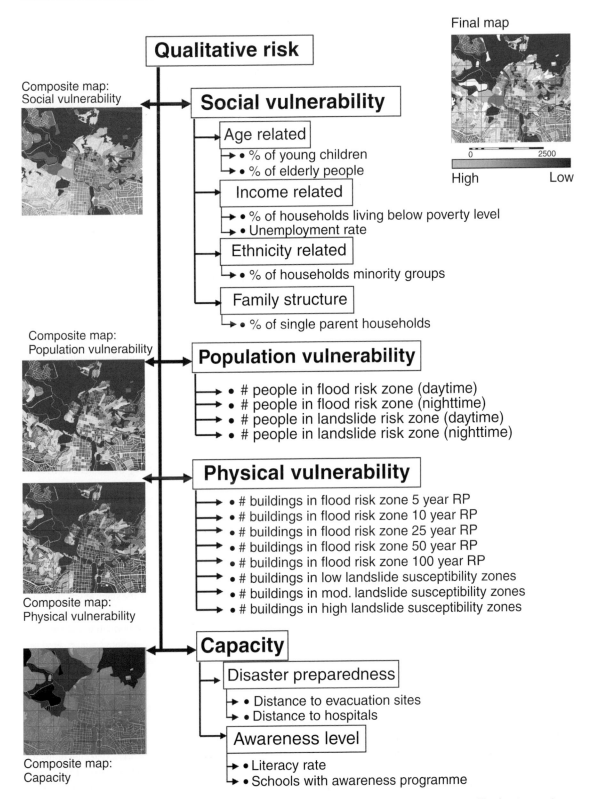

Final map

High   Low

Composite map:
Social vulnerability

Composite map:
Population vulnerability

Composite map:
Physical vulnerability

Composite map:
Capacity

Qualitative risk

Social vulnerability

Age related
→ • % of young children
→ • % of elderly people

Income related
→ • % of households living below poverty level
→ • Unemployment rate

Ethnicity related
→ • % of households minority groups

Family structure
→ • % of single parent households

Population vulnerability

→ • # people in flood risk zone (daytime)
→ • # people in flood risk zone (nighttime)
→ • # people in landslide risk zone (daytime)
→ • # people in landslide risk zone (nighttime)

Physical vulnerability

→ • # buildings in flood risk zone 5 year RP
→ • # buildings in flood risk zone 10 year RP
→ • # buildings in flood risk zone 25 year RP
→ • # buildings in flood risk zone 50 year RP
→ • # buildings in flood risk zone 100 year RP
→ • # buildings in low landslide susceptibility zones
→ • # buildings in mod. landslide susceptibility zones
→ • # buildings in high landslide susceptibility zones

Capacity

Disaster preparedness
→ • Distance to evacuation sites
→ • Distance to hospitals

Awareness level
→ • Literacy rate
→ • Schools with awareness programme

FIGURE 17.4. Overview of the SMCE procedure for qualitative multi-hazard vulnerability and risk assessment. The decision tree is indicated with the goal, sub-goals, and criteria. Also the composite maps of the sub-goals and goal are shown.

Landslides have a wide range of movement types, ranging from fast-impacting rockfalls to slow-moving deep-seated slides, and there is no unique indicator to use for the hazard intensity. Furthermore, damage databases for landslides are far less advanced than for flooding or earthquakes, which makes it difficult to derive vulnerability curves using an empirical approach (see Chapter 5). For the RiskCity case study, three GIS-based approaches were used to include landslide vulnerability in the risk calculations. First of all a qualitative vulnerability rating was made, using the population density and building density for each mapping unit, which was then combined with landslide susceptibility using a simple risk matrix. A second approach was to simply take the landslide vulnerability as 1 for individual buildings, assuming that if a building was hit by a landslide, the chance of destruction would be nearly 100 percent. The third approach tried to relate building vulnerability to the size of the building, and related the total floorspace to the degree of damage, assuming that single-storey small buildings would be destroyed completely, and taller buildings would suffer less damage.

17.4.3 Multi-hazard risk assessment

Multi-hazard risk assessment was carried out in the RiskCity case study for buildings and population. First attribute maps were generated that contain the number of buildings affected for each hazard type and hazard class (A_{Es} in Equation (17.2)) for each of the 1,306 mapping units in the urban area. Then the values were multiplied with values for vulnerability ($V_{(Es \mid Hs)}$) and with the temporal and spatial probability ($P_{(T:Hs)}$ and $P_{(L:Hs)}$) to convert them into annual risk values. The consequences were plotted against the annual probability and risk curves were generated (see Figure 17.5). For each hazard type, separate scenarios were made for each return period. In the case of flooding, this was done for return periods of 5, 10, 25, 50, and 100 years, based on the results of the HEC-RAS and SOBEK modeling. For landslides, scenarios were made with return periods of 50, 100, 200, 300, and 400 years, based on the landslide inventory, which included a considerable subjective component in defining the age of many of the large landslides in the area (see Figure 17.5). For flooding, the spatial probability ($P_{(L:Hs)}$ in Equation (17.2)) was taken as 1 since the individual hazard scenarios indicate the areas that will be flooded. For the landslide risk assessment the return period of the triggering event ($P_{(T:Hs)}$) was multiplied with the spatial probability of landslides occurring in the high, moderate, and low susceptibility classes ($P_{(L:Hs)}$ in Equation (17.2)), the vulnerability per land use type, and the number of

buildings located in each of the three zones. The resulting risk curves were plotted and the annual risk was calculated by integrating the area under the curve, and by using a simply graphical area calculated in Excel.

Population losses were estimated on the basis of building losses, based on the population vulnerability estimates for different injury levels indicated by HAZUS (FEMA, 2008). HAZUS uses four severity levels ranging from minor injuries (level 1) to deaths (level 4). For each of these four injury levels a relation is made with the building damage level. In a table linked to the mapping units the percentage of the population with a particular severity level was calculated. This calculation was done for the three temporal population scenarios indicated before. The population values were plotted using F-N curves to serve as a basis for defining the risk acceptability levels.

The calculation of direct economic losses due to flooding and landslides was restricted to building losses. The degree of loss to buildings was taken as input for this assessment. In order to evaluate the buildings in terms of unit replacement costs, standard values were linked to the urban land use types, and the floorspace for individual buildings. Unit costs (per square meter), based on literature review and evaluation of real estate values, were then applied per mapping unit for buildings and for contents of buildings. These were multiplied with the floorspace to get the total costs per mapping unit. After this attribute, maps were generated that contain the costs of buildings affected for each hazard type and hazard class, which were then used as the A_{Es} parameter in Equation (17.2), and combined with the vulnerability and probability information to generate risk curves with economic losses.

17.5 Risk management

The risk curves were subsequently used as the basis for risk evaluation. Risk evaluation is the stage at which values and judgements enter the decision process, explicitly or implicitly, by including consideration of the importance of the estimated risks and the associated social, environmental, and economic consequences, in order to identify a range of alternatives for managing the risks (IUGS, 1997).

For the RiskCity case study an evaluation of possible risk reduction options was made for both flood and landslide hazards, based on a cost/benefit analysis, in which the investments for risk reduction measures were compared with the benefits of the risk reduction. Two risk reduction options were evaluated for flooding and two for landslides. The first option for flooding involved the possible removal of residential buildings in the zone with a 10-year return period flood hazard. The option considered removal of

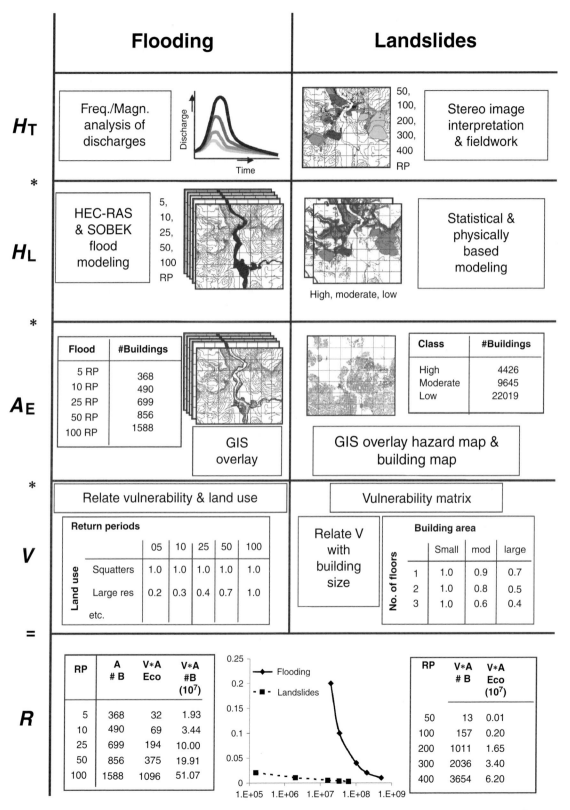

FIGURE 17.5. Procedure to produce the risk curves for flooding and landslide. The individual components of the risk equation (17.2) are given on the left-hand side. For each component some schematic analysis results are presented. See text for further explanation.

buildings, acquisition of land and construction of new buildings and infrastructure, and the conversion of the high hazard zone into a park. The plan also included the setting-up of a municipal department with a vigilance function to prevent illegal settlements in the area by squatters. The risk in the hazard zone with a 10-year return period will be reduced to 0, as a consequence of this risk reduction measure, and the expected losses for scenarios with higher return periods than 10 years will become lower. The second flood risk reduction option involves the construction of an upstream storage basin. The flood retention basin and drainage channel also need regular maintenance. In terms of flood risk reduction the retention basin will reduce the risk to 0 for events less than 10 years, and lower the effects for the higher return periods. The two options considered for landslides were the evacuation of buildings in the highest risk zones, and the construction of slope stabilization measures in the areas with highest landslide risk.

The GIS-based procedure for cost/benefit analysis started with the calculation of the annual risk in economic terms based on the current situation (described in Section 17.4.3). The reduction in risk was calculated for the four options, based on the assumptions described above. Next the investment costs were calculated. Standard values for buildings were used based on the urban land use type and floorspace, and investment costs were distributed over a number of years, which was considered to be greater for the options that involved relocation of buildings. Also the long-term costs were calculated (e.g. for the establishment of a department controlling illegal settlements and for maintenance and operation) taking into account inflation rates. Also for each option a project lifetime was defined, during which the investments of the project should be paid off. For instance, for the flood retention basin the project lifetime was taken as 40 years. Based on the annual risk reduction, investment costs, and the operation and maintenance costs, an annual incremental benefit was calculated for the project lifetime. Finally an evaluation was made of the options based on the so-called net present value (NPV), taking into account that the same amount of money in the future will probably be less valuable than today, the interest rate, and the internal rate of return (the discount rate/interest rate at which the NPV = 0). Based on the cost/benefit analysis, the risk reduction options involving the evacuation of existing buildings turned out to be much less beneficial than the other ones.

17.6 Conclusions

In the cost/benefit analysis only the economic aspects of the risk reduction scenarios were considered. There are of course many more socio-economic aspects to consider. For instance, communities living in the high-risk areas may not be willing to move, as they have historic ties with the place where they live, and depend on the location where they live for their livelihood, etc. These intangible aspects should also be taken into account apart from the purely economic ones. Therefore, a risk management should involve many more aspects, and should be based on a sound risk governance approach, involving all stakeholders in the procedure of risk assessment and risk evaluation. In this process, risk communication is also essential, and GIS can be a powerful tool in visualizing the risk for the stakeholders involved; for instance, using Web-GIS. Stakeholders, including local communities, should have an influence on the decisions that are taken for risk reduction and spatial planning. This could be done using spatial multi-criteria evaluation, which is designed to involve the evaluation of alternatives and the importance of many subjective criteria based on group decision-making.

One of the important aspects to consider in multi-hazard risk assessment is the degree of uncertainty of the risk. Since the analysis is based on a chain of operations, each one with a fair amount of uncertainty, this will have a large effect on the resulting risk levels. It is difficult to calculate the range of uncertainty of all input factors and represent risk as a range rather than a single value. When considering the components of Equation (17.2), most uncertainty is in the assumption of the temporal probability (particularly for landslides), and in the estimation of physical vulnerability. Most risk assessment studies are limited to a general estimation of the direct physical (building damage) and human-social (persons affected/injured/killed) losses. However, attempts should also be made to evaluate the direct and indirect economic, social, cultural, and environmental losses. Most of these cannot be evaluated quantitatively, due to lack of data or due to the difficulty of expressing these losses in quantitative terms (e.g. environmental or cultural losses). In this process of multi-dimensional direct and indirect risk assessment, spatial multi-criteria evaluation is also considered to be a very useful tool.

Another aspect that should be mentioned here is that a risk assessment cannot be a static procedure. The elements-at-risk change continuously, due to urban growth and changes in urban land use. Also the hazard footprints and the temporal frequency may change considerably in the short term due to human influences, such as deforestation, improper land use planning, etc., which may aggravate the hazard phenomena. But also on a longer term, due to climate change, the extreme events are expected to become more frequent in many places.

The purpose of this chapter was to illustrate the importance of GIS for multi-hazard risk assessment and risk management. There is a need to train disaster management experts and professionals working in many different disciplines that have an important disaster reduction component, such as planners, engineers, architects, etc.

The RiskCity training package is intended as a tool for this type of training and to demonstrate the utility and requirements of spatial data in urban multi-hazard risk assessment. The package is being extended and the plan is to incorporate more participatory GIS exercises, as well as to include more hazard modeling, using open source software, and to include a component of Web-GIS. It is offered as a distance education course, organized by ITC: http://www.itc.nl/education/courses/course_descriptions/C09-ESA-DED-01.aspx

Acknowledgements

We would like to thank Gonzalo Funes from Honduras for providing the initial datasets. The digital surface model and flood information were obtained from a study by the United States Geological Survey. The high-resolution image was obtained from a project funded by JICA. Ruben Vargas Franco, Dinand Alkema, Lorena Montoya, Michiel Damen, Nanette Kingma, Antonio Naverette, Jean Pascal Iannacone, Manzul Hazarika, and Norman Kerle are thanked for their contributions on various aspects of this case study. This work is part of the United Nations University – ITC School for Disaster GeoInformation Management (www.itc.nl/unu/dgim).

References

Alkema, D. (2007). *Simulating Floods; on the Application of a 2D Hydraulic Model for Flood Hazard and Risk Assessment.* ITC dissertation 147.

Aronoff, S. (1993). *Geographic Information Systems: A Management Perspective.* Ottawa: WDL Publications.

Birkmann, J. (ed.) (2006). *Measuring Vulnerability to Natural Hazards: Towards Disaster Resilient Societies.* Tokyo: UNU-EHS, United Nations University Press.

CEOS (2001). Committee on Earth Observation Satellites Disaster Management Support Group, available at www.ceos.org/pages/DMSG/.

Crozier, M. J. and Glade, T. (2005). Landslide hazard and risk: issues, concepts and approach. In T. Glade, M. Anderson and M. J. Crozier (eds.), *Landslide Hazard and Risk.* Chichester: John Wiley & Sons, Ltd., pp. 1–40.

EM-DAT (2008). EM-DAT, Emergency Events Database, available at: www.emdat.be/.

FEMA (2008). HAZUS, FEMA's software for estimating potential losses from disasters. Federal Emergency Management Agency, www.fema.gov/plan/prevent/hazus/.

Harp, E. L., Castaneda, M. and Held, M. D. (2002). *Landslides triggered by Hurricane Mitch in Tegucigalpa, Honduras.* USGS Open-File Report 02–0033. pubs.usgs.gov/of/2002/ofr-02-0033/.

IPCC (2007). *Climate Change 2007: Impacts, Adaptation and Vulnerability.* Contribution of Working Group II to the Fourth Assessment Report of the Intergovernmental Panel on Climate Change, available at: http://www.ipcc.ch/ipccreports/ar4-wg2.htm.

ITC (2001). ILWIS 3.0 Academic – User's Guide. ITC, Enschede, Netherlands, 520 pp, available at http://52north.org/index.php?option=com_content&task=view&id=149&Itemid= 127.

IUGS (1997). Quantitative risk assessment for slopes and landslides: the state of the art. Working Group on Landslides, Committee on Risk Assessment, 1997. In D. Cruden and R. Fell (eds.), *Landslide Risk Assessment.* Rotterdam: A. A. Balkema, pp. 3–12.

Mastin, M. C. and Olsen, T. D. (2002). *Fifty-year Flood-inundation Maps for Tegucigalpa, Honduras.* U.S. Geological Survey Open-File Report 02–261, pubs.usgs.gov/of/2002/ofr02261/.

PCRaster (2008). PCRaster Environmental Software, University of Utrecht, the Netherlands, available at: www.pcraster.nl.

Saaty, T. L. (1980). *The Analytic Hierarchy Process: Planning, Priority Setting, Resource Allocation.* New York: McGraw-Hill.

UN-ISDR (2004). Terminology of disaster risk reduction. UN-International Strategy for Disaster Reduction, available at www.unisdr.org/eng/library/lib-terminology-eng%20home.htm.

UN-ISDR (2005). Hyogo Framework for Action 2005–2015: Building the Resilience of Nations and Communities to Disasters, Kobe, Hyogo, Japan, available at www.unisdr.org/eng/hfa/hfa.htm.

USACE (2008). Hydrologic Engineering Centers River Analysis System (HEC- RAS) software, United States Army Corps of Engineers, available at www.hec.usace.army.mil/software/hec-ras/.

van Beek, L. P. H. and van Asch, T. W. (2004). Regional assessment of the effects of land-use change on landslide hazard by means of physically based modelling. *Natural Hazards*, **31**(1), 289–304.

WL/Delft Hydraulics (2008). SOBEK-Rural: Hydrodynamics. Available at delftsoftware.wldelft.nl.

18 Hazard assessment for risk analysis and risk management

Michael Crozier and Thomas Glade

18.1 Approach

The focus in this chapter is on the client – what is it that hazard and risk managers want from geomorphologists and what do geomorphologists believe that their science can constructively offer hazard and risk management? However, communicating skills and requirements can be difficult because scientists and practitioners come from different backgrounds and work within different constraints. On the one hand, the geomorphologist primarily needs to satisfy the research community, while the manager, on the other hand, has to deal with their client base and the public in general, often within a strict statutory, regulatory, policy and financial framework. Clearly, the basic information demands of hazard assessment, of *where* (location), *what* (type of event), *when* (how often) are fundamental to reducing risk but the manager might also legitimately ask 'which areas are free from hazard?', 'what type of mitigation might be appropriate?', 'what sort of monitoring should be undertaken?', 'what changes can we expect in the future?' and 'what is the cost effectiveness of different management options?'.

In post-event situations, geomorphologists may also be required for forensic investigation. In many cases this will be to establish the cause, apportion weight to the causative factors, and to determine the relative importance of human versus natural factors in creating both cause and consequences.

By understanding the geomorphic system, not only in space but also through time, the geomorphologist should be capable of predicting or at least indicating the hazardous characteristics of processes and places within the system, at a range of spatiotemporal scales. They should be able to identify the 'hotspots' in the system, the direction of system change, where and when the intrinsic or extrinsic thresholds might be met but more importantly how those

thresholds may change in time and space. For high-risk situations, there will always be a need for further detailed investigations often by a range of specialists. The geomorphologist, however, should be able to identify the important issues in the light of interconnections within the geosystem and consequently to guide those investigations, and pose the critical questions that need to be pursued in greater detail on site.

While this chapter is intentionally confined to the 'hazard assessment' component of the wider framework of risk management, it is important for the geomorphologist to recognise that authorities involved with hazards are ultimately concerned with assessment, evaluation and treatment of risk (that is the expected losses associated with hazard). Risk itself is not just a function of the hazard (probability of occurrence of a given magnitude of event in a given region/location and period) but also the elements of value exposed to the hazard and their vulnerability (Crozier and Glade, 2005). A general problem in any hazard or risk study is, however, that uncertainty is commonly not well addressed or communicated in hazard and risk assessments.

18.2 Basic concepts and issues

18.2.1 Behaviour of geomorphic processes: what makes them hazardous

The characteristics of geomorphic processes that make them hazardous include a wide range of parameters including: volume (mass), velocity, depth, mechanisms, duration, areal extent, and speed of onset. The relevance and damage capability (referred to here as *intensity*) that can be attached to these parameters varies with the type of hazard, the magnitude of these *intensity parameters* at the time of occurrence, and with the frequency with which

Geomorphological Hazards and Disaster Prevention, eds. Irasema Alcántara-Ayala and Andrew S. Goudie. Published by Cambridge University Press. © Cambridge University Press 2010.

they can be expected to occur in a given place. Unfortunately, only some of these critical parameters are recorded in historic data bases in a systematic way conducive to risk assessment. For example, floods may have a good record of peak discharge but little on the duration or residence time of flood water (which may be more important than peak discharge in causing economic loss). Frequency–magnitude analysis (Crozier, 1996, 1999; Crozier and Glade, 1999), which is at the heart of hazard assessment, is often restricted to conventional representations of magnitude, e.g. volume of landslide or snow avalanche, depth of precipitation, discharge of floods, rather than the parameters which more accurately represent the damage potential of the hazard. Additionally, there are three other hazard parameters of importance that are as much a function of the human condition as the physical process. These are predictability, controllability and lethality. Whereas they are not necessarily the domain of the geomorphologist, an appreciation of such parameters is essential to the hazard manager in choosing appropriate mitigation and risk reduction options.

The losses associated with the occurrence of a hazard are rarely confined solely to the event itself or to the specific locality of initial impact (Glade and Crozier, 2005a). It is important for management purposes to view hazard impact at a variety of scales and form (Glade and Crozier, 2005b). The impacts may be direct or indirect, acute (immediate) or chronic (delayed) or may lead to the development of consequential hazards. *Direct impacts* are those consequences incurred by direct physical contact with the hazard process itself. *Indirect impacts*, on the other hand, are changes brought about in the properties and behaviour of other natural systems as a result of hazard activity. Some of these induced changes may give rise to consequential hazards, e.g. a wave being generated by a landslide entering a reservoir, biological hazards arising from stagnant water left in the aftermath of flooding, or a flash flood resulting from a bursting lake formed by a surging glacier. Indirect impacts can be immediate or delayed, occur in the proximity of the initial hazard impact or at some distance from the impact site. For example, a tsunami consequent on submarine fault displacement may have its maximum impact delayed by many hours and manifest hundreds of kilometres away from the site of origin. Similarly, a large debris flow event caused by heavy precipitation in mountain ranges may cause extensive damage tens of kilometres away from the site of initiation (e.g. Lopez *et al.*, 2003). *Acute impacts* are short lived while *chronic impacts* may be manifest over a longer period of time, as for example economic losses attendant upon damaged infrastructure or loss of means of production.

Our understanding of the complexity of hazard and risk very often depends on analysis of existing records. However, one has to be cautious. The knowledge of past events and possible consequent damage is only available if these incidences have been reported. Thus, no information on former events does not automatically imply that there were no events in the past, it might just be an expression of missing records (Glade *et al.*, 2001). Although a trivial aspect, this is often disregarded in hazard and risk analysis and consequent interpretation.

18.2.2 Non-linearity and frequency–magnitude assessment

There is a variety of approaches used to assess magnitude and frequency including: the use of physically based modelling (Brooks *et al.*, 2004), analysis of the instrumental, historic, oral, secular or documentary record (Kemp, 2003), as well as the interpretation of geo archives such as sedimentary stratigraphy (Page *et al.*, 1994). It is in the last approach that geomorphologists, using a full range of increasingly sophisticated dating techniques (Walker, 2005; Gartner, 2007), make a distinctive contribution, particularly in the area of determining trends and shifts in the state of geosystem equilibria.

Our understanding of frequency–magnitude behaviour of physical processes has often been achieved indirectly by establishing the relationship between the behaviour of a forcing agent (triggering agent) and the associated geomorphic response (e.g. the relationship between river discharge and sediment movement, wind velocity and sediment entrainment, rainfall intensity and landsliding (Glade *et al.*, 2000), or earthquake shaking intensity and landsliding (Keefer, 1984; Keefer and Wilson, 1989)). Analysis of this relationship is designed to obtain the triggering (critical) threshold for a given response – thus allowing the record of the triggering agent (often much more complete than that of the hazard itself) to be examined in order to determine the frequency with which the critical threshold is likely to be equalled or exceeded. Many such studies (Wolman and Miller, 1960; Wolman and Gerson, 1978) have established frequency–magnitude relationships on the assumption of steady state or dynamic equilibrium conditions – assuming, for example, that process response is purely power constrained (Richards, 1999). However, in certain situations, the condition of the ambient environment can change sufficiently to alter the critical threshold, thus inducing changes in frequency and magnitude of hazard events, independent of the behaviour of the geomorphic agent. For instance, the relationships established between channel degradation, soil erosion

(Favis-Mortlock and Boardman, 1995) or wind erosion and the behaviour of the triggering agent may not be temporally stable and can readily break down with, for example, the development of bed armouring or lag deposits. In the case of landsliding, synergistic relationships derived from repeated episodes in a given place can increase thresholds and decrease event frequency in relatively short periods of time. For example, the removal of available material by repeated episodic shallow landsliding has been shown to increase overall catchment stability (referred to as *event resistance* (Crozier and Preston, 1999; Brooks *et al.*, 2002)). Similarly, repeated deep-seated bedrock landslides can also induce negative feedback by depleting the availability of susceptible sites – an effect referred to as *site exhaustion* (Cruden and Hu, 1993). Such synergistic changes need to be taken into account in landslide hazard mapping, which conventionally views the presence of landslides as indicators of future landslide susceptibility (referred to as the *precedence approach*).

Assessment of reactivation of existing landslides, on the other hand, may need to consider the possibility of the positive feedback conditions induced by initial movement. In certain cases, movement can reduce material strength from a peak to a weaker residual condition and poorly evacuated landslides can develop a morphology that inhibits drainage and enhances water entry into the slope, both lowering activation thresholds.

Non-linear relationships resulting from response-induced changes make frequency–magnitude assessment a difficult task, calling for an understanding of landform evolution at a range of temporal and spatial scales.

18.2.3 Vulnerability

Vulnerability (the expected degree of loss associated with a given level of hazard intensity) can be viewed as a function of both social and physical conditions. It can be expressed in terms of structural damage (damage ratios), or in terms of human, economic, cultural and environmental loss (Birkmann, 2006; Douglas, 2007; refer also in this book to Chapter 19 by Hufschmidt and Glade).

One of the least understood issues in hazard and risk assessment is the development of damage ratios and physical vulnerability indices with respect to different types of hazard and their intensity (e.g. Fuchs *et al.*, 2007). While damage ratios are well defined for different building materials and design with respect to earthquake shaking (which by definition incorporates different ground conditions (Dowrick, 1996)), there is little such information available for most other hazardous processes. Table 18.1, for example, treats physical vulnerability as a function of degree of exposure, and the nature of the physical process impact. However, the vulnerability ratios listed in this table are very tentative and could clearly be further qualified by degree of structural integrity and type of design. The degree of exposure considered in this table is static and, in order to calculate risk of a population likely to be affected, it is necessary to introduce a dynamic exposure term (e.g. the proportion of the time a person is likely to be in a location exposed to such a hazard).

TABLE 18.1. *Vulnerability of a person to landsliding under different degrees of exposure – the value 1.0 indicating a 100% probability of death*

Location	Description	Vulnerability of a person		
		Data range	Recommended value	Comment
Open space	Struck by rock fall	0.1–0.7	0.5	May be injured but unlikely to cause death
	Buried by debris	0.8–1	1	Death by asphyxia
	Not buried, but hit by debris	0.1–0.5	0.1	High chance of survival
Vehicle	Vehicle is buried/crushed	0.9–1	1	Death almost certain
	Vehicle is damaged only	0–0.3	0.3	High chance of survival
Building	Building collapse	0.9–1	1	Death almost certain
	Inundated building with debris and person is buried	0.8–1	1	Death is highly likely
	Inundated building with debris, but person is not buried	0–0.5	0.2	High chance of survival
	Debris strikes the building only	0–0.1	0.05	Virtually no danger

Modified by Glade (2003) after Wong *et al.* (1997)

Vulnerability as a function of social conditions is a complex concept involving aspects of coping capacity, adaptive capacity, and resilience, which in turn may be related to fundamental developmental and socio-political structural issues of the affected society (Alcántara-Ayala, 2002; Birkmann, 2006).

18.2.4 Hazard assessment

Hazard assessment (hazard analysis) in this context focuses on the physical behaviour of natural processes, particular at their extremes where the magnitude and frequency is such that those affected have been unwilling or unable to make the adaptations and adjustments that would allow nullification of the impacts. *Hazard*, by definition, is the probability of a damaging event and, expressed as such, has the implicit element of prediction and a requirement for exceedence of a notional damage threshold. The concept of probability, however, can be used in two different ways in hazard assessment. First, in the sense of *susceptibility*, i.e. the probability that the pre-conditions at a site will allow occurrence. Commonly, susceptibility assessments simply rank terrain on its likelihood of ever experiencing a hazard event, often in the form of a red (yes), orange (maybe) and green (never) zonation for different spatial expressions and resolutions (e.g. pixel by pixel or polygons, expressing slope segments or even catchments). Second, probability can refer to recurrence in time, i.e. the probability that a hazard event will recur. This latter characterisation of probability provides an expression of frequency that allows the statement of recurrence intervals (return periods). One main drawback in most temporal probability studies is that there is commonly little information on the spatial variation of temporal probability available. While hazard framed in these terms is much more useful for management purposes, it has more rigorous data requirements (van Westen *et al.*, 2006). Many of these issues have been recently addressed in a comprehensive set of international standards for defining, analysing and representing susceptibility and landslide hazard (Fell *et al.*, 2008).

It is critical that assessments of hazard provided to managers have taken into account future likely changes in frequency and magnitude. Such changes result from either culturally related reduction of the damage threshold or from physical causes related to changes in the variability, and/or magnitude of physical processes or from changes in environmental susceptibility (Crozier, 2008). Herein lies the danger of using historical data and empirical models and of treating hazard established at one point of time as a constant for a location. Hazard as well as risk is as dynamic

and evolutionary as the physical system itself (Hufschmidt and Crozier, 2008).

As indicated in Figure 18.1, hazard analysis is only one component of the risk management system and can be carried out at different levels of sophistication, depending on the scope and objectives of the project undertaken and the value of elements at risk that may be threatened. Assessments can range from regional to site scales or to specific object assessments such as buildings or life-line infrastructure (Glade and Crozier, 2005b). They may be required in areas where there is abundant evidence of former hazard events or, on occasions, in areas where there is no previous evidence of hazard occurrence. Hazard assessments may be a component of 'greenfields' planning projects, where considerable options are available for avoidance and mitigation or they may be conducted in high value, densely populated areas, where there are few opportunities for avoidance and technological treatments are the predominant options. Such considerations form part of the scoping phase and will strongly influence the detail and methodology of assessment as well as ultimately the choice of risk reduction solutions.

18.3 The contribution of geomorphology to hazard assessment

18.3.1 Location

Answering the '*where*' question in hazard assessment addresses the question of susceptibility and, while not providing a statement of hazard in itself, it is an essential first step of the assessment process. Geomorphic hazards can be loosely assigned to two groups, on the basis of locational preference. First, those where the required set of critical geomorphic conditions are repeatedly met at the same locality, thus hazard recurrence is expected in the same location (*location specific*), and second, those where the terrain requirements are less specific and can be met over a wide range of locations (*non-location specific*). For example, fault rupture (*location specific*) is confined to linear fault zones whereas ground shaking (*non-location specific*) may be manifest at a wide range of terrain types over a relatively large area. Non-location specific hazards, of course, are much more difficult to manage from a planning perspective, while location specific hazards represent geomorphic hazard 'hot spots' and can be targeted by a range of treatments. Although geomorphic hazard hot spots are well recognised by the geomorphic community, they are not always fully appreciated by planners, managers and the public at large. An obvious example of a hazard hotspot is represented by debris flow and alluvial fans. To the lay

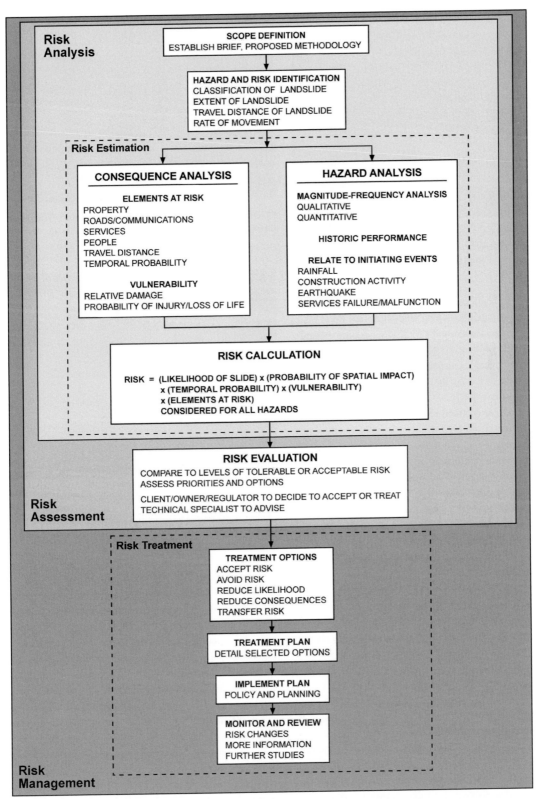

FIGURE 18.1. Flow chart showing all the stages involved in landslide risk management. (Based on Australian Geomechanics Society, 2000.)

FIGURE 18.2. Rakia River fan, New Zealand, showing avulsion pathways of different ages (Note: homestead on the true left of the active channel). (Photo: Jan Thompson.)

FIGURE 18.3. Former sea cliff still actively adjusting by landsliding, Orewa, New Zealand, 2006. (Photo: Graham Hancox.)

person, these landforms can appear to be the best building sites (relatively gently sloping land above the flood plain, below mountain slopes, and easy to use building ground – Figure 18.2), yet they are notorious for channel avulsion, gullying, and complex response (Davies and Korup, 2007), even on parts of the fan that have been dormant for many decades (Gartner, 2007).

Floodplains are perhaps more widely understood hazard 'hotspots' but are in high demand globally for their productivity, easy terrain and trafficability. The geomorphologist can provide information on their mode of formation, frequency of inundation and whether they are contemporary or relict (e.g. Pelletier *et al.*, 2005). In terms of volcanic hazards, lahars tend to be location specific, recurring within well-defined pathways (Lecointre *et al.*, 2004). Volcanic ash showers, on the other hand, are not specific to any particular location and the place affected depends on wind direction and strength at the time of the eruption.

Major hotspots for landslides are 'oversteepened' slopes. These constitute slopes that have been brought into a state of *marginal stability* (Crozier, 1986) through increase in height and/or slope angle, as a result of coastal, fluvial, glacial, tectonic or human action. At marginal stability, even small perturbations can trigger adjustment by slope failure and these highly susceptible conditions can persist for decades or even hundreds of years, until a stable angle of repose with respect to landsliding is reached and slower, more benign, slope process begin to assume dominance. While these slopes are easily recognisable when the erosional activity is still taking place (e.g. contemporary coastal cliffs), after they have been abandoned they are much less identifiable and yet the adjustment process will still continue to operate (Figure 18.3).

FIGURE 18.4. Multiple occurrence regional landslide event. These phenomena can occur almost anywhere in New Zealand hill country, depending on intensity and location of storm rainfall. Gisborne, New Zealand, 2002. (Photo: Michael Crozier.)

Multiple occurrence regional landslide events (Crozier, 2005) are common non-location specific landslide phenomena. They involve the essentially simultaneous occurrence of hundreds to thousands of rainfall- or earthquake-triggered landslides occurring over vast areas of varied terrain (Figure 18.4). The density of landslide occurrence is closely related to the intensity of the triggering agent. While broad terrain thresholds for their occurrence can be recognised, their location is determined by the passage of intense rainfall cells or the epicentre of earthquake energy release, the location of which at any one time is essentially random.

Site specific hazards have the potential of being avoided by the use of planning tools and regulation, including coastal and floodplain marginal set-back zones. Indeed, in some cases of fault rupture, avoidance is the only option, because of the magnitude of ground deformation (for example, the

Wairarapa fault in Wellington, New Zealand, has a characteristic displacement of 13 m horizontal and 2.7 m vertical), thus engineering solutions are simply not feasible (Grapes, 1999). From a management perspective, mitigation of non-location specific hazards is best addressed by generic tools such as education, building design, preparedness and warning. On the other hand, the fact that many hazards are associated with specific landforms indicates that appropriately constructed geomorphic maps can provide a very important role in susceptibility identification (van Westen et al., 2003).

18.3.2 Frequency–magnitude

By definition, frequency and magnitude analysis is the core of hazard assessment. This is conventionally established empirically from the inventory of occurrence or, less directly, by the frequency with which triggering agents are likely to surpass critical thresholds for hazard occurrence. Frequency–magnitude distributions can be established from a temporal record or from a spatial distribution of morphological or sedimentary evidence arising from one or more events (Hovius et al., 1997; Malamud et al., 2004; Guzzetti et al., 2008). For the purpose of hazard prediction it is sometimes assumed (questionably) that spatial frequency–magnitude distribution faithfully represents temporal probabilities. For example, if a spatial inventory of a multiple occurrence landslide event indicates that 10% of the landslides are of a magnitude sufficient to cause damage, then this ratio is transferred to established annual frequencies – an annual frequency of 10 landslides a year would thus suggest an occurrence of one damaging landside per year. As noted earlier, synergistic changes and consequent non-linearity in process–response relationship demand that a sound understanding of geomorphic system behaviour is required if empirical methodologies are to be employed. In particular, empirical approaches also need to factor in the impact of climate and land use change, both of which have the potential to dramatically affect the incidence of hazard occurrence (Haerbeli et al., 1993; Collison et al., 2000; Dehn et al., 2000; Ashmore and Church, 2001; Van Beek, 2002; Goudie, 2006; Crozier, 2008; Clague, 2009).

Hazard assessments may also be required in situations where historical inventories do not exist or where there is no evidence of hazards having occurred in the past. This situation commonly applies in the case of river impoundment and reservoir formation, which have the potential to induce shoreline landslides. The catastrophic consequences associated with large rapid landslides entering a reservoir are such that a planner requires a statement of hazard in order to evaluate the environmental and risk effects of the project. If there is no existing evidence, then the probability

of first-time failure needs to be addressed. Whereas some information may be gained from analogues from other reservoirs in similar terrain and rock condition, the only feasible approach is to use a theoretically based stability analysis, employing scenarios based on the expected changes in slope hydrology related to reservoir filling and proposed operating levels. Probabilities then are usually based on the variability of input parameters and their ability to bring the slope to a factor of safety of 1.0 or less.

While most geomorphic hazards are dynamic, some important ones are static and require a different form of frequency analysis for risk assessment. For example, areas of weak foundation material such as organic-rich deposits, bentonitic clays or sinkholes can represent a substantial hazard to building development. In such situations, frequency can be represented by area ratios or more realistically by *encounter probabilities*. The latter are commonly employed to determine the risk to motorists from site specific landslides and avalanches and involve such factors as the number of vehicles and their speed. Fitzharris and Owens (1980) used such factors in calculating avalanche hazard and risk as follows:

Encounter probability for moving traffic (P_m)

$$P_m = \frac{T \times (L + D) \times F}{V \times 3600 \times 24},$$

where:

P_m = number of moving vehicles hit by avalanches per annum

T = average daily traffic volume (vehicles/day)

V = average speed of traffic (m/s)

L = length of road covered by avalanche (width of avalanche)

D = stopping distance on a snow-covered road for a vehicle with speed V

F = frequency of avalanche occurrence (average number per year).

Encounter probability for stationary traffic (P_w)

$$P_w = Ps \times F \times N,$$

where:

P_w = number of stationary vehicles hit per annum

P_s = probability of another avalanche occurring at the same or adjacent site to one that has forced traffic to stop

N = number of vehicles in the avalanche track.

The *hazard index* (I) is then calculated as:

$$I_{X1}^{X4} = \sum W(P_m + P_w),$$

where W is a weight applied to each category of avalanche X1–X4. The weightings reflect the cost and consequences of an avalanche from a particular category, and in the case of the Milford Road, New Zealand, were chosen to be:

X1 (Powder snow) = 1
X2 (light snow) = 4
X3 (deep snow) = 10
X4 (plunging snow) = 12.

Indeed, additional possibilities to calculate the hazard and risk exist (e.g. McClung, 2005; Zischg et al., 2005: refer also in this book to Chapter 5 by Bründl et al.), however, the basic principle remains the same.

18.3.3 Geomorphology and reconstruction of hazard intensity parameters

Geomorphologists have used a variety of physically based techniques to reconstruct the intensity parameters associated with a range of different hazardous processes. For example, the velocity of rapid earth flows and debris flows can be reconstructed from super-elevation debris run-up on the outside of bends, by the following relationship (Takihashi 2007):

$$E_{max} = \frac{U^2}{2r_{co}g}(2mh_o + b),$$

where:

E_{max} = maximum super-elevation above elevation of channel midpoint
U = cross-sectional mean velocity
r_{co} = radius of curvature for channel centre line
g = acceleration due to gravity
h_o = is mean depth
b and m = constants, in the case of turbulent debris flows $b = 0$ and $m = 3$.

Similarly, Reneau and Dietrich (1987) have estimated debris flow velocity from the difference in elevation between the debris marks on the upstream and downstream faces of mid-channel obstacles.

Nott (1997, 2003) has developed a series of equations for estimating wave height required to move boulders in different positions based on boulder and water density, weight, and friction. These have been applied together with dating techniques to establish frequency and magnitudes of both tsunami and storm waves (Figure 18.5), which differ largely in terms of wave period and celerity (Kennedy et al., 2007). Similar approaches can be used to reconstruct velocity from the maximum size of particles in flood deposits using Hjulström or similar size/velocity entrainment relationships, or from bedform analysis (Simons and Richardson, 1966).

Flood discharges can be reconstructed by identification of the highest flood marks and rack deposit to define

FIGURE 18.5. Tsunami deposit of imbricated boulders used to determine size of emplacement wave. The overlying sand and loess deposit were dated to provide a minimum age for the event. Shag point, Otago, New Zealand. (Photo: David Kennedy.)

channel area and then application of considerations such as Manning's formula.

There are also well-established relationships between fault displacement, length of fault rupture, and earthquake magnitude. For example, Bonilla et al.'s (1984) relationship between surface wave magnitude (M_s) and fault rupture length (L) in kilometres is

$$M_s = 6.04 + 0.704(\log L);$$

whereas for maximum surface displacement in metres (D_{max}) per event the relationship with earthquake magnitude (M_s) is

$$M_s = 7.0 + 0.782(\log D_{max}).$$

Identification of displacement by successive disruption of dated fluvial or lacustrine deposits can then be used to establish the earthquake frequency and magnitude record for given faults.

Wilson and Keefer (1985) have also established relationships between earthquake magnitude and the ellipsoidal area within which earthquake-triggered landslides occur and also with the distance of the furthermost landslide or liquefaction feature from the earthquake epicentre. Hancox et al. (1997) using a similar approach have established that the magnitude of an earthquake can be determined by

$$M = 1.04 \log_{10} A + 3.85,$$

where M is Richter magnitude and A is area (km^2) affected by landslides.

In this way, identification of previous earthquake events through landslide evidence and other earth deformation features can assist in establishing the frequency–magnitude record of an area over long periods of time (Crozier et al. 1995).

The examples given here (although far from complete) have been derived from studies of geomorphic processes carried out to determine their behaviour, causative factors, and their role in landform evolution, and not necessarily with the aim of reconstructing magnitude and frequency for hazard assessment purposes. However, they serve to illustrate that geomorphology can play an important part in understanding hazard behaviour of natural systems.

18.4 Conclusions and perspectives

The previous examples have shown the importance of geomorphic hazard assessments for risk analysis and risk management. This contribution, in particular, focuses on 'hazard assessment' and demonstrates the need for detailed process studies viewed as part of the geosystem as a whole, as a basis for any risk decision. The advantages of the geomorphic approaches can be summarized as follows:

Within a geomorphic assessment, not only currently monitored process intensities are of importance, but also the record of former events determined by sedimentary archives or documentary archives is indispensable. Geomorphic studies are capable of extending the hazard record and thus allowing the full range of energy fluctuations and responses to be appreciated as well as helping to distinguish between variability and change within the system.

Temporal probabilities are often related to single catchments only, but respective spatial information is also required most importantly for spatially extrapolating these relationships and allowing the development of comprehensive planning strategies.

Accordingly, the investigation of interconnections and linkages between processes, earth materials, landforms and land use is a major geomorphic contribution to hazard assessment.

Additionally, a geomorphic assessment considers the current criticality status of the investigated system, thus addressing whether the system is in a stable state, or is near an exceedence threshold, requiring only small trigger magnitudes (e.g. fully saturated slopes).

The limitations of these assessments are in line with restrictions of other hazard and risk assessment approaches and include the following considerations:

Not enough information on former events is available, either for general occurrence or for detailed event characteristics (especially the critical intensity parameters). The associated uncertainty is commonly not addressed. This error includes not only the error (and its propagation) within the analysis, but also in the completeness and accuracy of the available data.

Although conceptually addressed, the linkage between different geomorphic systems (e.g. sediment routing from slopes to flood plains), the importance of triggers (e.g. required earthquake magnitude to cause landform response) as well as the threshold conditions are not sufficiently understood for the full range of terrain types and conditions.

Besides the already addressed advantages and limitations of geomorphic assessments in hazard analysis, it is evident that the social component needs much more integration into relevant hazard and risk research. A selection of major research issues required to address these concerns is represented in the following questions:

How much do humans influence the geomorphic systems directly (e.g. deforestation) and indirectly (e.g. climate change)? Can the ultimate drivers of these influences such as socio-economic conditions, population growth and urbanisation, be factored into future hazard and risk predictive models?

How does the rate of change in the human environment relate to the condition of the geomorphic environment; thus, how large are environmental buffer capacities, what is the human driving force, and how can this be influenced?

Can we predict hazard and risk with sufficient confidence in order to allow sustainable development?

How do we cope with errors, error propagation, and related uncertainty in hazard and risk assessments?

Geomorphology takes hazard assessment beyond the realm of a specific site or a specific moment in time. It uses a range of discipline tools, methodologies and concepts to explore variability and change in time and space. In so doing, it has established that robustness, accuracy and value of hazard assessments can only be met by addressing the interconnections between both physical and human systems at a range of spatial and temporal scales.

References

Alcántara-Ayala, I. (2002). Geomorphology, natural hazards, vulnerability and prevention of natural disasters in developing countries. *Geomorphology*, **47**, 107–124.

Ashmore, P. and Church, M. (2001). *The Impact of Climate Change on Rivers and River Processes*. Geological Survey of Canada Bulletin 555.

Australian Geomechanics Society (2000). Landslide risk management concepts and guidelines. *Australian Geomechanics*, **35**(1), 49–92.

Birkmann, J. (ed.) (2006). *Measuring Vulnerability to Natural Hazards*. Tokyo: United Nations University Press.

Bonilla, M. G., Mark, R. F. and Lienkaemper, J. J. (1984). Statistical relations among earthquake magnitude, surface rupture length, and surface fault displacement. *Bulletin of the Seismological Society of America*, **74**, 2379–2411.

Brooks, S. M., Crozier, M. J., Preston, N. J. and Anderson, M. G. (2002). Regolith evolution and the control of shallow translational hillslope failure: application of a 2-dimensional coupled soil hydrology-slope stability model, Hawke's Bay, New Zealand. *Geomorphology*, **45**(3–4), 165–179.

Brooks, S., Crozier, M. J., Glade, T. and Anderson, M. G. (2004). Towards establishing climatic thresholds for slope instability: use of a physically-based combined soil hydrology-slope stability model. *Pure and Applied Geophysics*, **161**(4), 881–905.

Clague, J. J. (2009). Climate change and slope instability. In K. Sassa and P. Canuti (eds.), *Landslides: Disaster Risk Reduction*. Heidelberg: Springer, pp. 557–572.

Collison A., Wade, S., Griffiths, J. and Dehn, M. (2000). Modelling the impact of predicted climate change on landslide frequency and magnitude. *Engineering Geology*, **55**, 205–218.

Crozier, M. J. (1986). *Landslides: Causes, Consequences and Environment*. London: Croom Helm.

Crozier, M. J. (1996). Magnitude/frequency issues in landslide hazard assessment. In R. Maeusbacher and A. Schulte (eds.), *Beitrage zur Physiogeographie: Barsch Festschrift*. Heidelberger Geographische Arbeiten 104, pp. 221–236.

Crozier, M. J. (1999). Frequency and magnitude of geomorphic processes. In M. J. Crozier and R. Maeusbacher (eds.), *Magnitude and Frequency in Geomorphology*. Zeitschrift für Geomorphologie Supplementband 115, pp. 35–50.

Crozier, M. J. (2005). Multiple-occurrence regional landslide events: hazard management perspectives. *Journal of the International Consortium on Landslides*, **2**(4), 245–256.

Crozier, M. J. (2008). Linking erosion with environmental and societal impacts in a rapidly changing environment. In *Sediment Dynamics in Changing Environments*, Proceedings of a Symposium held in Christchurch, New Zealand, December 2008. International Association of Hydrological Science (IAHS) Publication 325, pp. 469–476.

Crozier, M. J. and Glade, T. (1999). The frequency and magnitude of landsliding: fundamental research issues. In M. J. Crozier and R. Maeusbacher (eds.), *Magnitude and Frequency in Geomorphology*. Zeitschrift für Geomorphologie Supplementband 115, pp. 141–155.

Crozier, M. J. and Glade, T. (2005). Landslide hazard and risk: concepts and approach. In T. Glade, M. G. Anderson and M. J. Crozier (eds.), *Landslide Hazard and Risk*. London: Wiley, pp. 1–40.

Crozier, M. J. and Preston, N. J. (1999). Modelling changes in terrain resistance as a component of landform evolution in unstable hillcountry. In S. Hergarten and H. J. Neugebauer (eds.), *Process Modelling and Landform Evolution*. Lecture Notes in Earth Sciences 78. Berlin: Springer, pp. 267–284.

Crozier, M. J., Diemel, M. S. and Simon, J. S. (1995). Investigation of earthquake triggering for deep-seated landslides, Taranaki, New Zealand. *Quaternary International*, **25**, 65–73.

Cruden, D. M. and Hu, X. Q. (1993). Exhaustion and steady state models for predicting landslide hazards in the Canadian Rocky Mountains. *Geomorphology*, **8**, 279–285.

Davies, T. R. H. and Korup, O. (2007). Persistent alluvial fan-head trenching resulting from large, infrequent sediment inputs. *Earth Surface Processes and Landforms*, **32**, 725–742.

Dehn, M., Bürger G., Buma, J. and Gasparetto, P. (2000). Impact of climate on slope stability using expanded downscaling. *Engineering Geology*, **55**, 193–204.

Douglas, J. (2007). Physical vulnerability modelling in natural hazard risk assessment. *Natural Hazard and Earth System Science*, **7**, 283–288.

Dowrick, D. J. (1996). The Modified Mercalli earthquake intensity scale: revisions arising from recent studies of New Zealand earthquakes. *Bulletin of the New Zealand National Society for Earthquake Engineering*, **29**(2), 92–106.

Favis-Mortlock, D. and Boardman, J. (1995). Non linear responses of soil erosion to climate change: modelling study of the UK South Downs. *Catena*, **25**, 365–387.

Fell, R., Corominas, J., Bonnard, C. *et al.* (2008). Guidelines for landslide susceptibility, hazard and risk zoning for land use planning. *Engineering Geology*, **102**, 85–98.

Fitzharris, B. B. and Owens, I. F. (1980). *Avalanche Atlas of the Milford Road: An Assessment of the Hazard to Traffic*. New Zealand Mountain Safety Council, Avalanche Committee Report No. 4.

Fuchs, S., Heiss, K. and Hübl, J. (2007). Towards an empirical vulnerability function for use in debris flow risk assessment. *Natural Hazard and Earth System Science*, **7**, 495–506.

Gartner, H. (2007). Tree roots: methodological review and new development in dating and quantifying erosive processes. *Geomorphology*, **86**, 243–251.

Glade, T. (2003). Vulnerability assessment in landslide risk analysis. *Die Erde*, **134**(2), 121–138.

Glade, T. and Crozier, M. J. (2005a). The nature of landslide hazard impact. In T. Glade, M. G. Anderson and M. J. Crozier (eds.), *Landslide Hazard and Risk*. London: Wiley, pp. 43–74.

Glade, T. and Crozier, M. J. (2005b). A review of scale dependency in landslide hazard and risk analysis. In T. Glade, M. G. Anderson and M. J. Crozier (eds.), *Landslide Hazard and Risk*. London: Wiley, pp. 75–138.

Glade, T., Crozier, M. J. and Smith, P. (2000). Establishing landslide-triggering rainfall thresholds using an empirical antecedent daily rainfall model. *Journal of Pure and Applied Geophysics*, **157**, 1059–1079.

Glade, T., Albini, P. and Francés, F. (eds.) (2001). *The Use of Historical Data in Natural Hazard Assessments*. Advances in Natural and Technological Hazards Research, Dordrecht: Kluwer Academic Publishers.

Goudie, A. S. (2006). Global warming and fluvial geomorphology. *Geomorphology*, **79**, 384–394.

Grapes, R. H. (1999). Geomorphology of faulting: the Wairarapa Fault, New Zealand. *Zeitschrift für Geomorphologie*, **115**, 191–217.

Guzzetti, F., Peruccacci, S., Rossi, M. and Stark, C. P. (2008). The rainfall intensity-duration control of shallow landslides and debris flows: an update. *Landslides*, **5**(1), 3–18.

Haerbeli, W., Guodong, C., Gorbnov, A. P. and Harris, S. A. (1993). Mountain permafrost and climate change. *Periglacial Processes*, **4**, 165–174.

Hancox, G. T., Perrin, N. D. and Dellow, G. D. (1997). *Earthquake-Induced Landsliding in New Zealand and Implications for MM Intensity and Seismic Hazard Assessment*. Wellington: Institute of Geological and Nuclear Sciences Ltd.

Hovius, N., Stark, C. P. and Allen, P. A. (1997). Sediment flux from a mountain belt derived by landslide mapping. *Geology*, **25**, 231–234.

Hufschmidt, G. and Crozier, M. J. (2008). Evolution of natural risk: analysing changing landslide hazard in Wellington, Aotearoa/New Zealand. *Natural Hazards*, **45**, 255–276.

Keefer, D. K. (1984). Landslides caused by earthquakes. *Geological Society of America Bulletin*, **95**(4), 406–421.

Keefer, D. K. and Wilson, R. C. (1989). Predicting earthquake-induced landslides with emphasis on arid and semi-arid environments. *Publication of the Inland Geological Society*, **2**, 118–149.

Kemp, J. (2003). Documentary flood records from a remote valley in the Scottish Highlands: the River Beuly, UK. In V. R. Thorndycraft, G. Benito, M. C. Llasat and M. Barriendos (eds.), *Paleofloods, Historical Data and Climatic Variability: Applications in Flood Risk Assessment*. Madrid: CSIC, pp. 113–118.

Kennedy, D. M., Tannock, K. L., Crozier, M. J. and Reiser, U. (2007). Boulders of MIS 5 age deposited by a tsunami on the coast of Otago, New Zealand. *Journal of Sedimentary Geology*, **200**(3&4), 222–231.

Lecointre, J., Hodgson, K., Neall, V. and Cronin, S. (2004). Lahar-triggering mechanisms and hazard at Ruapehu Volcano, New Zealand. *Natural Hazards*, **31**(1), 85–109.

Lopez, J. L., Perez, D. and Garcia, R. (2003). Hydrologic and geomorphologic evaluation of the 1999 debris-flow event in Venezuela. In D. Rickenmann and C.-L. Chen (eds.), *Debris-flow Hazards Mitigation: Mechanics, Prediction, and Assessment*, 10–12 September 2003, Davos, Switzerland. Rotterdam: Millpress, pp. 989–1000.

Malamud, B. D., Turcotte, D. L., Guzzetti, F. and Reichenbach, P. (2004). Landslide inventories and their statistical properties. *Earth Surface Processes and Landforms*, **29**(6), 687–711.

McClung, D. (2005). Risk-based definition of zones for land-use planning in snow avalanche terrain. *Canadian Geotechnical Journal*, **42**, 1030–1038.

Nott, J. (1997). Extremely high-energy wave deposits inside the great Barrier Reef, Australia; determining the cause: tsunami or tropical cyclone. *Marine Geology*, **141**, 193–207.

Nott, J. (2003). Waves, coastal boulder deposits and the importance of the pre-transport setting. *Earth and Planetary Letters*, **210**, 269–276.

Page, M. J., Trustrum, N. A. and DeRose, R. C. (1994). A high resolution record of storm-induced erosion from lake sediments, New Zealand. *Journal of Palaeolimnology*, **11**, 333–348.

Pelletier, J. D., Mayer, L., Pearthree, P. A. *et al.* (2005). An integrated approach to flood hazard assessment on alluvial fans using numerical modeling, field mapping, and remote sensing, *Geological Society of America Bulletin*, **117**(9), 1167–1180.

Reneau, S. L. and Dietrich, W. E. (1987). The importance of hollows in debris flow studies: examples from Marin County, California. In J. E. Costa and G. F. Wieczorek (eds.), *Debris Flows/Avalanches: Process, Recognition, and Mitigation*. Reviews in Engineering Geology 7, Boulder: Geological Society of America, pp. 165–180.

Richards, K. (1999). The magnitude-frequency concept in fluvial geomorphology: a component of a degenerating research programme? *Zeitschrift für Geomorphologie Supplementband*, **115**, 1–18.

Simons, D. B. and Richardson E. V. (1966). *Resistance to Flow in Alluvial Channels*. United States Geological Survey Professional Paper 422J.

Takahashi, T. (2007). *Debris Flow Mechanics, Prediction and Countermeasures*. London: Routledge.

Van Beek, R. (2002). *Assessment of the Influence of Changes in Land Use and Climate on Landslide Activity in a Mediterranean Environment*. Nederlandse Geografische Studies 294, Universiteit Utrecht, the Netherlands.

van Westen, C. J., Rengers, N. and Soeters, R. (2003). Use of geomorphological information in indirect landslide susceptibility assessment. *Natural Hazards*, **30**(3), 399–419.

van Westen, C. J., van Asch, T. and Soeters, R. (2006). Landslide hazard and risk zonation: why is it still so difficult? *Bulletin of Engineering Geology and Environment*, **65**(2), 167–184.

Walker, M. (2005). *Quaternary Dating Methods*. Chichester: John Wiley and Sons.

Wilson, R. C. and Keefer, D. K. (1985). Predicting areal limits of earthquake-induced landsliding. In J. I. Ziony (ed.), *Evaluating Earthquake Hazards in the Los Angeles Region*. United States Geological Survey Professional Paper 1360, pp. 317–493

Wolman, M. G. and Gerson, R. (1978). Relative scales of time and effectiveness of climate in watershed geomorphology. *Earth Surface Processes*, **3**, 189–208.

Wolman, M. G. and Miller, J. P. (1960). Magnitude and frequency of forces in geomorphic processes. *Journal of Geology*, **68**(1), 54–74.

Wong, H. N., Ho, K. K. S. and Chan, Y. C. (1997). Assessment of consequences of landslides. In D. M. Cruden and R. Fell (eds.), *Landslide Risk Assessment*. Proceedings of the Workshop on Landslide Risk Assessment, Honolulu, Hawaii, USA, 19–21 February 1997. Rotterdam: A. A. Balkema, pp. 111–149.

Zischg, A., Fuchs, S., Keiler, M. and Stötter, J. (2005). Temporal variability of damage potential on roads as a conceptual contribution towards a short term avalanche risk simulation. *Natural Hazards and Earth System Sciences*, **5**, 235–242.

19 Vulnerability analysis in geomorphic risk assessment

Gabi Hufschmidt and Thomas Glade

19.1 Rationale

The application of vulnerability concepts within the discipline of geomorphology is relatively new. Due to the lack of a theory of its own, the usage of these concepts for geomorphic risk assessments is not without pitfalls. A first step towards avoiding these pitfalls is to recognise and appreciate the theoretical backgrounds and meanings in other fields of research. Consequently, it is fundamental to review these concepts carefully and to link them – where possible – to our own discipline of geomorphology. Indeed, some of the reviewed concepts are distant to geomorphic applications. However, it is important to acknowledge their meanings in order to distinguish different aspects linked to various scientific roots. The major aim of this contribution is to review these scientific roots and their vulnerability concepts, to provide ideas for applying these concepts within geomorphology and to give some examples to illustrate advances, but also pitfalls.

The first part of this chapter outlines the series of key developments in vulnerability research, and then builds the (historical) framework for vulnerability analysis in geomorphic risk reduction strategies. The common approach and methodology of vulnerability analysis from a geomorphic (i.e. natural science) perspective is reviewed and discussed in the context presented.

19.2 Different vulnerability approaches towards risk reduction

Research in the fields of natural hazard and risk research has diversified substantially during its relatively short history. A series of key developments can be identified:

1. the notion of risk changed from expressing the likelihood of geophysical processes occurring, such as a landslide, to a concept that includes the possible overall adverse effects on people and their (built) environment;
2. the shift from defining 'hazard' as a natural process only towards a concept that includes the frequency–magnitude relationship of the process;
3. the recognition that measures of loss reduction based on achievements in the field of science and technology only are insufficient for effective and sustainable loss reduction;
4. the importance of human adjustment;
5. the emergence of the vulnerability concept and its development into a research field of its own, where multiple dimensions (space, time, cultural, environmental, political, economic) for understanding and measuring the type and degree of damage inflicted on people, societies or economies are included.

Geomorphic risk reduction only reluctantly acknowledges and responds to the role of this 'social' vulnerability (i.e. living environment of social groups, e.g. Wisner *et al.*, 2004) as opposed to 'physical' vulnerability (i.e. structures of the built environment, e.g. Quarantelli, 2003).

19.3 Science and technology

In the early half of the twentieth century, the Western world became slowly but increasingly aware of the damage and resulting financial losses following the exploitation of their own environmental resources. For example, in the late nineteenth century, US agriculturalists, supported by technological innovation as well as the railway and government policy, had pushed the margin of production into a region of highly variable climate. As a result, great droughts occurred in the subsequent decades, vividly remembered in the images of the 1930s dust bowls

Geomorphological Hazards and Disaster Prevention, eds. Irasema Alcántara-Ayala and Andrew S. Goudie. Published by Cambridge University Press. © Cambridge University Press 2010.

FIGURE 19.1. Location of a petrol station on a bridge over a currently not active debris flow channel in Bildudalur, Iceland. (Photograph: Rainer Bell.) It is obvious that the vulnerability of the 'elements at risk' (i.e. bridge, road, petrol station, power supply, etc.) towards large debris flows is very high.

fall predominantly into the realm of science and technology. In 1972, the United Nations stated:

It is believed that not only the causes of … disasters fall within the province of science and technology, but also, in some cases their prevention, as well as the organizational arrangements made for forecasting them and reducing their impact when they occur. (United Nations Department of Economics and Social Affairs, 1972, p. 1).

Consequently, the United Nations Advisory Committee on the Application of Science and Technology to Development promoted research in the fields of physical processes, forecasting, technological measures for protection and warning systems (Burton et al., 1978). This research stream continues today, and against the background of recent disasters has intensified on an international scale.

From the perspective of this research stream, damage is predominantly seen as proportional only to the magnitude, frequency and type of the natural process (Hewitt, 1983; UNDP, 2004). What is more, in the 1970s and 80s 'risk' was used by natural scientists to express the likelihood of an event occurring (frequency and magnitude). The focus was very much on the natural process itself, without including damage or loss estimates (Timmerman, 1981; Cardona, 2004). The process-focused understanding of risk has now mostly shifted towards including the probability of damage and loss; this means adverse consequences for humans and their environment. Some examples include studies on landslides (Ragozin and Tikhvinsky, 2000; Glade, 2003; Alexander, 2005), debris flows (Fuchs et al., 2007; Lu et al., 2007), volcanic eruptions (Aceves-Quesada et al., 2007), or snow avalanches (Fuchs and Bründl, 2005) (refer also to Chapter 5 by Bründl et al. in this book), to name a few examples and exemplary references only.

Additionally, 'hazard' applied today usually implies that someone or something is threatened (e.g. Glade et al., 2005), while in the 1980s, 'hazard' was frequently used to describe the natural process only (Hewitt, 1983). This shift in terminology reflects the development of natural hazard and risk research, as will be seen in the following.

19.4 The human ecology approach

In his dissertation, published in 1945, Gilbert F. White clearly placed the responsibility for flood damage into the realm of human action, not 'nature' (White, 1945). White, who had worked with H. Barrows, questioned the efficacy of solutions based on science and technology only, such as large flood-control measure expenditures for dams, channel modification and levees (White et al., 1958; White,

covering the Great Plains of North America (Warrick, 1983). Additionally, major floods which occurred in the USA of the 1940s and 1950s clearly demonstrated that somehow not the flood as a natural process, but society with inappropriate development and adaption strategies is the main driver for the consequences suffered following these 'natural' events (Burton et al., 1968; Kates, 1970). Even today, more recent examples of inappropriate development are ubiquitous (Figure 19.1).

Attempts at controlling and predicting natural processes such as soil erosion and floods, usually with engineering measures, dominated within the scientific community. Simultaneously, the pressure to utilise natural resources increased, and so did the amount of money potentially assigned for public engineering works (Mitchell, 1990; Smith, 2004). Nationally and internationally, solutions for reducing losses incurred from natural hazards continued to

1961). This questioning can indeed be transferred to various hazardous processes such as landslides and snow avalanches.

White and colleagues concluded that reducing loss by technology-driven strategies only proved to generate even more rather than less damage. Avoiding smaller losses in the short term turned out to increase losses in the long term (Burton *et al.*, 1968, 1978; Kates, 1970; Burton and Hewitt, 1974). While, for example, the construction of dams reduces flood damage of a specific magnitude, 'protected' areas are still exposed to higher-magnitude events. Furthermore, by constructing levees human settlement is encouraged in a supposedly flood-protected area, which creates a false sense of security and increases the damage potential. Urban growth outside the range of levees raises the damage potential further (Hewitt and Burton, 1971; White, 1974).

Against the background of a common aim to optimise the cost–benefit ratio of mitigation measures (e.g. engineering flood protection works), White (1973, 1974) criticised that the knowledge of physical processes and their behaviour is 'imperfect' and therefore mitigation measures were not as successful as envisaged. Today, the problem of uncertainty in prediction has still not been resolved for many hazards, especially for landslide hazards as Crozier and Glade (2005) pointed out.

Hewitt and Burton (1971) advocated White's human ecology perspective on hazards. They identified beneficial outcomes of the dualism of 'man' and 'nature' i.e. resources and goods, and negative outcomes, such as hazard and risk (see also Kates, 1970, and Burton *et al.*, 1993). White (1974, p. 3) emphasised that 'natural' hazards are not generated by 'nature': 'By definition, no hazard exists apart from human adjustment to it. It always involves human initiative and choice. Floods would not be hazards were not man tempted to occupy floodplains …'. This statement implies that humans can take action to reduce these losses, and therefore bears a strong possibilistic, not a deterministic, connotation.

Human adjustment to natural hazards is the central topic of the human ecologist school, which is also sometimes labelled the 'Chicago School' (Tobin and Montz, 1997) or the 'behavioural paradigm' (Pelling, 2003; Smith, 2004). White (1974, p. 4) defined adjustment as 'a human activity intended to reduce the negative impact of the event'. It is when the impact of a natural process exceeds the adjustments in place that damage and loss unfold. This entails 'a continuing effort to make the human use system less vulnerable to the vagaries of nature' (Kates, 1970, p. 1).

From a human ecology perspective, a combination of the magnitude of a geophysical variable, like water discharge,

and human adjustment delineates a zone where damage is not significant, i.e. does not cross a threshold above which the positive effects of resource utilisation flip into adverse effects creating 'negative resources' (hazards) (Hewitt and Burton, 1971; Burton and Hewitt, 1974). This damage threshold is closely interlinked with human adjustment strategies which can alter the threshold, and hence widen or lessen the zone of insignificant damage. The zone of insignificant damage is adjusted to buffer 'normal' events of a certain frequency–magnitude relation. 'Extreme events' exceeding the damage threshold, however, are not covered and damage is incurred. It is important to recognise that such damage thresholds vary in time. This variation is related not only to changes in the geo(morphological) system, but also within the social realm, for example the implementation of new urban development strategies or changes in land use.

The zone of insignificant damage is what the climate change and global environmental change community calls the 'coping range' (Smit *et al.*, 2000; Ford, 2004), which can be extended depending on people's 'adaptive capacity' (Smit and Pilifosova, 2003; Adger, 2006; Smit and Wandel, 2006).

The Chicago School proposes a combination of adjustments such as (1) '*modify the cause*', i.e. keep the hazard away from the population (for example by structural measures such as constructing levees, avalanche fences, or by operations such as snow melting, slope stabilisation); (2) '*modify the loss*' by keeping the population away from the hazard (for example warning systems, building design, land use planning); and (3) '*distribute and adjust to losses*' such as purchasing insurance (Burton *et al.*, 1968, 1993; Kates, 1970; White, 1973; Mitchell, 1990). For a variety of adjustments in relation to different hazard types, see Burton and Hewitt (1974). Especially when 'modifying the cause' is not possible, non-structural management options are rated as increasingly relevant. Therefore, common strategies of loss reduction based on science and technology are not dismissed as such, but rated to be insufficient by themselves and only effective if in combination with other strategies.

Choices of adjustment are seen as mirroring a limited human rationality, or 'bounded rationality' (Kates, 1970; White, 1973; Burton *et al.*, 1978, 1993). Rational behaviour overruled by political or economic power is only marginally recognised within the classical human ecology perspective, which is identified as a major deficit by the emerging criticism of the 'structuralist paradigm'.

One component of a human ecologist approach to loss reduction is to include the socio-economic causes and effects of risk. The human ecology school defines vulnerability as the '*capacity to be wounded*' (Kates, 1985, p. 9). The most vulnerable element, i.e. the one most susceptible to be wounded, is relative. This means it varies with hazard type and community type. Acknowledging the partially context-specific nature of vulnerability is regarded as especially important.

White and Haas (1975) suggested that three interacting elements have to be analysed in order to estimate losses: the 'natural event generator' (for example frequency and magnitude of earthquakes and storms), the 'population-at-risk in each area' (density or distribution of people and buildings), and the 'vulnerability of population-at-risk to loss for a given severity of an event' (p. 123). This work initiated the concept of risk as it is predominantly understood today and synthesised by the United Nations Disaster Relief Office (UNDRO, 1982).

In summary, the emerging human ecology school shifted the focus from human *control* of nature to human *adjustment* to nature. The 'naturalness' of hazards is questioned, as well as the limited effectiveness of purely scientific and engineering approaches to loss reduction. What is more, the human ecology school paved the way for the concept of vulnerability that today is a key to understanding the magnitude of damage induced from natural processes.

In the 1980s, when a *Zeitgeist* of science-scepticism emerged and whole societies perceived themselves as threatened in various ways – labelled 'risk society' by Ulrich Beck (1986) – vulnerability research developed into two different ways of tackling the challenge of reducing losses from natural hazards: one as a continuous development of the human ecology school ('applied sciences'), the other being the 'structuralist paradigm' triggered by criticism of the former and the 'science and technology' approach.

19.5 Vulnerability and the applied sciences

Applied sciences, such as engineering, economics, politics, geography and environmental studies, are increasingly inspired by the human ecologist school of natural hazards. At the end of the 1970s, the vulnerability concept was fostered and implemented in guidelines for future research in the fields of energy, risk management and climate impact assessment. Models of social collapse and ecology were combined under the vulnerability umbrella (Timmerman, 1981).

Applied sciences define vulnerability as the degree of loss, which can be expressed as a damage ratio (for example from 0 to 1). This understanding is based on definitions suggested by the United Nations Disaster Relief Organisation (UNDRO, 1982). Hollenstein *et al.* (2002) presented a comprehensive compilation and evaluation of approaches to vulnerability within the applied sciences,

covering a range of natural hazard types. Within these, Hollenstein (2005) detected a dominance of earthquake- and wind-related vulnerability models.

Analysing vulnerability can be done qualitatively, semi-quantitatively or quantitatively. Whatever the way, the construct is expressed as a condition. In the overall context of risk assessment, qualitative descriptions are often used as a first assessment to identify different vulnerability aspects, or when numerical data are not available (AS/NZS, 2004). Wisner (2006) for instance discussed and exemplified the advantages of participatory approaches that are qualitative self-assessments. Semi-quantitative methods assign values to qualitative ranks in order to introduce a more expanded scale. These values are, however, not 'real' values and are usually expressed on an ordinal scale that bears limited mathematical possibilities. Villagran (2006) described such an analysis where structural characteristics of buildings (e.g. material of the roof and walls) are associated with classes of low, medium and high vulnerability. These are assigned values of one, three and five, respectively, and combined into one figure. Thirdly, quantitative analysis relies on numerical values based on metric variables on an interval/ratio scale allowing for mathematical operations. Differences between variables can be quantified as true numeric magnitudes. Metric variables can be used as indicators and aggregated into one index. Examples of an index-based methodology are given by Briguglio (1995), Davidson (1997), Davidson and Shah (1998), Cutter *et al.* (2000), Davidson and Lambert (2001), Cutter (2003), Boruff *et al.* (2005), Cardona (2005, 2006), Bollin and Hidajat (2006) and Plate (2006). Challenges encountered when developing an index-based approach of vulnerability assessment are manifold and include issues such as data availability, subjectivity and opportunities to manipulate results, and overall the lack of validation options. Consequently, uncertainty and sensitivity analysis are mandatory for maximising methodological transparency and soundness, and hence the acceptance of research findings (Gall, 2007). Despite this demand, both analyses are often missing in vulnerability assessments.

The following exemplifies the rather classical way of incorporating vulnerability in the context of landslide risk studies. Scales covered in such landslide vulnerability and risk studies range from the very detailed local scale to the regional scale and the more abstract scale of society when calculating 'societal' vulnerability and risk. Vulnerability to landslides is usually considered for individuals in relation to buildings, for buildings themselves and roads in the potential path of the landslide (Glade, 2003; Alexander,

2005). Generally, when analysing the vulnerability to land-slides, the following factors are taken into account:

1. the location of the element at risk in respect to the land-slide (e.g. uphill, on the landslide, or downhill);
2. the temporal component (e.g. day/night, weekend/workweek);
3. the impact of the landslide, assuming that the level of vulnerability changes with the level of impact, which depends on:
   a. the velocity of the landslide,
   b. the depth of the landslide (e.g. magnitude: volume/spatial extent);
4. the characteristics of the element at risk; for structures this is well documented, e.g. in order to design structures that can resist the impact (IUGS, 1997). See Corominas et al. (2005) for an example of such a conceptualisation. Leone et al. (1996) developed a damage matrix to classify the potential damage according to the building structure. Another approach is to assemble historical data on damages, as done by Remondo et al. (2005) for the built environment within a catchment in northern Spain. The problem with historical data is their availability, and even if historical damage data are available, the level of detail is often not sufficient to capture landslide type and magnitude (Glade, 2003).

The landslide risk assessment carried out for the Geoscience Australia 'Cities Project' with respect to urban communities in Cairns is a (modified) example for an approach as described above. The vulnerability of people, buildings and roads is expressed as the probability of a fatal injury or destruction, expressed on a scale of 0 to 1 (Michael-Leiba et al., 2003, 2005).

The need to understand vulnerability in all its multi-dimensionality on the one hand (often done qualitatively), and the need for relatively simple tools of analysis on the other hand (often required quantitatively), can be conflicting. This dilemma was observed by Davidson (1997), who reviewed earthquake risk assessment models developed from the two camps of social sciences and engineering. For example, clear and readily available vulnerability indicators are often what hazard and disaster managers seek – before, during or after an emergency (Birkmann, 2006; Queste and Lauwe, 2006). Vulnerability indicators or indices simplify reality, and the degree of simplification depends on the target audiences (Karlsson et al., 2007; Moldan and Dahl, 2007; Stanners et al., 2007). Hence there is no general preference for qualitative, semi-quantitative or quantitative approaches. As a general rule the approach best suited to meet the defined goal should guide the methodology applied (AS/NZS, 2004).

Besides hazard-specific studies, multi-hazard approaches emerge. A multi-hazard perspective was initiated by the human ecologist approach of Hewitt and Burton (1971), who presented a case study that includes 'all hazards at a place'. Based on this work, Cutter developed the 'hazards-of-a-place-model of vulnerability' (Cutter, 1996; Cutter et al., 2000, 2003). The progress from single-hazard dominated research is particularly important considering the combined occurrence of many hazards, such as a hurricane followed by floods and landslides, or an earthquake triggering landslides. Some concepts of addressing multi-hazard risks within natural sciences have already been suggested (e.g. van Westen et al., 2002; von Elverfeldt and Glade, 2008). However, constructing a 'multi-hazard' assessment of vulnerability still appears to be extremely challenging (UNDP, 2004).

Although the importance of social aspects is recognised, the concept of 'physical vulnerability', meaning the susceptibility of physical elements such as houses or infrastructures, dominates the applied sciences (Mueller-Mahn, 2005). The common catalogue of risk reduction measures is based on the adjustment strategies identified by the human ecology school, as for example presented by Dai et al. (2002).

With respect to landslides, their prediction and control, especially for first-time failures, is limited. Consequently, the assessment of vulnerability and the identification of loss reduction strategies are especially important. As Liu et al. (2002) and Liu and Lei (2003) demonstrated, this is highly relevant for debris flows, in which case the modification of the human system (removing people and assets from hazardous areas) is the most feasible strategy to reduce landslide loss. Alexander (2005) underlined that vulnerability can determine the extent of losses to a greater extent than the landslide process itself. Overall, the importance of the vulnerability concept in landslide risk studies has been and is increasingly acknowledged. However, more cooperation between the 'physical' and 'social' vulnerability approaches, and more synergies with perspectives and solutions offered by the 'structuralist paradigm', are needed in order to maximise the effectiveness of loss reduction strategies.

19.6 Vulnerability and the structuralist paradigm

In the 1970s, criticism of the increasingly dominant research approach to natural hazards and disasters, the combination of science, technology and human ecology, mounted ('dominant paradigm'). This criticism came to be known as the 'structuralist paradigm' (Smith, 2004). The structuralist alternative focuses not so much on people's perception and their subsequent choice of adjustment, as on people's individual socio-economic and demographic characteristics within a specific social, cultural, economic, political and environmental fabric. It is not the choice as such, but the ability to choose between adjustment options that is the nucleus of this vulnerability research.

Similarly to White and colleagues, one can argue against labelling disasters as 'natural' (O'Keefe et al., 1976). Richards (1975) stressed the interaction of natural and social processes. He highlighted that factors 'such as economic development can affect natural systems "causing" famine and soil erosion for example. This should make us think again about the term "natural" disaster' (cited in Timmerman, 1981, p. 11).

It is criticised that within the dominant view, 'The sense of causality or the direction of explanation still runs from the physical environment to its social impacts' (Hewitt, 1983, p. 5). The usefulness of better process understanding and improved forecasting is not doubted per se. However, the emerging paradigm questions the view that mainly improved predictions will reduce damage and loss, while social aspects are ignored (Hewitt, 1983). In addition to predictions, the trust in security provided by the existence of structural measures is seen as most dominant in society.

Especially in less developed countries the unsuccessful strategies of loss reduction conceptualised by the dominant view featured as an impetus for the structuralist paradigm (Smith, 2004). Increasingly, social scientists active mainly in the less developed countries of Latin America and Asia could not sufficiently decipher the rising number of disasters using the characteristics of the natural process alone (van Westen et al., 2005). Focusing on hazard as a specialised problem, which can only be cured by scientific expertise and technology transfer, is identified as part of the problem, not the solution. The usual approach is seen as more of a technical monologue rather than a dialogue with 'grass roots' knowledge (Copans, 1983; Hewitt, 1983). It is increasingly recognised that under the pressure of daily threat, local people have developed their own successful strategies to cope with hazards and disasters (Bankoff, 2004; Heijmans, 2004).

The understanding of vulnerability from a structuralist perspective incorporates a wider appreciation of the social, economic, cultural and political context people live in, as well as their day-to-day personal socio-economic situation (Blaikie et al., 1994; Wisner et al., 2004). Wisner et al. (2004, p. 11) identified a range of variables determining vulnerability as 'class (including differences in wealth), occupation, caste, ethnicity, gender, disability and health status, age and immigration status ("legal" or "illegal") and the nature and extent of social networks'. Sometimes poverty is identified as the main cause of vulnerability, since

very often the poor are those who suffer the most (Cuny, 1983). However, equalling vulnerability with poverty is an approach that is too simplistic (Wisner et al., 2004). Wisner (1993) rated the model of marginalisation as one of the most useful of disaster occurrence anchored in social theory. Examples of marginalisation are informal settlements in dangerous areas prone to landslides as underlined by Smyth and Royle (2000) or by Anderson and Holcombe (2006).

Pronounced economic, social, environmental and political marginalisation can generally render people vulnerable – no matter if they are threatened by a flood, a landslide or an earthquake. In this context, Briguglio (2003) and Cardona (2005, p. 12) used the term 'inherent' vulnerability. Allen (2003, p. 170) referred to this phenomenon as 'underlying vulnerability', which she interpreted as a 'contextual weakness or susceptibility underpinning daily life'. Wisner (1993) and Wisner et al. (2004) preferred the term 'generalised' vulnerability. In the context of community response to disasters, Quarantelli (1997) differentiated between 'agent-specific' (hazard-specific) and 'generic' factors.

A comprehensive model summarising the structuralist perspective on multi-causal vulnerability combining far-reaching political, economic and cultural processes is the 'pressure and release' model (PAR). The complementary 'access model' focuses on the household scale and identifies access to resources, such as capital, land, or relief, as the drivers of vulnerability (Blaikie et al., 1994; Wisner et al., 2004). Research outcomes, such as the PAR-model, can deliver insight and increase the understanding of how risk is created in the developed world, as demonstrated recently by the events during and following Hurricane Katrina.

The new and challenging interpretation of disasters leads to an ideological battle that rejects technology-based approaches and the 'behavioural' paradigm of the human ecologist school radically. The dominant view is flagged as 'naïve determinism' and 'technocratic optimism'. Political responsibility, capitalism and the resulting marginal situation of many are viewed within a Marxist context: 'Acts of God become Acts of Capital' (Waddell, 1983, p. 38). Hence the structuralist paradigm carries neo-Marxist implications (Pelling, 2003).

Predominantly, people are perceived as victims without the ability to choose where they live or how they earn a livelihood, restricted by political and economical power structures. The 'bounded rationality' of the human ecologist school is replaced by a reality where potentially rational behaviour is suppressed by political, economic and cultural forces. It should be noted that White (1961), too, identified social, political and economic constraints influencing and potentially limiting people's perception and hence choice of adjustments. From White's perspective, however, the responsibility to undertake adjustments lies within the individual realm.

A field of research influencing the structuralist perspective on social vulnerability during the 1980s and later is the 'sustainable livelihood' approach (Carney, 1998). Also Sen's (1981) 'entitlement' approach showed that major hunger crises in India in the mid twentieth century were rooted within the societal structures with different entitlements (access) to resources. A vulnerability model containing human ecological (for example effects of land use, desertification) and political-economic elements (household income, access to markets, price development) influenced by Sen was developed by Watts and Bohle (1993) and Bohle *et al.* (1994). The economy-based entitlement approach developed by Sen can be seen as a third field besides the human ecology school and the structuralist paradigm (Pelling, 2003).

19.7 Summary and perspectives

The path of vulnerability as a key concept within hazard and risk studies was paved by the human ecology school, and increasingly explored by several other disciplines including geomorphology. Its interpretation and application is advanced by the applied sciences. Importantly, the vulnerability concept within applied natural hazard and risk research, e.g. to describe the susceptibility of built structures and areas under agricultural production, has been broadened to encompass the demographic and socio-economic characteristics of people and topics such as risk preparedness. From a structuralist perspective, globalisation and increasing socio-economic and ecologic marginalisation are identified as root causes for people's vulnerability, especially in the less developed countries. The structuralist paradigm opened up the field of hazard and disaster research for socio-economic, cultural and political aspects within risk reduction strategies. In comparison with the classical human ecology school, the recognition that some social groups are simply not able to adjust to hazards, or cannot choose between different adjustment options – offered by traditional geomorphic hazard and risk studies, which are also described in this book – is a major contribution towards reducing loss in developing and developed countries alike.

The key synergy between the applied sciences, the (traditionally opposing) classical human ecology school and the structuralist paradigm is the underpinning that natural

hazards and disaster are not just 'natural' but also social phenomena. Having this in mind, the consequence should be that any hazard and risk study including vulnerability assessment has to address both the natural science approach and the related social dimension. Hewitt (1997) developed an explanation of risk out of the human ecologist school and key elements of the structuralist approach to vulnerability. Hence human agency (behaviour/adjustment) and societal structures potentially restricting human agency are combined.

It is this kind of synergy that is necessary to reduce loss induced by natural processes in a sustainable way, recognising the 'inherent' as well as the 'context-specific' nature of vulnerability, in the less developed and the developed world alike. Vulnerability analysis with respect to geomorphic hazards is traditionally associated with the science and technology domain, and only slowly responds to the developments within the field of vulnerability research. Therefore, targeting a sustainable development of a given region or area demands not only the treatment of hazard and risk with modelling and simulation techniques, it also requires social research. The key element herein is not to carry out both analyses independently. Rather a coupling of both approaches is necessary to allow and ensure sustainable development.

Furthermore, it should be increasingly recognised that vulnerability is not static, whether in social, in technological or in natural science approaches. Refer to Hufschmidt *et al.* (2005) for a summary and discussion of temporal variability in the context of risk analysis (including hazard and vulnerability), with a focus on geomorphic processes such as landslides. A key point is that ongoing global change has major implications for designs of any technical structures, and also for modelling, simulation and predictions because the boundary variables change continuously, and at different rates. Thus, any planning procedure is strongly influenced by high uncertainties as a result of the unknown future. Frequently, this is not addressed in respective research or applications.

But the challenge is even greater. Engineering structures protecting environment and society are most important in highly endangered areas and locations. However, it is often not realised that by building the structure, the problem is not solved. Any structure has a given lifetime – and if no or not enough resources are allocated to maintain these structures, they might fail in the future and might cause even greater damage. For example, if snow avalanche fences are not maintained, they artificially collect more snow than would accumulate under natural conditions. If they fail due to neglected maintenance, the consequences are much worse than ever experienced before – and we read in the literature 'Nature fights back'. This problem is also evident for other processes (e.g. landslides, river dams etc.).

There is indeed a high demand for further research on hazard and risk in the specific traditional natural, engineering and social sciences. However, there is an even greater demand to couple these approaches. Herein, vulnerability analysis can play a predominant role. The respective innovation potential is incredibly high and has to be explored.

References

Aceves-Quesada, J. F., Díaz-Salgado, J. and López-Blanco, J. (2007). Vulnerability assessment in a volcanic risk evaluation in Central Mexico through a multi-criteria-GIS approach. *Natural Hazards*, **40**(2), 239–256.

Adger, N. (2006). Vulnerability. *Global Environmental Change*, **16**, 268–281.

Alexander, D. (2005). Vulnerability to landslides. In T. Glade, M. Anderson and M. J. Crozier (eds.), *Landslide Hazard and Risk*. Chichester: John Wiley & Sons, pp. 175–198.

Allen, K. (2003). Vulnerability reduction and the community-based approach. In M. Pelling (ed.), *Natural Disasters and Development in a Globalizing World*. London, New York: Routledge, pp. 170–184.

Anderson, M. and Holcombe, L. (2006). Purpose-driven public sector reform: the need for within-government capacity build for the management of slope stability in communities in the Caribbean. *Environmental Management*, **37**(1), 15–29.

AS/NZS (2004). *Risk Management Guidelines, AS/NZS 4360:2004*. Sydney, Wellington: Standards Australia International Ltd. and Standards New Zealand.

Bankoff, G. (2004). The historical geography of disaster: 'vulnerability' and 'local knowledge'. In G. Bankoff, G. Frerks and D. Hilhorst (eds.), *Mapping Vulnerability: Disasters, Development and People*. London: Earthscan, pp. 25–36.

Beck, U. (1986). *Risikogesellschaft*. Frankfurt/M: Suhrkamp.

Birkmann, J. (2006). Indicators and criteria for measuring vulnerability: theoretical bases and requirements. In J. Birkmann (ed.), *Measuring Vulnerability to Natural Hazards Towards Disaster Resilient Societies*. Tokyo: United Nations University, pp. 55–77.

Blaikie, P., Cannon, T., Davis, I. and Wisner, B. (1994a). *At Risk: Natural Hazards, People's Vulnerability, and Disasters*. London: Routledge.

Bohle, H. G., Downing, T. E. and Watts, M. J. (1994). Climate-change and social vulnerability: toward a sociology and geography of food insecurity. *Global Environmental Change: Human and Policy Dimensions*, **4**(1), 37–48.

Bollin, C. and Hidajat, R. (2006). Community-based disaster risk index: pilot implementation in Indonesia. In J. Birkmann (ed.), *Measuring Vulnerability to Natural Hazards Towards Disaster Resilient Societies*. Tokyo: United Nations University, pp. 271–289.

Boruff, B. J., Emrich, C. and Cutter, S. L. (2005). Erosion hazard vulnerability of U.S. coastal counties. *Journal of Coastal Research*, **21**(5), 932–942.

Briguglio, L. (1995). Small island developing states and their economic vulnerabilities. *World Development*, **23**(9), 1615–1632.

Briguglio, L. (2003). *Methodological and Practical Considerations for Constructing Socioeconomic Indicators to Evaluate Disaster Risk*. IDB/IDEA Program of indicators for disaster risk management, National University of Colombia: Manizales.

Burton, I. and Hewitt, K. (1974). Ecological dimensions of environmental hazards. In F. Sargent II (ed.), *Human Ecology*. Amsterdam: North-Holland Publishing Company, pp. 253–283.

Burton, I., Kates, R. W. and White, G. F. (1968). *The Human Ecology of Extreme Geophysical Events*, Toronto: Department of Geography, University of Toronto.

Burton, I., Kates, R. W. and White, G. F. (1978). *The Environment as Hazard*. Oxford: Oxford University Press.

Burton, I., Kates, R. W. and White, G. F. (1993). *The Environment as Hazard*. New York, London: The Guilford Press.

Cardona, O. (2004). Curriculum adaptation and disaster prevention in Colombia. In J. Stoltman, J. Lidstone and L. Dechano (eds.), *International Perspectives on Natural Disasters: Occurrence, Mitigation, and Consequences*. Dordrecht: Kluwer, pp. 397–408.

Cardona, O. D. (2005). *Indicators of Disaster Risk and Risk Management. Program for Latin America and the Caribbean, Summary Report*. Washington, D.C.: Inter-American Development Bank (IADB), Sustainable Development Department.

Cardona, O. D. (2006). A system of indicators for disaster risk management in the Americas. In J. Birkmann (ed.), *Measuring Vulnerability to Natural Hazards*. Tokyo: United Nations University Press: pp. 189–209.

Carney, D. (ed.) (1998). Sustainable rural livelihoods: what contribution can we make? Paper presented at the Department for International Development's Natural Resources Advisers Conference: London.

Copans, J. (1983). The Sahelian drought: social sciences and the political economy of underdevelopment. In K. Hewitt (ed.), *Interpretations of Calamity From the Viewpoint of Human Ecology*. Winchester: Allen & Unwin Inc., pp. 83–97.

Corominas, J., Copons, R., Moya, J. *et al.* (2005). Quantitative assessment of the residual risk in a rockfall protected area. *Landslides*, **2**, 343–357.

Crozier, M. J. and Glade, T. (2005). Landslide hazard and risk: issues, concepts, and approach. In T. Glade, M. G. Anderson and M. J. Crozier (eds.), *Landslide Hazard and Risk*. Chichester: Wiley, pp. 1–38.

Cuny, F. C. (1983). *Disasters and Development*. New York: Oxford University Press.

Cutter, S. L. (1996). Vulnerability to environmental hazards. *Progress in Human Geography*, **20**(4), 529–539.

Cutter, S. L. (2003). The vulnerability of science and the science of vulnerability. *Annals of the Association of American Geographers*, **93**(1), 1–12.

Cutter, S. L., Mitchell, J. T. and Scott, M. S. (2000). Revealing the vulnerability of people and places: a case study of Georgetown County, South Carolina. *Annals of the Association of American Geographers*, **90**(4), 713–737.

Cutter, S. L., Boruff, B. J. and Shirley, W. L. (2003). Social vulnerability to environmental hazards. *Social Science Quarterly*, **84**(2), 242–261.

Dai, F. C., Lee, C. F. and Ngai, Y. Y. (2002). Landslide risk assessment and management: an overview. *Engineering Geology*, **64**(1), 65–87.

Davidson, R. (1997). *An Urban Earthquake Disaster Risk Index*. John A. Blume Earthquake Engineering Centre, Stanford University, Stanford, California.

Davidson, R. and Lambert, K. B. (2001). Comparing the hurricane disaster risk of U.S. coastal counties. *Natural Hazards Review*, **2**(3), 132–142.

Davidson, R. and Shah, H. C. (1998). *Evaluation and Use of the Earthquake Disaster Risk Index*. Understanding Urban Seismic Risk Around the World Project, Stanford University, Stanford, California.

Ford, J. (2004). Inuit adaptive strategies and environmental conditions. Nunavut Research Institute conference, 26 July 2004.

Fuchs, S. and Bründl, M. (2005). Damage potential and losses resulting from snow avalanches in settlements of the canton of Grisons, Switzerland. *Natural Hazards*, **34**(1), 53–69.

Fuchs, S., Heiss, K. and Hübl, J. (2007). Towards an empirical vulnerability function for use in debris flow risk assessment. *Natural Hazard and Earth System Science*, **7**, 495–506.

Gall, M. (2007). *Indices of Social Vulnerability to Natural Hazards: A Comparative Evaluation*. University of South Carolina.

Glade, T. (2003). Vulnerability assessment in landslide risk analysis. *Die Erde*, **134**(2), 121–138.

Glade, T., Anderson, M. G. and Crozier, M. J. (eds.) (2005). *Landslide Hazard and Risk*. Chichester: Wiley.

Heijmans, A. (2004). From vulnerability to empowerment. In G. Bankoff, G. Frerks and D. Hilhorst (eds.), *Mapping Vulnerability: Disasters, Development and People*. London: Earthscan, pp. 114–127.

Hewitt, K. (1983). The idea of calamity in a technocratic age. In K. Hewitt (ed.), *Interpretations of Calamity From the Viewpoint of Human Ecology*. Winchester: Allen & Unwin Inc, pp. 3–32.

Hewitt, K. (1997). *Regions of Risk: A Geographical Introduction to Disasters*. Harlow, Essex: Addison Wesley Longman Limited.

Hewitt, K. and Burton, I. (1971). *The Hazardousness of a Place: A Regional Ecology of Damaging Events*. Research Publication 6. Toronto: University of Toronto Press.

Hollenstein, K. (2005). Reconsidering the risk assessment concept: standardizing the impact description as a building block for vulnerability assessment. *Natural Hazard and Earth System Science*, **5**, 301–307.

Hollenstein, K., Bieri, O. and Stueckelberger, J. (2002). *Modellierung der Vulnerability von Schadensobjekten gegenueber Naturgefahrenprozessen*. Zürich: ETHZ.

Hufschmidt, G., Crozier, M. J. and Glade, T. (2005). Evolution of natural risk: research framework and perspectives. *Natural Hazards and Earth System Sciences*, **5**, 375–387.

IUGS Working Group on Landslides: Committee on Risk Assessment (1997). Quantitative assessment for slopes and landslides: The state of the art. In D. M. Cruden and R. Fell (eds.), *Proceedings of the Workshop on Landslide Risk Assessment, Honolulu, Hawaii, USA, 19–21 February 1997*. Rotterdam: A.A. Balkema, pp. 3–12.

Karlsson, S., Dahl, L., Biggs, A. L. *et al.* (2007). Conceptual challenges. In T. Hak, B. Moldan and L. Dahl (eds.), *Sustainability Indicators*. Washington, D.C.: Island Press, pp. 27–48.

Kates, R. W. (1970). *Natural Hazard in Human Ecological Perspective: Hypotheses and Models*. Working Paper 14. Department of Geography, University of Toronto, Toronto.

Kates, R. W. (1985). The interaction of climate and society. In R. W. Kates, J. H. Ausubel and M. Berberian (eds.), *Climate Impact Assessment*. New York: Wiley, pp. 3–36.

Leone, F., Asté, J. P. and Leroi, E. (1996). Vulnerability assessment of elements exposed to mass-movement: working toward a better risk perception. In K. Senneset (ed.), *Landslides*. Rotterdam: A. A. Balkema, pp. 263–270.

Liu, X. L. and Lei, J. Z. (2003). A method for assessing regional debris flow risk: an application in Zhaotong of Yunnan province (SW China). *Geomorphology*, **52**(3–4), 181–191.

Liu, X. L., Yue, Z. Q., Tham, L. G. and Lee, C. F. (2002). Empirical assessment of debris flow risk on a regional scale in Yunnan province, southwestern China. *Environmental Management*, **30**(2), 249–264.

Lu, G., Chiu, L. and Wong, D. (2007). Vulnerability assessment of rainfall-induced debris flows in Taiwan. *Natural Hazards*, **43**(2), 223–244.

Michael-Leiba, M., Baynes, F., Scott, G. and Granger, K. (2003). Regional landslide risk to the Cairns community. *Natural Hazards*, **30**, 233–249.

Michael-Leiba, M., Baynes, F., Scott, G. and Granger, K. (2005). Quantitative landslide risk assessment of Cairns, Australia. In T. Glade, M. B. Anderson and M. J. Crozier (eds.), *Landslide Hazard and Risk*. Chichester: John Wiley & Sons, pp. 621–642.

Mitchell, K. (1990). Human dimensions of environmental hazards. In A. Kirby (ed.), *Nothing to Fear*. Tuscon: University of Arizona Press, pp. 131–175.

Moldan, B. and Dahl, L. (2007). Challenges to sustainability indicators. In T. Hak, B. Moldan and L. Dahl (eds.), *Sustainability Indicators*. Washington, D.C.: Island Press, pp. 1–24.

Mueller-Mahn, D. (2005). Von 'Naturkatastrophen' zu 'Complex Emergencies': Die Entwicklung integrativer Forschungsansaetze im Dialog mit der Praxis. In D. Mueller-Mahn and U. Wardenga (eds.), *Moeglichkeiten und Grenzen integrativer Forschungsansaetze in Physischer Geographie und Humangeographie*. Forum IFL, Leipzig: Leibniz-Institut fuer Laenderkunde e.V., pp. 69–77.

O'Keefe, P., Westgate, K. and Wisner, B. (1976). Taking the naturalness out of natural disasters. *Nature*, **260**, 566–567.

Pelling, M. (2003). Paradigms of risk. In M. Pelling (ed.), *Natural Disasters and Development in a Globalizing World*. London: Routledge, pp. 3–16.

Plate, E. J. (2006). A human security index. In J. Birkmann (ed.), *Measuring Vulnerability to Natural Hazards*. Tokyo: United Nations University Press, pp. 246–267.

Quarantelli, E. L. (1997). Ten criteria for evaluating the management of community disasters. *Disasters*, **21**(1), 39–56.

Quarantelli, E. (2003). Urban vulnerability to disasters in developing countries: managing risks. In A. Kreimer, M. Arnold and A. Carlin (eds.), *Building Safer Cities; The Future of Disaster Risk*. Disaster Risk Management Series. Washington D.C.: The World Bank, pp. 211–232.

Queste, A. and Lauwe, P. (2006). User needs: why we need indicators. In J. Birkmann (ed.), *Measuring Vulnerability to Natural Hazards: Towards Disaster Resilient Societies*. Tokyo: United Nations University, pp. 103–114.

Ragozin, A. L. and Tikhvinsky, I. O. (2000). Landslide hazard, vulnerability and risk assessment. In E. Bromhead, N. Dixon and M.-L. Ibsen (eds.), *Landslides in Research, Theory and Practice*. Cardiff: Thomas Telford, pp.1257–1262.

Remondo, J., Soto, J. S., Gonzalez-Diez, A., de Teran, J. R. D. and Cendrero, A. (2005). Human impact on geomorphic processes and hazards in mountain areas in northern Spain. *Geomorphology*, **66**(1–4), 69–84.

Richards, P. (1975). *African Environment: Problems and Perspectives*. London: International African Institute.

Sen, A. (1981). *Famines and Poverty*. Oxford: Clarendon Press.

Smit, B. and Pilifosova, O. (2003). From adaptation to adaptive capacity and vulnerability reduction. In J.B. Smith, R. J. T. Klein and S. Huq (eds.), *Climate Change, Adaptive Capacity and Development*. London: Imperial College Press, pp. 9–28.

Smit, B. and Wandel, J. (2006). Adaptation, adaptive capacity and vulnerability. *Global Environmental Change*, **16**, 282–292.

Smit, B., Burton, I. and Klein, R. J. T. (2000). An anatomy of adaptation to climate change and variability. *Climate Change*, **45**, 223–251.

Smith, K. (2004). *Environmental Hazards: Assessing Risk and Reducing Disaster*. London, New York: Routledge.

Smyth, C. G. and Royle, S. A. (2000). Urban landslide hazards: incidence and causative factors in Niteroi, Rio de Janeiro State, Brazil. *Applied Geography*, **20**(2), 95–117.

Stanners, D., Bosch, P., Dom, A. *et al.* (2007). Frameworks for environmental assessment and indicators at the EEA. In

T. Hak, B. Moldan and L. Dahl (eds.), *Sustainability Indicators*. Washington, D.C.: Island Press, pp. 127–144.

Timmerman, P. (1981). *Vulnerability, Resilience and the Collapse of Society: A Review of Models and Possible Climatic Applications*. Environmental Monograph 1. Institute for Environmental Studies, University of Toronto, Toronto.

Tobin, G. A. and Montz, B. E. (1997). *Natural Hazards: Explanation and Integration*. New York: The Guilford Press.

UNDP (2004). *Reducing Disaster Risk: A Challenge for Development. A Global Report*. New York: United Nations.

UNDRO (1982). *Natural Disasters and Vulnerability Analysis*. Geneva: United Nations Disaster Relief Organisation.

United Nations Department of Economics and Social Affairs (1972). *The Role of Science and Technology in Reducing the Impacts of Natural Disasters on Mankind*. New York: United Nations.

van Westen, C. J., Montoya, A. L., Boerboom, L. G. J. and Badilla Coto, E. (2002). Multi-hazard risk assessment using GIS in urban areas: a case study for the city of Turrialba, Costa Rica. In *Proceedings of the Regional Workshop on Best Practices in Disaster Mitigation*, Bali, pp. 120–136.

van Westen, C. J., Kumar Piya, B. and Guragain, J. (2005). Geo-information for urban risk assessment in developing countries: the SLARIM project. In P. J. M. van Oosterom, S. Zlatanova and M. Elfriede (eds.), *Proceedings of the 1st International Symposium on Geo-information for Disaster Management, 21–23 March 2005*, Delft, the Netherlands. Berlin: Springer, pp. 379–392.

Villagran, J. C. (2006). Vulnerability assessment: the sectoral approach. In J. Birkmann (ed.), *Measuring Vulnerability to Natural Hazards*. Tokyo: United Nations University Press, pp. 300–315.

von Elverfeldt, K. and Glade, T. (2008). Development of a multihazard and multirisk concept. In M. Mikos and J. Huebl (eds.), *11th Congress INTERPRAEVENT*, Dornbirn, Vorarlberg, Austria. International Research Society INTERPRAEVENT, pp. 422–423.

Waddell, E. (1983). Coping with frosts, governments and disaster experts: some reflections based on New Guinea experience and a perusal of the relevant literature. In K. Hewitt (ed.), *Interpretations of Calamity From the Viewpoint of Human Ecology*. Winchester: Allen & Unwin Inc., pp. 33–43.

Warrick, R. A. (1983). Drought in the US Great Plains: shifting social consequences? In K. Hewitt (ed.), *Interpretations of Calamity From the Viewpoint of Human Ecology*. Winchester: Allen & Unwin Inc., pp. 67–82.

Watts, M. J. and Bohle, H.-G. (1993). The space of vulnerability: the causal structure of hunger and famine. *Progress in Human Geography*, **17**(1), 43–67.

White, G. F. (1945). *Human Adjustments to Floods*. Chicago: University of Chicago Department of Geography Research Paper 29.

White, G. F. (1961). The choice of resource management. *Natural Resources Journal*, **23**, 23–40.

White, G. F. (1973). Natural hazards research. In R. J. Chorley (ed.), *Directions in Geography*. London: Methuen & Co Ltd., pp. 193–216.

White, G. F. (ed.) (1974). *Natural Hazards: Local, National, Global*. New York: Oxford University Press.

White, G. F. and Haas, J. E. (1975). *Assessment of Research on Natural Hazards*. Cambridge, MA: MIT Press Environmental Studies Series.

White, G. F., Calef, W. C., Hudson, J. W. *et al.* (1958). *Changes in Urban Occupance of Flood Plains in the United States.*

Chicago: University of Chicago Department of Geography Research Paper 57.

Wisner, B. (1993). Disaster vulnerability: scale, power and daily life. *GeoJournal*, **30**(2), 127–140.

Wisner, B. (2006). Self-assessment of coping capacity: participatory, proactive and qualitative engagement of communities in their own risk management. In J. Birkmann (ed.), *Measuring Vulnerability to Natural Hazards*. Tokyo: United Nations University Press, pp. 316–328.

Wisner, B., Blaikie, P. M., Cannon, T. and Davis, I. (2004). *At Risk: Natural Hazards, People's Vulnerability and Disasters*. London, New York: Routledge.

20 Geomorphological hazards and global climate change

Andrew S. Goudie

20.1 Introduction

It is likely that global climate will change substantially in coming decades (IPCC, 2007) and will have a series of impacts on the operation of geomorphological hazards as a result of changes in temperatures, precipitation amounts and intensities, and soil moisture conditions. Some environments will change more than others – 'geomorphological hotspots' (Goudie, 1996, 2006a), especially when crucial thresholds are crossed. Ice caps and glaciers will melt, permafrost will thin and retreat, shorelines will be subject to inundation by rising sea-levels, extreme hydrological events (both floods and droughts) may become more frequent, and dune and dust activity may change.

There are four types of reasons why some geomorphological processes, hazards and landform assemblages will show substantial modification as climate changes.

20.1.1 Threshold reliance

Some landforms and landforming processes are prone to change across crucial thresholds of temperature and precipitation. For example, the melting of components of the cryosphere is strongly temperature dependent and permafrost can only exist where mean annual temperatures are negative. Thus, as temperatures rise, permafrost will move polewards and/or upwards in altitude and the depth of summer thaw will change (Couture and Pollard, 2007). The mass balance of glaciers is largely controlled by the relative significance of ablation and snow nourishment and these in turn depend on temperatures and precipitation amounts. Likewise, stream flow, especially in dry regions, can vary greatly with modest changes in moisture caused by changes in evapotranspiration. The Holocene history of valley fills in the American south west has shown how abruptly and greatly stream systems can switch between incision (*arroyo* cutting) and aggradation. Aeolian activity is also strongly dependent on wind energy and the nature and extent of vegetation cover. If the latter falls below a certain level wind action is sharply intensified. Another example of threshold dependence of a landform type is the coral reef. These are highly sensitive to any changes in cyclone activity but also to coral bleaching caused by elevated sea-surface temperatures, and to sea-level rise itself. Terminal lake basins are other landforms that have shown very rapid and substantial variations in response to Holocene and twentieth-century climate changes.

20.1.2 Compound effects

There are numerous examples of landform change being promoted by a combination of climate changes and other human pressures. The USA Dust Bowl of the 1930s, for example, was caused not only by a run of dry, hot years, but also coincided with a phase of land use intensification that led to the busting of the sod and the ploughing up of prairie grasslands by tractor-drawn ploughs. Indeed, desertification is a phenomenon that is often at its most intense when climatic and human pressures coincide. In coastal regions, the effects of rising sea-levels are compounded by local subsidence caused by fluid abstraction, while beaches, marshes and deltas starved of sediment by the damming of rivers and the construction of coastal defences will be especially prone to erosion and inundation as global sea-levels rise.

20.1.3 Susceptible environments

Some landforms are robust, while others are not. As we saw earlier, alluvial channels such as the arroyos of the south west USA, are intrinsically weak, being formed of relatively easily eroded material. The same is true of

Geomorphological Hazards and Disaster Prevention, eds. Irasema Alcántara-Ayala and Andrew S. Goudie. Published by Cambridge University Press. © Cambridge University Press 2010.

FIGURE 20.1. Erosional scars (dongas) produced by erosion of susceptible colluvium in Swaziland, southern Africa.

FIGURE 20.2. Low-lying, soft coastlines such as this coastal settlement in Kota Kinabalu, Sabah, Malaysia, will be especially prone to the effects of sea-level rise and coastal inundation.

dongas developed in erodible colluvial aprons in southern Africa (Figure 20.1). Likewise, once their protective vegetation cover is removed, sand dunes in arid areas are easily reactivated. Muddy and sandy coastlines will plainly be more prone to erosion than hard-rock coastlines, and slopes, shorelines and river banks glued by permafrost will be more resistant than those features in the absence of permafrost. At the other extreme, there are some landscapes that appear to have remained relatively unperturbed by environmental change over millions of years (e.g. low-relief surfaces characteristic of shield areas in the interior of Australia).

20.1.4 Severity of climate change

The severity of climate change, and thus its likely impact, will vary spatially. For example, the degree of temperature increase will be particularly great in high northern latitudes (IPCC, 2007, p. 15) (e.g. northern Canada) so that the cryosphere in such regions may be subjected to particular pressures. Similarly, reductions in soil moisture and stream flows may be especially great in some areas that are currently relatively dry, so that stream networks will shrink (Goudie, 2006b). Warming will have an especially strong impact on river behaviour in areas where winter precipitation currently falls as snow. Giorgi (2006) developed a regional climate change index (RCCI) based on four variables: changes in mean regional precipitation and inter-annual precipitation variability, and changes in regional surface air temperatures and their inter-annual variability. On this basis certain climate change hotspots were identified, including the Mediterranean and north eastern European regions, Central America, southern Equatorial Africa and the Sahara, and eastern North America. Similarly, Sheffield and Wood (2008)

recognised some potential future long-term drought hot spots, including the Mediterranean, West Africa, Central Asia and Central America. The likely incidence of more severe drought conditions in the Mediterranean area has also been suggested by Gao and Giorgi (2008).

Having considered some of the general factors that will determine which areas may prove to be geomorphological hotspots (Goudie, 1996), let us now consider some specific examples of how geomorphological hazards may be affected by future climate change.

20.2 Coastal hazards

If temperatures climb then so will sea-levels. This is partly because of the steric effect but also because of melting of the cryosphere. Rising sea-levels will have substantial geomorphological consequences for the world's coastlines and will impact on a large proportion of the Earth's human population (Leatherman, 2001; FitzGerald et al. 2008) (Figure 20.2). As Viles and Spencer (1995) pointed out, about half of the population in the industrialised world lives within one kilometre of a coast, and two-thirds of the world's cities with populations of over 2.5 million people are near estuaries. Thirteen of the world's twenty largest cities are located on coasts. Areas at less than 10 m in altitude (the so-called low elevation coastal zone – the LECZ) account for c. 2 per cent of the world's land area but contain 10% of the population. While only 13 per cent of urban settlements with populations of less than 100,000 occur in the LECZ, the figure rises to over 65 per cent among cities of 5 million or more.

There has been a considerable diversity of views about how much sea-level rise is likely to occur by 2100. In general, however, estimates have tended to be revised downward through time (Pirazzoli, 1996) and have now settled at best estimates of just under 50 centimetres by 2100. This implies rates of sea-level rise of around 5 millimetres per year, which compares with a rate of about 1.5 to 2.0 millimetres during the twentieth century (Miller and Douglas, 2004). However, should the Greenland ice sheet melt at a faster rate than is currently predicted, then the amount of rise will be greater.

The effects of global sea-level rise will be compounded in those areas that suffer from local subsidence resulting from tectonic movements, isostatic adjustments and extraction of solids and liquids. Areas where land is rising because of isostasy (e.g. Fennoscandia, Scotland and the Canadian Shield) or because of tectonic uplift (e.g. much of the Pacific coast of the Americas) will be less at risk than subsiding regions. Sectors of the world's coastline that have been subsiding in recent decades include a large tract of the eastern seaboard of the USA, south eastern England, and some of the world's great river deltas (e.g. Indus, Ganges, Nile, Mississippi, Mekong, Tigris-Euphrates and Zambezi).

Other areas of appreciable subsidence include some ocean islands. Tide gauge records from the Hawaiian Ridge demonstrate ongoing subsidence of 1,500 millimetres per thousand years for Oahu and 3,500 millimetres per thousand years for Hawaii. The causes of this are a matter of debate (see Lambeck, 1988, pp. 506–509) but some of it may be caused by the loading of volcanic material onto the crust, while some may be due to a gradual contraction of the seafloor as the ocean lithosphere moves away from either the ridge or the hotspot that led to the initial formation of the volcanoes. The presence of guyots and seamounts in the Pacific Ocean attest to the fact that such subsidence has occurred.

Some high and mid-latitude areas are subject to subsidence because of ongoing adjustment to the application and removal of ice loadings to the crust (glacio-isostasy). During Pleistocene glacials, areas directly under the weight of ice caps were depressed, whereas areas adjacent to them popped up by way of compensation, giving the so-called peripheral bulge. Conversely, during the Holocene, following removal of the ice load, the formerly glaciated areas have rebounded whereas the marginal areas have foundered. A good example of this is the Laurentide area of North America. It is also one of the reasons for the sinking of south east England and its greater susceptibility to flooding by storm surges.

Elsewhere, human actions promote subsidence: the withdrawal of solids (coal, salt, sulphur etc.) or fluids (groundwater, oil and gas); the hydro-compaction of sediments; the oxidation and shrinkage of organic deposits such as peats and humus-rich soils; the melting of permafrost; and the catastrophic development of sinkholes in karstic terrain.

Coastal salt marshes that may be highly prone to sea-level rise include areas of deltaic sedimentation where, because of sediment trapping by dams upstream or because of cyclic changes in the location of centres of deposition, rates of sediment supply are low. Such areas may also, as we have seen, be areas with high rates of subsidence. A classic example of this is portions of the Mississippi delta.

In arid areas, such as the Arabian Gulf in the Middle East, extensive tracts of coastline are fringed by low-level salt-plains called *sabkhas*. These are generally regarded as equilibrium forms that are produced by a combination of depositional and erosional processes (e.g. wind erosion and storm surge effects). They tend to occur at or about high tide level. A large proportion of the industrial and urban infrastructure of the United Arab Emirates, particularly around Abu Dhabi, is located on or in close proximity to sabkhas.

Deltaic coasts and their environs are also home to large numbers of people. They are likely to be threatened by submergence as sea-levels rise, especially where prospects of compensating sediment accretion are not evident. Many deltas are currently zones of subsidence because of the isostatic effects of the sedimentation that caused them to form. Holocene subsidence rates for the Mississippi delta are *c.* 15 millimetres per annum (Fairbridge, 1983), for the Yangtze 1.6–4.4 (Stanley and Chen, 1993), for the Rhone 0.5–4.5 (L'Homer, 1992), and for the Nile *c.* 4.7 (Sherif and Singh, 1999). This will compound the effects of eustatic sea-level rise (Milliman and Haq, 1996). The future of our large deltas looks grim (Ericcson *et al.* 2006). Broadus *et al.* (1986), for example, calculated that were the sea-level to rise by just 1 metre in 100 years, 12–15 per cent of Egypt's arable land would be lost and 16 per cent of the population would have to be relocated. With a 3 metres rise the figures would be a 20 per cent loss of arable land and a need to relocate 21 per cent of the population. Alexandria, Rosetta and Port Said are at particular risk and even a sea-level rise of 50 centimetres could mean that 2 million people would have to abandon their homes (El-Raey, 1997). In Bangladesh, a 1 metre rise would inundate 11.5 per cent of the total land area of the state and affect 9 per cent of the population directly, while a 3 metres rise in sea-level would inundate 29 per cent of the land area and affect 21 per cent of the population. It is sobering to remember that at the present time approximately one-half of Bangladesh's rice production is in the area that is less than 1 metre above sea-level. Many of the world's major conurbations might be flooded in whole or in part, sewers and drains rendered inoperative (Kuo, 1986), and peri-urban agricultural productivity reduced by saltwater incursion (Chen and Yong, 1999).

Even without accelerating sea-level rise, the Nile delta has suffered accelerated recession because of sediment retention by dams. The Nile sediments, on reaching the sea, formerly generated sandbars and dunes which contributed to delta accretion. About a century ago an inverse process was initiated and the delta began to retreat. For example, the Rosetta mouth lost about 1.6 kilometres of its length from 1898 to 1954. The imbalance between sedimentation and erosion appears to have started with the delta barrages (1861) and then been continued by later works, culminating with the Aswan High Dam. In addition, large amounts of sediment are retained in an extremely dense network of irrigation and drainage channels that has been developed in the Nile delta itself (Stanley, 1996). Much of the Egyptian coast is now 'undernourished' with sediment and, as a result of this overall erosion of the shoreline, the sandbars bordering Lake Manzala and Lake Burullus on the seaward side are eroded and likely to collapse. If this were to happen, the lakes would be converted into marine bays, so that saline water would come into direct contact with low-lying cultivated land and freshwater aquifers.

It needs to be remembered, however, that deltas will not solely be affected by sea-level changes. The delta lands of Bangladesh (Warrick and Ahmad, 1996), for example, receive very heavy sediment loads from the rivers (like the Ganges) that feed them, so that it is the relative rates of accretion and inundation that will be crucial (Milliman *et al.*, 1989). Land use changes upstream, such as deforestation, could increase rates of sediment supply. Deltas could also be affected by changing tropical cyclone activity (see below).

Bangkok is an example of a city that is being threatened by a combination of accelerated subsidence and accelerated sea-level rise (Nutalaya *et al.*, 1996). Between 1960 and 1988, 20 to 160 centimetres of depression of the land surface occurred. The situation is critical because Bangkok is situated on a very flat, low-lying area, where the ground-level elevations range from only 0 to 1.5 metres above mean sea-level.

Barrier islands, such as those that line the eastern seaboard of the United States and the southern North Sea, are dynamic and often densely settled landforms that will tend to migrate inland with rising sea-levels and increased intensity of overtopping by waves (Eitner, 1996). If sea-level rise is not too rapid, and if they are not constrained by human activities (e.g. engineering structures and erosion control measures), they are moved inland by wash over – a process similar to rolling up a rug (Titus, 1990); as the island rolls landward, it builds landwards and remains above sea-level. As sea-level rises, they will be exposed to higher storm surges and greater flooding. In the New York area, Gornitz *et al.* (2002) have calculated that by the 2080s the return period of the 100-year storm flood could be reduced to between 4 and 60 years (depending on location). As the marshes behind barriers are converted to open water, tidal exchange through barrier inlets will increase, which will lead to the sequestration of sand in tidal deltas and the subsequent erosion of the adjacent barrier island shorelines (Fitzgerald, 2008).

Coral reefs, because of their low-lying nature, may be subject to over-washing and inundation as sea-level rises, particularly if, because of pollution, they are unable to grow upwards at an adequate rate to keep up with sea-level rise. They may also suffer from increasing ocean acidification. A proportion of the extra carbon dioxide being released into the atmosphere by the burning of fossil fuels and biomass is absorbed by sea water. As carbon dioxide combines with water it produces carbonic acid. An increase in carbonic acid in sea water will cause it to become more acidic (i.e. it will have a lower pH than now). Several centuries from now, if we continue to add carbon dioxide to the atmosphere, ocean pH will be lower than at any time in the past 300 million years (Doney, 2006). This will be harmful to those organisms like corals that depend on the presence of carbonate ions to build their hard parts out of calcium bicarbonate (Orr *et al.*, 2005).

Reefs play an important role in coastal protection against storms so that 'we can anticipate that decreasing rates of reef accretion, increasing rates of bioerosion, rising sea levels, and intensifying storms may combine to jeopardize a wide range of coastal barriers' (Hoegh-Guldberg *et al.*, 2007, p. 1742).

20.3 Hydrological hazards

Turning to fluvial and hydrological hazards, rainfall intensity is a major factor in controlling such phenomena as flooding, rates of soil erosion and mass movements (Sidle and Dhakal, 2002). Under increased greenhouse gas concentrations some general circulation models (GCMs) exhibit enhanced mid-latitude and global precipitation intensity and shortened return periods of extreme events (McGuffie *et al.*, 1999; Jones and Reid, 2001; New *et al.*, 2001). There is some evidence of increased rainfall events in various countries over recent warming decades, which lends some support to this notion. Examples are known

FIGURE 20.3. The possibility of increasing hurricane intensities and frequencies has implications for infrastructure, as is illustrated by the destruction of this bridge across the Usutu River in Swaziland as a result of Cyclone Domoina, 1984.

from North America (Francis and Hengeveld, 1998; Karl and Knight, 1998), Australia (Suppiah and Hennessy, 1998), Japan (Iwashima and Yamamoto, 1993), South Africa (Mason *et al.*, 1999) and Europe (Forland *et al.*, 1998). In the UK there has been an upward trend in the heaviest winter rainfall events (Osborn *et al.*, 2000). In their analysis of flood records for 29 river basins from high and low latitudes with areas greater than 200,000 square kilometres, Milly *et al.* (2002) found that the frequency of great floods had increased substantially during the twentieth century, particularly during its warmer later decades.

It is possible, though by no means certain, that as the oceans warm up, so the geographical spread and frequency of hurricanes will increase. These are a major hazard for human communities (Figure 20.3) and also have a whole series of hazardous consequences, including accentuated river flooding, greater wave heights (Komar and Allan, 2008), coastal surges, the triggering of landslides, and accelerated land erosion and siltation. The possibility that their frequency, intensity and geographical spread might increase in a warmer world became a particular concern after the ravages of Hurricane Katrina on the Gulf Coast of the USA in August 2005. This storm killed at least 1,836 people and caused damage estimated at $81.2 billion. Roughly 80 per cent of New Orleans was flooded. Furthermore, it is also likely that the intensity of these storms will be magnified. Emanuel (1987) used a GCM which predicted that with a doubling of present atmospheric concentrations of carbon dioxide there will be an increase of 40–50 per cent in the destructive potential of hurricanes. More recently Knutson and Tuleya (1999) simulated hurricane activity for a sea-surface temperature warming of 2.2 °C and found that this yielded hurricanes

that were more intense by 3–7 metres per second for wind speed, an increase of 5–12 per cent. The relationship between sea-surface temperature increase and increasing global hurricane activity has been confirmed by Hoyos *et al.* (2006) and Saunders and Lea (2008). Santer *et al.* (2006) have demonstrated that human factors have caused the increase in sea-surface temperatures and cyclogenesis in both the Atlantic and Pacific regions. The Intergovernmental Panel on Climate Change (IPCC, 2007, p. 239) has argued, 'Globally, estimates of the potential destructiveness of hurricanes show a significant upward trend since the mid-1970s, with a trend towards longer lifetimes and greater storm intensity...' On the other hand, some models have indicated the possibility that there could be a reduction in Atlantic hurricane frequency under twenty-first-century warming (Knutson *et al.* 2008).

Changes in precipitation amounts and evapotranspiration rates will modify river flow regimes (Goudie, 2006b). Probabilistic analysis of GCMs by Palmer and Räisänen (2002), applied to Western Europe and the Asian Monsoon region, shows under global warming a clear increase in extreme winter precipitation for the former and for extreme summer precipitation for the latter. Increased monsoonal rainfall events would have potentially grave implications for flooding in Bangladesh.

Let us also consider the Rhine. It stretches from the Swiss Alps to the Dutch coast and its catchment covers 185,000 square kilometres (Shabalova *et al.*, 2003). Models suggest that by the end of the century its discharge will become markedly more seasonal with mean discharge decreases of about 30 per cent in summer and increases of about 30 per cent in winter. The increase in winter discharge will be caused by a combination of increased precipitation, reduced snow storage and increased early melt. The decrease in the summer discharge is related mainly to a predicted decrease in precipitation combined with increases in evapotranspiration. Glacier melting in the Alps also contributes to the flow of the Rhine. Increased rates of glacier melting may, for a period of years, cause an increased incidence of summer meltwater floods, but when the glaciers have disappeared, river flow volumes may be drastically reduced (Braun *et al.*, 2000), leading to severe water shortages.

Lakes will respond to the temperature and precipitation changes that may result from global warming. An example of such responses is provided by models that have been developed for the Great Lakes of North America by Croley (1990) and Hartmann (1990). They suggest that for a doubling of carbon dioxide there may be a 23–51 per cent reduction in net basin supplies of water to all the Great Lakes, and that as a result levels will fall at rates ranging from 13 millimetres per decade (for Lake Superior) to 93 millimetres per decade (for Lake Ontario). Major falls of up to 9 metres by the end of the present century have been predicted for the Caspian (Elguindi and Giorgi, 2006).

20.4 Mass movement and soil erosion hazards

Mass movements will be impacted upon by changes in hydrological conditions. In particular, slope stability and landslide activity are greatly influenced by groundwater levels and pore-water pressure fluctuations, though Collison et al. (2000) have argued that other factors, such as land use change and human activity, would be likely to have a greater impact than climate change in some situations.

Various attempts have been made to model potential slope responses. For example, in the Italian Dolomites, Dehn et al. (2000) suggested that future landslide activity would be reduced because there would be less storage of precipitation as snow. Therefore, the release of meltwater, which under present conditions contributes to high ground-water levels and strong landslide displacement in early spring, would be significantly diminished. However, because of the differences between GCMs and problems of downscaling, there are still great problems in modelling future landslide activity, and Dehn and Buma (1999) found that the use of three different GCM experiments for the assessment of the activity of a small landslide in south east France did not show a consistent picture of future landslide frequencies.

Geomorphologists have modelled changes in soil erosion that may occur as a consequence of rainfall changes (Sun et al., 2002; Yang et al., 2003) though it is difficult to determine the likely effects of climate change compared to future land management practices (Wilby et al., 1997). For Brazil, Favis-Mortlock and Guerra (1999) used the Hadley Centre HADCM2 GCM and an erosion model (WEPP – Water Erosion Predictions Project). They found that by 2050 the increase in mean annual sediment yield in their area in the Mato Grosso would be 27 per cent. For the south east of the UK, where winter rainfall is predicted to increase modestly, Favis-Mortlock and Boardman (1995) recognised that changes in rainfall not only impact upon erosion rates directly, but also through their effects on rates of crop growth and on soil properties. Nonetheless, they showed that erosion rates were likely to rise, particularly in wet years. Decreases in soil moisture levels may increase fire frequencies, and this could have implications for soil erosion and debris flow generation in post-fire rainfall events, as in Chapparral shrublands in California. An

analysis by Flannigan et al. (2000) suggests that future fire severity could increase over much of North America. They anticipate increases in the area burned in the USA of 25–50 per cent by the middle of the twenty-first century, with most of the increases occurring in Alaska and the south east United States.

20.5 Glacial and permafrost hazards

Changes in the cryosphere have already taken place in recent decades, and may intensify in the future (Li et al., 2008). In many parts of the world glaciers are retreating and this is likely to be a trend that accelerates with global warming (Figure 20.4). The fast retreat of glaciers can have a series of adverse geomorphological effects. Glacial lakes in areas such as the Himalayas of Nepal and Bhutan are rapidly expanding as they are fed by increasing amounts of meltwater. Outburst floods from such lakes are extremely hazardous. Likewise, slopes deprived of the buttressing effects of glaciers (Holm et al., 2004) can become unstable,

FIGURE 20.4. In a warmer world, valley glaciers, such as the Bossons Glacier in the French Alps, will retreat, with implications for downstream runoff regimes and for local slope stability.

generating a risk of increased landsliding and debris ava-lanches (Kirkbride and Warren, 1999; Haeberli and Burn, 2002). The shrinkage of hanging glaciers in steep rock walls may uncover large expanses of bedrock, lead to changed temperature and stress fields in the rock, expose the uncovered rock to mechanical weathering and thermal erosion, and permit the wedging effects of ice formation caused by water percolation into cracks and fissures (Fischer *et al.* 2006).

Degradation of permafrost may lead to an increasing scale and frequency of slope failures, particularly in moun-tainous areas (Gruber and Haeberli, 2007). Thawing reduces the strength of both ice-rich sediments and frozen jointed bedrock (Davies *et al.*, 2001). Ice-rich soils undergo thaw consolidation during melting, with resulting elevated pore-water pressures, so that formerly sediment-mantled slopes may become unstable. Equally, bedrock slopes may be destabilised if warming reduces the strength of ice-bonded open joints or leads to groundwater movements that cause pore pressures to rise (Harris *et al.*, 2001). Increases in the thickness of the active layer may make more material available for debris flows.

Coastal bluffs may be subject to increased rates of erosion if they suffer from thermal erosion caused by permafrost decay. This would be accelerated still further if sea ice were to be less prevalent, for sea ice can protect coasts from wave erosion and debris removal (Carter *et al.*, 1987; Lantuit and Pollard, 2008). Local coastal losses to erosion of as much as 40 metres per year have been observed in some locations in both Siberia and Canada in recent years, while erosive losses of up to 600 metres over the past few decades have occurred in Alaska (Parson *et al.*, 2001).

One of the severest consequences of permafrost degra-dation is ground subsidence and the formation of thermo-karst (Nelson *et al.*, 2001; Nelson, 2002). This is likely to be a particular problem for engineering structures in the zones of relatively warm permafrost (the discontinuous and sporadic zones), and where the permafrost is rich in ice (Woo *et al.*, 1992). With thawing, ice-rich areas, such as ice-cored palsas and pingos, will settle more than those with lesser ice contents, producing irregular hummocks and depressions. Water released from ice melt may accu-mulate in such depressions. The depressions may then enlarge into thaw lakes as a result of thermal erosion (Harris, 2002), which occurs because summer heat is trans-mitted efficiently through the water body into surrounding ice-rich material (Yoshikawa and Hinzman, 2003). Once they have started, thaw lakes can continue to enlarge for decades to centuries because of wave action and continued thermal erosion of the banks. However, if thawing eventually penetrates the permafrost, drainage occurs that leads to ponds drying up.

20.6 Aeolian hazards

Changes in climate could affect wind erosion either through their impact on erosivity or through their effect on erodibil-ity. Erosivity is controlled by a range of wind variables including velocity, frequency, duration, magnitude, shear and turbulence. Unfortunately, GCMs as yet give little indi-cation of how wind characteristics might be modified in a warmer world, so that prediction of future changes in wind erosivity is problematic. Erodibility is largely controlled by vegetation cover and surface type, both of which can be influenced markedly by climate. In general, vegetation cover, which protects the ground surface and modifies the wind regime, decreases as conditions become more arid. Likewise climate affects surface materials by controlling their moisture content, the nature and amount of clay mineral content (cohesiveness) and organic levels. Soils that are dry, have a low clay content and little binding humus are highly susceptible to wind erosion. However, modelling the response of wind erosion to climatic variables on agricultural land is vastly complex, not least because of the variability of soil characteristics, topographic variation, the state of plant growth and residue decomposition, and the existence of windbreaks. To this needs to be added the temporal varia-bility of aeolian processes and moisture condition and the effects of different land management practices (Leys, 1999), which may themselves change with climate change.

If soil moisture levels decline as a result of changes in precipitation and/or temperature, there is the possibility that dust storm activity could increase in a warmer world (Wheaton, 1990). A comparison between the Dust Bowl years of the 1930s and model predictions of precipitation and temperature for the Great Plains of Kansas and Nebraska indicates that mean conditions could be similar to those of the 1930s under enhanced greenhouse condi-tions (Smith and Tirpak, 1990), or even worse (Rosenzweig and Hillel, 1993).

If dust storm activity were to increase as a response to global warming it is possible that this could have a feed-back effect on precipitation that would lead to further decreases in soil moisture (Tegen *et al.*, 1996; Miller and Tegen, 1998). However, the impact and occurrence of dust storms will depend a great deal on land management prac-tices, and recent decreases in dust storm activity in North Dakota have resulted from conservation measures (Todhunter and Cihacek, 1999).

Sand dunes, because of the crucial relationships between vegetation cover and sand movement, are highly susceptible to changes of climate. Some areas, such as the south west Kalahari (Stokes *et al.*, 1997) or portions of the High Plains of the USA (Gaylord, 1990) may have been especially prone to changes in precipitation and/or wind velocity because of their location in climatic zones that are close to a climatic threshold between dune stability and activity.

The explosive development in the use of luminescence dating of sand grains and studies of explorers' accounts (Muhs and Holliday, 1995) has led to the realisation that such marginal dune fields have undergone repeated phases of change at decadal and century time scales in response to extended drought events during the course of the Holocene (Arbogast, 1996; Muhs *et al.*, 1997; Stokes and Swinehart, 1997; Thomas *et al.*, 1997).

The mobility of desert dunes (M) is directly proportional to the sand-moving power of the wind, but indirectly proportional to their vegetation cover (Lancaster, 1995, p. 238). An index of the wind's sand-moving power is given by the percentage of the time (W) the wind blows above the threshold velocity (4.5 metres per second) for sand transport. Vegetation cover is a function of the ratio between annual rainfall (P) and potential evapotranspiration (PE). Thus, $M = W/(P/PE)$. Empirical observations in the United States and southern Africa indicate that dunes are completely stabilised by vegetation for $M < 50$, and are fully active for $M > 200$.

Muhs and Maat (1993) used the output from GCMs combined with this dune mobility index to show that many dunes and sand sheets on the Great Plains of the USA are likely to become reactivated, particularly if the frequencies of wind speeds above the threshold velocity increase by even a moderate amount. For Washington State, Stetler and Gaylord (1996) suggested that with a 4 °C warming vegetation would be greatly reduced and that as a consequence sand dune mobility would increase by over 400 per cent. For the Canadian Prairies, Wolfe (1997) found that while most dunes were currently inactive or only had active crests, under conditions of increased future drought most dunes would become more active.

Perhaps the most detailed scenarios for dune remobilisation by global warming have been developed for the Kalahari in southern Africa (Thomas *et al.*, 2005) (Figure 20.5). Much of the mega-Kalahari is currently vegetated and stable, but GCMs suggest that by 2099 all dune fields, from South Africa in the south to Zambia and Angola in the north will be reactivated. This could disrupt pastoral and agricultural systems. Indeed, the consequences of dune encroachment and reactivation could be serious and might lead to a loss of agricultural land, the overwhelming of buildings, roads, canals, runways and the like, abrasion

FIGURE 20.5. The interior of southern Africa, the Kalahari, has large expanses of now stable sand dunes which may become reactivated in a drier, warmer world. These vegetated linear dunes are in the south east of Angola. Scale bar is 10 km. (Courtesy NASA.)

of structures and equipment, damage to crops, and the impoverishment of soil structure. However, the methods used to estimate future dune field mobility are still full of problems and much more research is needed before we can have confidence in them (Knight *et al.* 2004).

20.7 Conclusions

It is evident that future climate change will have an impact on the frequency, distribution and intensity of a wide range of geomorphological hazards. As human population levels rise and cities become ever larger, any increase in hazard severity would have a growing impact. Some hazards will undoubtedly become more serious and others less so, but a great deal depends on the exact nature of the geomorphological environment in which human populations dwell.

References

Arbogast, A. F. (1996). Stratigraphic evidence for late-Holocene aeolian sand mobilization and soil formation in south-central Kansas, USA. *Journal of Arid Environments*, **34**, 403–14.

Braun, L. N., Weber, M. and Schulz, M. (2000). Consequences of climate change for runoff from alpine regions. *Annals of Glaciology*, **31**, 19–25.

Broadus, J., Milliman, J., Edwards, S., Aubrey, D. and Gable, F. (1986). Rising sea level and damming of rivers: possible effects in Egypt and Bangladesh. In J. G. Titus (ed.), *Effects of Changes in Stratospheric Ozone and Global Climate*. Washington D.C.: UNEP/USEPA, pp. 165–189.

Carter, L. D. (1987). Arctic lowlands: introduction. In W. L. Graf (ed.), *Geomorphic Systems of North America*. Boulder: Geological Society of America, Centennial Special Volume, 2, pp. 583–615.

Chen, X. and Zong, Y. (1999). Major impacts of sea-level rise on agriculture in the Yangtze Delta area around Shanghai. *Applied Geography*, **19**, 69–84.

Collison, A., Wade, S., Griffiths, J. and Dehn, M. (2000). Modelling the impact of predicted climate change on landslide frequency and magnitude in S. E. England. *Engineering Geology*, **55**, 205–218.

Couture N. J. and Pollard, W. H. (2007). Modelling geomorphic response to climatic change. *Climatic Change*, **85**, 407–431.

Croley, T. E. (1990). Laurentian Great Lakes double-CO_2 climate change hydrological impacts. *Climatic Change*, **17**, 27–47.

Davies, M. C. R., Hamza, O. and Harris, C. (2001). The effect of rise in mean annual temperature on the stability of rock slopes containing ice-filled discontinuities. *Permafrost and Periglacial Processes*, **12**, 137–144.

Dehn, M. and Buma, J. (1999). Modelling future landslide activity based on general circulation models. *Geomorphology*, **30**, 175–187.

Dehn, M., Gurger, G., Buma, J. and Gasparetto, P. (2000). Impact of climate change on slope stability using expanded downscaling. *Engineering Geology*, **55**, 193–204.

Doney, S. C. (2006). The dangers of ocean acidification. *Scientific American*, **294**(3), 38–45.

Eitner, V. (1996). Geomorphological response to the East Frisan barrier islands to sea level rise: an investigation of past and future evolution. *Geomorphology*, **15**, 57–65.

Elguindi, N. and Giorgi, F. (2006). Projected changes in the Caspian Sea level for the 21st century based on the latest AOGCM simulations. *Geophysical Research Letters*, **33**, L08706, doi:10.1029/2006GL025943.

El-Raey, M. (1997). Vulnerability assessment of the coastal zone of the Nile delta of Egypt to the impact of sea level rise. *Ocean and Coastal Management*, **37**, 29–40.

Emanuel, K. A. (1987). The dependence of hurricane intensity on climate. *Nature*, **326**, 483–485.

Ericcson, J. P, Vorosmarty, C. J, Dingham, L., Ward, L. G. and Meybeck, M. (2006). Effective sea-level rise and deltas: causes of change and human dimension implications. *Global and Planetary Change*, **50**, 63–82.

Fairbridge, R. W. (1983). Isostasy and eustasy. In D. E. Smith and A. G. Dawson (eds.), *Shorelines and Isostasy*. London: Academic Press, pp. 3–25.

Favis-Mortlock, D. and Boardman, J. (1995). Non linear responses of soil erosion to climate change: modelling study on the UK South Downs. *Catena*, **25**, 365–387.

Favis-Mortlock, D. T. and Guerra, A. J. T. (1999). The implications of general circulation model estimates of rainfall for future erosion: a case study from Brazil. *Catena*, **37**, 329–354.

Fischer, L., Kääb, A., Huggel, C. and Noetzli, J. (2006). Geology, glacier retreat and permafrost degradation as controlling factors of slope instabilities in a high-mountain rock wall: the Monte Rosa east face. *Natural Hazards and Earth System Sciences*, **6**, 761–772.

FitzGerald, D. M., Fenster, M. S., Argow, B. A. and Buynevich, I. V. (2008). Coastal impacts due to sea-level rise. *Annual Review of Earth and Planetary Sciences*, **36**, 601–647.

Flannigan, M. D., Stocks, B. J. and Wotton, B. M. (2000). Climate change and forest fires. *The Science of the Total Environment*, **262**, 221–230.

Forland, E. J., Alexandersson, H., Drebs, A. et al. (1998). *Trends in Maximum 1-day Precipitation in the Nordic Region*. DNMI report 14/98, Klima. Oslo: Norwegian Meteorological Institute, 1–55.

Francis, D. and Hengeveld, H. (1998). *Extreme Weather and Climate Change*. Downsview, Ontario: Environment Canada.

Gao X. and Giorgi, F. (2008). Increased aridity in the Mediterranean region under greenhouse gas forcing estimated from high resolution simulations with a regional climate model. *Global and Planetary Change*, **62**, 195–209.

Gaylord, D. R. (1990). Holocene palaeoclimatic fluctuations revealed from dune and interdune strata in Wyoming. *Journal of Arid Environments*, **18**, 123–138.

Giorgi F. (2006). Climate change hot-spots. *Geophysical Research Letters*, **33**, L08707, doi:10.1029/2006GL025734.

Gornitz, V., Couch, S. and Hartig, E. K. (2002). Impacts of sea level rise in the New York City metropolitan area. *Global and Planetary Change*, **32**, 61–88.

Goudie, A. S. (1996). Geomorphological 'hotspots' and global warming. *Interdisciplinary Science Reviews*, **21**(3), 253–259.

Goudie A. S. (2006a). *The Human Impact on the Natural Environment*, 6th edition. Oxford: Blackwell.

Goudie A. S. (2006b). Global warming and fluvial geomorphology. *Geomorphology*, **79**, 384–394.

Gruber, S. and Haeberli, W. (2007). Permafrost in steep bedrock slopes and its temperature-related destabilization following climate change. *Journal of Geophysical Research*, **112**, F02S18, doi:10.1029/2006FJ000547.

Haeberli, W. and Burn, C. R. (2002). Natural hazards in forests: glacier and permafrost effects as related to climate change. In R. C. Sidle (ed.), *Environmental Change and Geomorphic Effects in Forests*. Wallingford: CABI, pp. 167–202.

Harris, C., Davies, M. C. R. and Etzelmüller, B. (2001). The assessment of prudential geotechnical hazards associated with mountain permafrost in a warming global climate. *Permafrost and Periglacial Processes*, **12**, 145–156.

Harris, S. A. (2002). Causes and consequences of rapid thermokarst development in permafrost or glacial terrain. *Permafrost and Periglacial Processes*, **13**, 237–242.

Hartmann, H. C. (1990). Climate change impacts on Laurentian Great Lakes levels. *Climatic Change*, **17**, 49–67.

Hoegh-Guldberg, O., Mumby, P. J., Hooten, A. J. et al. (2007). Coral reefs under rapid climate change and ocean acidification. *Science*, **318**, 1737–1742.

Holm, K., Bovis, M. and Jacob, M. (2004). The landslide response of alpine basins to post-Little Ice Age glacial thinning and retreat in southwestern British Columbia. *Geomorphology*, **57**, 201–216.

Hoyos, C. D., Agudelo, P. A., Webster, P. J. and Curry, J. A. (2006). Deconvolution of the factors contributing to the increase in global hurricane intensity. *Science*, **312**, 94–97.

IPCC (2007). *Climate Change 2007: The Physical Science Basis. Contribution of Working Group I to the Fourth Assessment Report of the Intergovernmental Panel on Climate Change* (edited by S. Solomon *et al.*). Cambridge and New York: Cambridge University Press.

Iwashima, T. and Yamamoto, R. (1993). A statistical analysis of the extreme events: long-term trend of heavy daily precipitation. *Journal of the Meteorological Society of Japan*, **71**, 637–640.

Jones, P. D. and Reid, P. A. (2001). Assessing future changes in extreme precipitation over Britain using regional climate model integrations. *International Journal of Climatology*, **21**, 1337–1356.

Karl, T. R. and Knight, R. W. (1998). Secular trends of precipitation amount, frequency and intensity in the United States. *Bulletin of the American Meteorological Society*, **79**, 1413–1449.

Kirkbride, M. P. and Warren, C. R. (1999). Tasman Glacier, New Zealand: 20th century thinning and predicted calving retreat. *Global and Planetary Change*, **22**, 11–28.

Knight, M., Thomas, D. S. G. and Wiggs, G. F. S. (2004). Challenges of calculating dunefield mobility over the 21st century. *Geomorphology*, **59**, 197–213.

Knutson, T. R. and Tuleya, R. E. (1999). Increased hurricane intensities with CO_2-induced warming as simulated using the GFDL hurricane prediction system. *Climate Dynamics*, **15**, 503–519.

Knutson, T. R., Sirutis, J. J., Garner, S. T., Vecchi, G. A. and Held, I. M. (2008). Simulated reduction in Atlantic hurricane frequency under twenty-first-century warming conditions. *Nature Geoscience*, doi:10.1038/ngeo202.

Komar, P. D. and Allan, J. C. (2008). Increasing hurricane-generated wave heights along the U.S. East Coast and their climate controls. *Journal of Coastal Research*, **24**, 479–488.

Kuo, C. (1986). Flooding in Taipeh, Taiwan and coastal drainage. In J. G. Titus (ed.), *Effects of Changes in Stratospheric Ozone and Global Climate*. Washington D.C.: UNEP/USEPA, pp. 37–46.

L'Homer, A. (1992). Sea level changes and impact on the Rhône Delta coastal lowlands. In M. J. Tooley and S. Jelgersma (eds.), *Impacts of Sea Level Rise on European Coastal Lowlands*. Oxford: Blackwell, pp. 136–152.

Lambeck, K. (1988). *Geological Geodesy*. Oxford: Clarendon Press.

Lancaster, N. (1995). *Geomorphology of Desert Dunes*. London: Routledge.

Lantuit, H. and Pollard, W. H. (2008). Fifty years of coastal erosion and retrogressive thaw slump activity on Herschel Island, southern Beaufort Sea, Yukon Territory, Canada. *Geomorphology*, **95**, 84–102.

Leatherman, S. P. (2001). Social and economic costs of sea level rise. In B. C. Douglas, M. S. Kearney and S. P. Leatherman (eds.), *Sea Level Rise: History and Consequences*. San Diego: Academic Press, pp. 181–223.

Leys, J. (1999). Wind erosion on agricultural land. In A. S. Goudie, I. Livingstone and S. Stokes (eds.), *Aeolian Environments, Sediments and Landforms*. Chichester: Wiley, pp. 143–166.

Li, X., Cheng, G., Jin, H. *et al.* (2008). Cryospheric change in China. *Global and Planetary Change*, **62**, 210–218.

Mason, S. J., Waylen, P. R., Mimmack, G. M. Rajaratnam, B. and Harrison, J. M. (1999). Changes in extreme rainfall events in South Africa. *Climatic Change*, **41**, 249–257.

McGuffie, K., Henderson-Sellers, A., Holbrook, N. *et al.* (1999). Assessing simulations of daily temperature and precipitation variability with global climate models for present and enhanced greenhouse climates. *International Journal of Climatology*, **19**, 1–26.

Miller, L. and Douglas, B. C. (2004). Mass and volume contribution to twentieth-century global sea level rise. *Nature*, **428**, 406–409.

Miller, R. L. and Tegen, I. (1998). Climate response to soil dust aerosols. *Journal of Climate*, **11**, 3247–3267.

Milliman, J. D. and Haq, B. U. (eds.), (1996). *Sea Level Rise and Coastal Subsidence*. Dordrecht: Kluwer.

Milliman, J. D., Broadus, J. M. and Gable, F. (1989). Environmental and economic impacts of rising sea level and subsiding deltas: the Nile and Bengal examples. *Ambio*, **18**, 340–345.

Milly, P. C. D., Wetherald, R. T., Dunne, K. A. and Delworth, T. L. (2002). Increasing risk of great floods in a changing climate. *Nature*, **415**, 514–517.

Muhs, D. R. and Holliday, V. T. (1995). Evidence of active dune sand in the Great Plains in the 19th century from accounts of early explorers. *Quaternary Research*, **43**, 198–208.

Muhs, D. R. and Maat, P. B. (1993). The potential response of eolian sands to greenhouse warming and precipitation reduction on the Great Plains of the United States. *Journal of Arid Environments*, **25**, 351–361.

Muhs, D. R., Stafford, T. W., Swinehart, J. B. *et al.* (1997). Late Holocene eolian activity in the mineralogically mature Nebraska Sand Hills. *Quaternary Research*, **48**, 162–176.

Nelson, F. E. (2002). Climate change and hazard zonation in the Circum-Arctic Permafrost regions. *Natural Hazards*, **26**, 203–225.

Nelson, F. E., Anisimov, O. A. and Shiklomanov, N. I. (2001). Subsidence risk from thawing permafrost. *Nature*, **410**, 889–890.

New, M., Todd, M., Hulme, M. and Jones, P. (2001). Precipitation measurements and trends in the twentieth century. *International Journal of Climatology*, **21**, 1899–1922.

Nutalaya, P., Yong, R. N., Chumnankit, T. and Buapeng, S. (1996). Land subsidence in Bangkok during 1978–1988. In J. D. Milliman and B. U. Haq (eds.), *Sea Level Rise and Coastal Subsidence*. Dordrecht: Kluwer, pp. 105–130.

Orr, J. C, Pantoja, S. and Pörtner, H.-O. (2005). Introduction to special selection: the ocean in a high-CO_2 world. *Journal of Geophysical Research*, **110** (C), doi: 10 1029/2005 JC 003086.

Osborn, T. J., Hulme, M., Jones, P. D. and Basnett, T. A. (2000). Observed trends in the daily intensity of United Kingdom precipitation. *International Journal of Climatology*, **20**, 347–364.

Palmer, T. N. and Räisänen, J. (2002). Quantifying the risk of extreme seasonal precipitation events in a changing climate. *Nature*, **415**, 512–514.

Parson, E. A., Carter, L., Anderson, P., Wang, B. and Weller, G. (2001). Potential consequences of climate variability and change for Alaska. In National Assessment Synthesis Team, *Climate Change Impacts on the United States: The Potential Consequences of Climate Variability and Change*. Cambridge: Cambridge University Press, pp. 283–312.

Pirazzoli, P. A. (1996). *Sea Level Changes: The Last 20,000 Years*. Chichester: Wiley.

Rosenzweig, C. and Hillel, D. (1993). The dust bowl of the 1930s: analog of greenhouse effect in the Great Plains? *Journal of Environmental Quality*, **22**, 9–22.

Santer, B. D., Wigley, T. M. L., Gleckler, P. J. *et al.* (2006). Forced and unforced ocean temperature changes in Atlantic and Pacific tropical cyclogenesis regions. *Proceedings of the National Academy of Sciences*, **103**, 13 905–13 910.

Saunders, M. A. and Lee, A. S. (2008). Large contribution of sea surface warming to recent increase in Atlantic hurricane activity. *Nature*, **451**, 557–560.

Shabalova, M. V., van Deursen, W. P. A. and Buishand, T. A. (2003). Assessing future discharge of the river Rhine using regional climate model integrations and a hydrological model. *Climate Research*, **23**, 233–246.

Sheffield, J. and Wood, E. F. (2008). Projected changes in drought occurrence under future global warming from multi-model, multi-scenario, IPCC AR4 simulations. *Climate Dynamics*, **31**, 79–105.

Sherif, M. M. and Singh, V. P. (1999). Effect of climate change on sea water intrusion in coastal aquifers. *Hydrological Processes*, **13**, 1277–1287.

Sidle, R. C. and Dhakal, A. S. (2002). Potential effect of environmental change on landslide hazards in forest environments. In R. C. Sidle (ed.), *Environmental Change and Geomorphic Hazards in Forests*. Wallingford: CABI, pp. 123–165.

Smith, J. B. and Tirpak, D. A. (eds.) (1990). *The Potential Effects of Global Climate Change on the United States*. New York: Hemisphere.

Stanley, D. J. (1996). Nile delta: extreme case of sediment entrapment on a delta plain and consequent coastal land loss. *Marine Geology*, **129**, 189–195.

Stanley, D. J. and Chen, Z. (1993). Yangtze delta, eastern China: I. Geometry and subsidence of Holocene depocenter. *Marine Geology*, **112**, 1–11.

Stetler, L. L. and Gaylord, D. R. (1996). Evaluating eolian-climate interactions using a regional climate model from Hanford, Washington (USA). *Geomorphology*, **17**, 99–113.

Stokes, S. and Swinehart, J. B. (1997). Middle- and late-Holocene dune reactivation on the Nebraska Sand Hills, USA. *The Holocene*, **7**, 272–281.

Stokes, S., Thomas, D. S. G. and Washington, R. (1997). Multiple episodes of aridity in southern Africa since the last interglacial period. *Nature*, **388**, 154–158.

Sun, G. E., McNulty, S. G., Moore, J., Bunch, C. and Ni, J. (2002). Potential impacts of climate change on rainfall erosivity and water availability in China in the next 100 years. *Proceedings of the 12th International Soil Conservation Conference*. Beijing: Tsinghua University Press.

Suppiah, R. and Hennessy, K. J. (1998). Trends in total rainfall, heavy rain events and number of dry days in Australia. *International Journal of Climatology*, **18**, 1141–1164.

Tegen, I., Lacis, A. A. and Fung, I. (1996). The influence on climate forcing of mineral aerosols from disturbed soils. *Nature*, **380**, 419–422.

Thomas, D. S. G., Stokes, S. and Shaw, P. A. (1997). Holocene aeolian activity in the south-western Kalahari Desert, southern Africa: significance and relationships to late-Pleistocene dune-building events. *The Holocene*, **7**, 273–281.

Thomas, D. S. G., Knight, M. and Wiggs, G. F. S. (2005). Remobilization of southern African desert dune systems by twenty-first century global warming. *Nature*, **435**, 1218–1221.

Titus, J. G. (1990). Greenhouse effect, sea level rise, and barrier islands: case study of Long Beach Island, New Jersey. *Coastal Management*, **18**, 65–90.

Todhunter, P. E. and Chihacek, L. J. (1999). Historical reduction of airborne dust in the Red River Valley of the North. *Journal of Soil and Water Conservation*, **54**, 543–551.

Viles, H. A. and Spencer, T. (1995). *Coastal Problems: Geomorphology, Ecology and Society at the Coast*. London: Edward Arnold.

Warrick, R. A. and Ahmad, Q. K. (eds) (1996). *The Implications of Climate and Sea Level Change for Bangladesh*. Dordrecht: Kluwer.

Wheaton, E. E. (1990). Frequency and severity of drought and dust storms. *Canadian Journal of Agricultural Economics*, **38**, 695–700.

Wilby R. L., Dalgleish, H. Y. and Foster, I. D. L. (1997). The impact of weather patterns on historic and contemporary catchment sediment yields. *Earth Surface Processes and Landforms*, **22**, 353–363.

Wolfe, S. A. (1997). Impact of increased aridity on sand dune activity in the Canadian Prairies. *Journal of Arid Environments*, **36**, 412–432.

Woo, M.-K., Lewkowicz, A. G. and Rouse, W. R. (1992). Response of the Canadian permafrost environment to climate change. *Physical Geography*, **13**, 287–317.

Yang, D., Kanae, S., Oki, T., Koike, T. and Musiake, K. (2003). Global potential soil erosion with reference to land use and climate changes. *Hydrological Processes*, **17**, 2913–2928.

Yoshikawa, K. and Hinzman, L. D. (2003). Shrinking thermokarst ponds and groundwater dynamics in discontinuous permafrost near Council, Alaska. *Permafrost and Periglacial Processes*, **14**, 151–160.

21 Geomorphic hazards and sustainable development

David Higgitt

21.1 Introduction

Intuitively, there should be a close link between the topics of environmental hazards and sustainability. Both issues feature prominently in the contemporary language of policy makers, journalists and academics. Planning for an environmentally sustainable future clearly requires identification, assessment and management of risks and vulnerability. The vulnerability of humans to a range of environmental hazards affects the sustainability of societies. Vulnerability has both economic and social dimensions dictating that a complete study of geophysical risks requires the work of scientists and social scientists to be integrated (Beer, 2004). The aim of this chapter is to consider some aspects of the interdisciplinary debate on the interactions between hazard, risk and sustainability, from which the potential for geomorphology to perform at this intersection can be gauged. Some examples from Asian seismic and flooding hazards are included to illustrate these points.

It could be argued that the sustainability agenda is a relatively recent arrival in the domain of hazards research. The World Commission on Environment and Development (WCED, 1987) catapulted the concept of sustainable development into the limelight with its definition as 'development that meets the needs of the present without compromising the ability of future generations to meet their own needs'. While the hallmarks of sustainability thinking are implicit in much of the earlier hazards research, a shift towards a sustainability agenda demands adjustment to the paradigms of hazards research. If a comparison is made with a third prominent issue readily tripping from the tongues of policy makers – climate change – the association between climate change and natural hazards and between climate change and sustainability seem to have generated far greater policy attention than a hazards–sustainability synergy. Nevertheless, new approaches to conceptualizing the three-way interactions of climate change, hazard and sustainability, are emerging. What relevance, threats and opportunities does this agenda hold for the practice of geomorphology?

Though the goals of hazard management and sustainability appear closely aligned, the relationship is complex. Andres and Strappazzon (2007) question the link from three aspects: epistemological, technical and communicational. From the epistemological context, environmental scientists (including geomorphologists) draw understanding of risk and sustainability from ecological concepts of resilience, stability, thresholds and adaptation, each grounded in general systems theory. A systems perspective underpins an understanding of stability as a function of organization and connectivity. As sustainability is fundamentally about the maintenance of system function, risk can therefore be regarded as one aspect of sustainability (Beer, 2004).

From the technical context, planning regulations and institutional frameworks are increasingly associating the two concepts of risk and sustainability, but in most countries sustainable development has only become an explicit focus of public policy within the last 10–20 years. Hence the technical expertise involved in the tasks of managing risk and in sustainable planning is likely to be compartmentalized. The regulatory framework poses additional challenges for integrating planning and development decisions at different scales of administration. Limitations of national environmental policies in the 1970s and 1980s, constrained by the imperative of economic growth, have led to intensive critique of environmental regulatory regimes and their ability to address hazard management within an environmental sustainability framework (e.g. May *et al.*, 1996). It is not possible to explore that literature here in any detail. Suffice it to say that the link between risk and sustainability at a practical level of policy implementation remains in its infancy.

Geomorphological Hazards and Disaster Prevention, eds. Irasema Alcántara-Ayala and Andrew S. Goudie. Published by Cambridge University Press. © Cambridge University Press 2010.

From the communications context, the terms risk and sustainable development have readily become buzzwords, adopted by experts, media and the general public, infusing every field of public policy and almost every education curriculum. Sustainable development, in particular, suffers from (mis-)appropriation by different actors to serve a multitude of purposes. The link between risk and sustainability therefore suffers from linguistic confusion, a problem that diffuses across the hazards research domain. In convening a special issue for the International Human Dimension Programme on Global Environmental Change (IHDP), Janssen and Ostrom (2006), note the 'Tower of Babel' experience in hearing contributors interpret the core concepts of resilience, vulnerability and adaptation in diverse ways. This may not be surprising. After all, the use of these terms reflects intellectual histories originating from different disciplines. The momentum towards interdisciplinary research in the hazards arena should gradually erode some of these communication problems.

From a geomorphological perspective, the debate about the link between risk and sustainability may seem abstract. One is reminded of Richard Chorley's (1978) frequently quoted comment about a geomorphologist, when confronted by discussion of theory, instinctively reaching for his soil auger to seek solace in the field. To some, the term 'sustainability' may have a similar effect. While many geomorphologists will continue to contribute to hazard management through the careful collection of field data and the development and improvement of measuring techniques and assessment protocols, many prominent scientists in hazard research have acknowledged the need for greater engagement between scientists and social scientists. For example, the IUGG Commission on Geophysical Risk and Sustainability (GeoRisk) calls for scientific studies to provide hazard data and information to emergency managers, policy makers, scientists and the general public. The Budapest Manifesto drafted by the Commission calls for scientists to apply experience and expertise to the mitigation of urgent societal problems and to:

… go beyond traditional hazard mapping and monitoring. We must involve the community in extensive campaigns of knowledge exchange and communication. Risk evaluation must rely heavily, but not exclusively, on modelling and visualization of physical, biological and social processes and their implications. The results need to be easily grasped by emergency planners, the insurance industry, policy makers, and the public. We also need a deeper understanding, based on work across disciplines, of all of the processes that are involved. (IUGG GeoRisk, 2002)

In this brief chapter, the development of some of the key themes in hazard research is charted by linking the engineering-dominated agenda with recent developments in the social sciences. In so doing it illustrates opportunities for geomorphology to contribute to the sustainability agenda. As former President of the International Association of Geomorphologists, Olav Slaymaker (writing with Tom Spencer from the disciplinary perspective of physical geography), has commented: 'A discipline that does not have sustainability as a goal at all scales of analysis may itself be unsustainable' (Slaymaker and Spencer, 1998).

21.2 Challenges to the dominant paradigm of natural hazards

A common aspect of some of the classical definitions of natural hazards is their emphasis on the externality of the hazard event. For example, Burton and Kates (1964) defined natural hazards as 'elements in the physical environment harmful to man and caused by extraneous forces'. The American Geological Institute (1984) refers to 'naturally occurring geologic condition that presents a risk or potential danger to life or property'. Traditionally, natural hazards have often been regarded as 'Acts of God' (Smith, 1992) both in terms of folklore and legal systems, which has unhelpfully cultivated the idea that humans have no part to play in creating hazards. Though the role of humans in exacerbating risk has been widely recognized, the association of hazard with an external trigger is fundamental in the academic development of hazards research and remains central in engineering-led approaches. The premise that disasters are primarily associated with geophysical extremes logically suggests that the prediction and control of extreme natural events will provide effective protection.

This dominant paradigm is reflected and reinforced by media reporting of disasters, which tends to focus initially on quantifying the scale of the hazard event and proceeds to discussion of the anatomy of the event, the emergency relief effort and engineering solutions for future hazard protection. Occasionally there will be much interest in the monitoring of environmental conditions using new technologies such as satellite remote sensing. After the Indian Ocean Tsunami in 2004, for example, there was much media coverage of the technology and design plans for an early warning system. Media attention is less likely to focus on issues of pre-event vulnerability (whether physical, economic or social) and will, of course, attend to rapid-onset events (such as earthquakes, volcanic eruptions and storms) rather than chronic events (such as droughts or extreme land degradation). As such, the emphasis on rapid-onset events reinforces the notion of an external trigger as the formative action in the unfolding of a *natural* disaster.

In the last 20–30 years an alternative view of hazards and disasters has emerged within the social sciences. Primarily associated with action research in less developed countries (LDCs), the approach represents a shifting perspective from viewing disasters as environmentally controlled to one more grounded in social theory. Emphasis is given to the constraints placed on individuals and communities by socio-economic conditions and institutional forces. A disaster is seen in this context as a non-routine event which distinguishes it as unusual and different from everyday concerns (Kreps, 2001), though many in the field of development studies would regard the separation of people at risk from natural hazards and the many dangers of everyday life as artificial (e.g. Wisner *et al.*, 2004).

This is not to suggest that the social sciences were unconcerned with hazards until relatively recently. There is a long tradition of interest in the interactions of environment, hazard and society in the field that became known as human ecology (principally associated with the University of Chicago) in the first half of the twentieth century. From this school, Gilbert F. White became internationally renowned as a pioneer of natural hazards research and management, viewing hazards as the result of interacting natural and social forces. His work on flood hazard (White, 1945; reprised in White *et al.*, 1997) explored human responses to flood hazard, initiating important questions of why certain adjustments to perceived hazards were preferred over others and why, despite technological fixes, costs associated with hazards continued to increase. This behavioural approach to examining adjustments to hazard has continued to shape the agenda of hazards research in the United States and has complemented the engineering-dominated practice with its emphasis on monitoring, modelling and prediction of events, allied to a strong commitment to physical and managerial control. The relationship between the engineering and social science approaches was, to some extent, formalized when Gilbert White and Eugene Haas, a prominent sociologist, were invited to prepare the first national assessment of natural hazards and disasters in the United States. The report (White and Haas, 1975) represented an important step towards encouraging and setting the agenda for interdisciplinary research. The legacy of the behavioural approach is apparent in the central notion that individuals and groups choose how to react (adjust) to the risk of extreme events in their environment, albeit acknowledging that decision-making is constrained by limited knowledge and social factors. A second national assessment was conducted in the 1990s under the leadership of Dennis Mileti. The summary volume (Mileti,

1999) begins with an appraisal of progress since White and Haas (1975) before developing a sustainability framework for analyzing natural and technological hazards.

Mileti (1999) argues that the four-stage model of human adjustment to disaster – preparedness, response, recovery, mitigation – has served well as a framework for accumulating knowledge about specific hazards and the effectiveness of decision-making. However, in the face of ever-increasing losses from hazards (Figure 21.1), improvements are required and here sustainability thinking is informative. Firstly, there is concern that mitigation measures, such as flood control structures, have encouraged migration into hazardous areas and rather than eliminating damage, merely postpone it. Similarly, engineered solutions to earthquakes may reduce losses in moderate-sized seismic events but indirectly cause greater losses in larger events (Mileti, 1999). From this basis Mileti (1999) advocates adoption of a global systems perspective, acceptance of responsibility for hazards as human-induced, anticipation of change, rejection of short-term thinking, a broader view of social forces and the principles of sustainable development as the six tenets for more creative thinking about hazards in the twenty-first century. Essentially this is a call for a broader perspective capable of taking greater complexity in socio-ecological systems into account, of linking mitigation with the broader goals of environmental stewardship (Schneider, 2002).

More recently, the National Research Council (2006) has published an appraisal of the challenges facing social sciences hazards and disaster research and the opportunities for forging interdisciplinary collaborations with the natural sciences. The report includes an assessment of the accomplishments of social sciences research undertaken within the National Earthquake Hazards Reduction Program (NEHRP). Formed in 1977, NEHRP was mandated to include social sciences research within a broader programme dominated by earth science and engineering. The influence of the White and Haas report was no doubt important in opening the door. Both the National Assessment Reports and the report of the National Research Council (2006) reinforce the pedigree of American-based behavioural/adjustment paradigm. The contrast with alternative approaches emerging from LDCs remains marked, but the development of the sustainability agenda provides momentum for the two paradigms to evolve, mature and explore synergies.

21.3 Vulnerability and resilience: Asian earthquakes

Two core concepts in the emerging debate of human dimensions of environmental change and hazards are vulnerability and resilience. Both terms have long and contested

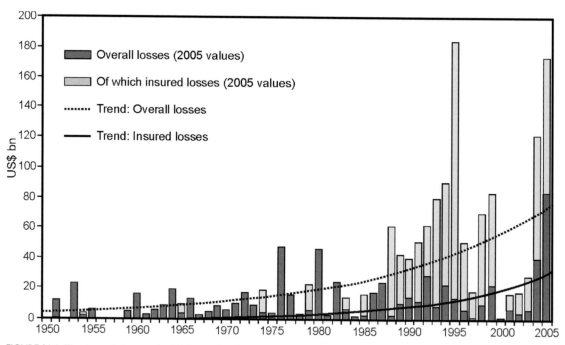

FIGURE 21.1. The dramatic increase in global overall and insured loss during the latter part of the twentieth century. (Data from Munich Re (2006) in 2005 values.)

histories reflecting different disciplinary emphases. Vulnerability is taken to mean 'the characteristics of a person or group and their situation that influence their capacity to anticipate, cope with, resist and recover from the impact of a natural hazard' (Wisner *et al.*, 2004).

The succession of large earthquakes in Asia during the first decade of this century emphatically demonstrates the vulnerability of human populations to extreme events. During this short period, more than half a million people in Asia have been killed by earthquakes, with the five largest events being in Bhuj, India, in 2001 (20,000 fatalities), Bam, Iran, in 2003 (40,000), Sumatra in 2004 (250,000 fatalities – mostly from the Indian Ocean Tsunami), Muzzaffarabad, Kashmir, in 2006 (80,000) and Sichuan, China, in 2008 (70,000) (estimates from Jackson, 2008). An analysis of historic earthquake data constructed for the last thousand years (Jackson, 2008) indicates that the number of earthquakes killing more than ten thousand people ran at a frequency of around 5 per century from AD 1000 to 1600, around 15 per century from 1600 to 1900 and about 30 in the twentieth century. Despite the technological advances, Asian populations are more vulnerable to earthquake hazards than ever before. There is no evidence to suggest that earthquake magnitude–frequency dynamics have changed over time. The increased vulnerability can

be explained by the massive increase in population located in areas of seismic activity and the manner in which people live, but geological and geomorphological factors played an important role in initiating the pattern of settlement. In arid Central Asia, early settlements were distributed in isolated oases and served an important function to trade routes that followed the edge of mountain fronts. The location of these early settlements was clearly governed by the need to reduce vulnerability to water stress and the opportunity to sustain some agriculture. Unfortunately, mountain fronts and the springs that sustain desert oases are characteristic of active fault scarps. As small settlements developed into towns and cities, so population has become concentrated in extremely hazardous locations. Somewhat controversially, Halvorson and Hamilton (2007) argue that vulnerability is enhanced in Central Asia by a diminishing 'seismic culture', which is constrained by a lack of public access to earthquake information, gender issues and migration, reducing indigenous knowledge of hazard reduction.

Janssen and Ostrom (2006) contend that the concept of vulnerability has its disciplinary origins in Geography and has largely been approached through the comparative analysis of case studies such that attempts to measure vulnerability quantitatively have been fairly limited in scope. In mainstream hazards research (i.e. the behavioural paradigm) most emphasis has been given to the physical elements of exposure and to the probability of hazards occurring. Coupled with ideas of landscape sensitivity, the concept of exposure has resonance for geomorphologists. The sensitivity of systems to external stresses is characterized by factors such as magnitude, frequency, duration and areal extent (Burton et al., 1993), which determine places vulnerable to hazard and hence exposure to people inhabiting those areas. Analysis of floodplain inundation risk would be a classic example. However, vulnerability goes beyond the sensitivity of the geophysical system. Cutter et al. (2003) identify vulnerability as exposure as one of three main traditions, the second viewing vulnerability as a social condition and the third linking exposure and societal response as represented in specific places. Adger (2006), in a comprehensive review and discussion about the emerging concept of vulnerability, draws attention to the contrasting views of vulnerability as outcome and vulnerability as process. Whereas engineering traditions will be focused towards methods to recognize the status of vulnerability (outcome), the social science traditions pay more attention to the political and structural causes of vulnerability within society (process), which result in some households, particularly the poor, living in riskier locations.

Some of these concerns can be illustrated by consideration of the aftermath of the Indian Ocean Tsunami. Generated by a magnitude 9.3 earthquake off the west coast of Sumatra (the second largest earthquake ever recorded), the displaced water reached heights of up to 30 m coming ashore in Sumatra, Malaysia, Thailand, Myanmar, Sri Lanka, India and the east coast of Africa. An estimated 250,000 people were killed and up to 1.5 million displaced. Figure 21.2 depicts a scene of devastation on the western shore of Sri Lanka, in the vicinity of Hikkaduwa. The physical vulnerability of the location to the force of the tsunami is immediately apparent. Geomorphological and engineering research can assess exposure to vulnerability at specific locations and provide recommendations to adjust patterns of human occupation and land use in the coastal zone. However, faced with the aftermath of a major disaster the questions of economic and social vulnerability take centre stage. Many of the fishing villages along the coast were established in contradiction of planning regulations requiring developments to be set back from the coast at least 100 m. Resettling these communities further inland is rarely possible because of existing occupation of land and because the livelihoods are dependent on ready access to the sea. Hence a dilemma arises between actions required to maximize protection from a future tsunami or storm surge and the humanitarian prerogative to

compensate and re-establish sustainable livelihoods for impacted communities as quickly as possible. In the aftermath of the Indian Ocean Tsunami, some geomorphologists found themselves working on the ground with teams of social scientists to make sense of the dimensions of the disaster and the resilience of the local communities recovering from the impact. Bird et al. (2007) and Horton et al. (2008), for example, report results of science-based and social science-based surveys from the Malaysian islands of Langkawi and Penang, respectively. The scientific field surveys conducted measurements of topography, flow depth and flow direction, complemented by eye witness accounts, in order to reconstruct the dimensions of inundation from perishable data, while the social science surveys relied on interviews with key informants.

A major contribution to the linking of physical and social traditions towards vulnerability was provided by the pressure and release model of Blaikie et al. (1994), more recently revamped by Wisner et al. (2004). Vulnerability is represented as a progression from root causes that lie in limited access to power and resources coupled with the constraints of economic and political systems. These root causes recognize that people who are marginalized

FIGURE 21.2. Tsunami destruction near Hikkaduwa, Sri Lanka. This picture was taken six months after the tsunami struck.

economically (e.g. the rural poor) or who inhabit marginal environments (floodplains, low-lying coastlines, unstable slopes) are also of marginal importance to the holders of economic and political power (Blaikie *et al.*, 1994). The plight of communities in the Irrawaddy Delta, Myanmar, following Cyclone Nargis in May 2008, is an example of this idea. In turn the root causes generate dynamic pressures (such as migration, rapid urbanization or land use change) that translate into unsafe conditions. Unsafe conditions, such as having to live in hazardous locations or having to engage in dangerous livelihoods, are the specific manifestations of vulnerability. The occurrence of a disaster is then dependent on the interaction between the vulnerable population and the realization of a hazard event. Thus, in the discussion of fishing villages on the coast of Sri Lanka, the root causes of poverty and lack of access to power generate a situation where communities are required to obtain their livelihood in an inherently hazardous location. In the absence of the tsunami-generating earthquake, there is no disaster. Likewise, a tsunami cannot induce disaster if vulnerability is close to zero, as, for example, in situations where the coastline is protected by tsunami-proof sea walls, where mangrove forests have been retained and are able to dissipate tsunami energy, where warning systems enable orderly evacuation, or in cases where the tsunami inundates uninhabited coastlines.

If vulnerability is perceived as a set of negative influences describing the combination of circumstances that contribute to the extent to which lives, livelihood, property or assets are exposed to risk by an identifiable hazard, the corollary is the concept of resilience – the capacity of the system to withstand recurrent disturbances. Janssen and Ostrom (2006) argue that the resilience concept in hazards research originates from ecological sciences through analysis of population ecology. As such its disciplinary legacy is linked to theoretical and mathematical models of the persistence of relationships within systems, in contrast to the more descriptive metaphor of vulnerability. Applied to socio-ecological systems, the key concern is the capacity of systems to retain essential structures and processes in the face of disturbance (Berkes, 2007) and to identify the role of humans in reducing the resilience of ecosystems. Resilience is thus an expression of society's ability to cope with hazard events. Resilience is widely seen as a desirable property of socio-ecological systems, but suffers equally from having become an umbrella concept used in multiple ways. Carpenter *et al.* (2001) consider resilience to have three components: the amount of disturbance a system can absorb and still remain in the same state; the degree to which the system is capable of self-organization; and the degree to which the system can build and develop self-learning. Adopted and developed by international agencies, this three-pronged approach has broadened the scope of resilience but arguably made it no closer to operational definition as a policy and management tool (Klein *et al.*, 2003).

Though definitions of vulnerability and resilience display considerable variability and there remain significant limitations in data and methods, the concepts have provided a rational basis for hazards researchers to provide information to decision makers and policy writers. Some progress has been made in developing metrics for assessing a wide range of stressors and analyzing the components of vulnerability in bounded systems at particular scales (Kates, 2001; Turner *et al.*, 2003; Haque and Burton, 2005). Pelling (2006) has reviewed approaches used to identify urban vulnerability, calling for improved communication between practitioners working on urban disaster risk management and those in the existing urban development community. Social survey and participatory methods tend to vary in focus between those concerned with vulnerability of places and those with sustainable livelihoods. In the search for science to underpin sustainability, measurement of vulnerability and resilience is a key objective. As Adger (2006) concludes:

The challenges for human dimensions research include those of measuring vulnerability within a robust conceptual framework, addressing perceptions of vulnerability and risk and of governance. All of these challenges are common to the domains of vulnerability, adaptation and resilience. They relate fundamentally to the relationship between vulnerability and both social resistance and the resilience of the ecosystems on which human well-being ultimately depends.

21.4 Geomorphology, hazards and sustainability

Having summarized some of the recent debate concerning the integration of science, engineering and social science interests around a sustainability framework, it is important to consider the implications for geomorphology. Much of the preceding discussion may seem peripheral to the practice of geomorphology in the field. It is possible to visualize the potential contribution of geomorphology to the hazards and sustainability agenda by considering its relevance in the various components of hazard and disaster research. Figure 21.3 is based on the conceptual model used by the National Research Council (2006) to explore social science contributions to the research field. The figure, partly derived from Cutter (1996), identifies the five principal realms of hazard and disaster research, which have mainly evolved under separate disciplinary routes: hazard

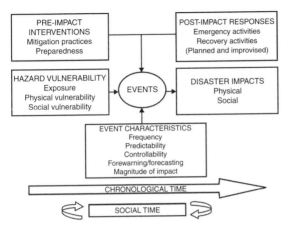

FIGURE 21.3. The Hazard and Disaster Management System. (Modified from National Research Council, 2006.)

vulnerability, hazard mitigation, disaster preparedness, emergency response and disaster recovery. Represented in the centre of the diagram, the events button is the causal link between pre-impact preparedness and mitigation and post-impact responses. These are the events that engender action before, during and after their realization. Geomorphology has contributed to all of these realms, both in pre-event analysis of risk and vulnerability and in post-mortem mode to reconstruct events from field data and assess physical impact (e.g. tsunami dimensions, flood extents, anatomy of landslides). Disaster preparedness derives from understanding 'event characteristics'. Key topics here include determination of the magnitude–frequency characteristics of events, a common theme in geomorphology, developments in forecasting, identification of precursors, and prediction. Ideally, geomorphological expertise can be used to identify the characteristics of potential events before they occur so that appropriate mitigation measures can be adopted and adjustments made to reduce socio-economic impact. However, geomorphological theory has advanced significantly through understanding accumulated from the analysis of actual extreme events – for example, the way that Wolman and Miller's (1960) theory of formative events in river channels was augmented from observations following extreme events (e.g. Baker, 1977; Wolman and Gerson, 1978).

An interesting aspect of Figure 21.3, with potential yet unexplored implications for geomorphology, is the distinction between chronological time, which is unidirectional and linear, and social time, which is nonlinear and multidimensional. The National Research Council (2006) argued that this attribute had received very little mention in the hazard and disaster research community but could be valuable. In chronological time, flowing from left to right in

Figure 21.3, vulnerability assessments influence mitigation strategies. Time is a taken-for-granted but essential tool for determining event characteristics. Social time, on the other hand, recognizes that time may be experienced differently by individuals or communities, both in the way that the past is reconstructed from the present and the way the present is constructed from the past.

Disasters can act as significant markers in individual, community or societal consciousness and feature prominently in reconstructions of history. Shared memory of past disasters may strongly influence cultural practices. These ideas have been explored for the hazard-prone Philippines by Bankoff (2002), who has also delved into the historical geography of hazards (Bankoff, 2004). Over longer time-scales, communal memories of disasters enter oral tradition or even assume myth-like status – 'geomythology'. This may be accompanied by cultural practices akin to mitigation strategies, such as forbidden zones around volcanoes. Cashman and Cronin (2008) have recently reviewed traditions associated with volcanic eruptions from the Pacific Northwest, Caribbean and New Zealand. The earlier example of earthquake hazard in Central Asia (Halvorson and Hamilton, 2007) argued that erosion of seismic culture played an important role in increasing vulnerability.

The second aspect of social time is the way that the past is re-evaluated to contextualize the present. This is particularly relevant to chronic (or slow-onset) hazards such as droughts or land degradation. At what point in time does a succession of dry seasons constitute a drought disaster, global climate change become a crisis or, for that matter, sustainability become the vital agenda of the global community? Far from being rhetorical questions, these constructions of priorities influence political and academic agendas, perhaps guiding the trajectory of research programmes and directing the availability of funding for particular activities, including geomorphology.

Having considered the potential input of geomorphology into the five realms of hazard and disaster research, the synergy between geomorphological research and the aspirations of the sustainability agenda can be briefly examined. The Second National Assessment of natural hazards in the United States (Mileti, 1999) identified a series of priorities for fostering environmental sustainability and mitigating natural hazards. These are summarized in Table 21.1 as a means of identifying areas of relevance for geomorphology. The mitigation strategies on the right-hand side are the traditional staple of natural hazard management. While the development of building codes and warning systems will more likely fall within the domain of engineering, the other headings have witnessed considerable geomorphological input. Gillespie et al. (2007) have

TABLE 21.1. *Key goals of fostering sustainability and mitigation strategies*

Fostering sustainability	Mitigation strategies
Enhancing environmental quality	Land use
Maintaining quality of life	Warnings
Local resiliency and responsibility: (person, community, society)	Building codes
Vibrant local economies	Insurance
Equity	Appropriate technology
	• GIS
	• Remote sensing
	• Numerical modeling
Consensus building	

Developed from ideas by Mileti, 1999; Andres and Strappazzon, 2007.

recently reviewed the new opportunities available from space-borne remote sensing to improve characterization of the Earth's surface as well as quantify the movement of water (floods, tsunamis and storms). High-resolution passive sensors are capable of producing highly accurate assessments of flood damage. Furthermore, launch plans for the next few years will increase resolution and reduce revisit times, improving the potential to assess and predict hazards from space (Gillespie *et al.*, 2007). Observational data are the first step in the integration of hazard prediction. GIS provides a tool for visualizing data as well as a platform for numerical modelling. Writing about volcanic hazards, Valentine (2003) notes the role of numerical modelling in making specific predictions about different eruption phenomena which can be overlaid with vulnerability data in a GIS framework. The generation of vulnerability data itself is unlikely to be in the domain expertise of the geoscientist, emphasizing the importance of interdisciplinary collaboration. The same principles guide the considerable application of GIS in sustainable development (Campagna, 2006).

Assessment and prediction of the areal extent of impacts (such as flood damage or landslide risk) form the basis for land use controls. The link between insurance and geomorphology has been less frequently explored but again the applications of geotechnologies to develop high-resolution maps of damage and risk are very attractive to insurance companies, especially given the sharp rise in insured losses from hazard events in the last 15 years (Figure 21.1). In Britain and Europe recent floods have precipitated a debate about the sustainability of engineered flood defences and the role of insurance (Treby *et al.*, 2006; Lamond and Proverbs, 2008). Ironically, lack of insurance

cover may be a factor in promoting sustainable use of floodplains by discouraging inappropriate development, but the withdrawal of insurance cover from existing properties is an emotive and politically fraught issue.

If the right-hand column of Table 21.1 represents management of natural hazard, the left-hand side represents a transition towards sustainable management of areas subject to hazard. At first glance the headings here may appear well beyond the scope of geomorphology. The remit is to realign from considering hazards as isolated incidents to support the broader goals of a sustainable society that has the capacity to co-exist with hazard. The emphasis of sustainable development moves towards maintaining or improving the quality of life, with an emphasis on equity, empowerment of individuals and vibrant local economies. Sustainable societies seek to improve environmental quality and implicitly this infers a retreat from hard-engineering towards management tools that are more sympathetic towards natural ecosystem functions. For the holistically minded geomorphologist, the link between sustainability and hazards should provide opportunities to integrate mitigation with enhancement, to experiment and to encourage public participation. Consensus building is another key feature of the sustainability agenda and again is a topic that may seem unrelated to the main concerns of science-based hazard researchers. However, there is increasing awareness of the responsibility on scientists to communicate their data and ideas and, where possible, to involve local communities in projects. There are clear parallels here to the experience of geomorphologists working in river restoration/rehabilitation, where focus groups and participatory processes have become commonplace. A key aspect of communication in river management has been explanation of uncertainty (Hillman and Brierley, 2008). Though data gathering and knowledge generation can improve understanding of complex systems, it is important to recognize that some uncertainty is irreducible. The complexity of social-ecological systems is such that the sustainable management of hazards should not be based on the expectation of predicting adjustments, but should embrace uncertainty, bolster resilience and build adaptive capacity. Some elements of this agenda are now considered in relation to flood hazard management in Southeast Asia.

21.5 Flood hazards in Southeast Asia: links with sustainable management

Flood hazard is prevalent in Asia due to a combination of high rainfall, large flow volumes and dense populations inhabiting floodplain areas and low-lying coasts. UNESCO estimates suggest that the annual global death toll caused by floods varies between 5,000 and 15,000 people but 70% of victims are in Asia (Douben and Ratnayake, 2006). Similarly, the vast majority of the average 125 million people per year impacted by floods are resident in Asia. Such data reinforce the conventional view of floods as trigger events where hazard is realized when magnitude exceeds a given threshold. However, from the viewpoint of sustainable livelihoods, it should be recognized that societies have evolved in monsoon-dominated river basins in Southeast Asia to exploit the benefits of the annual flood cycle. Parts of the Mekong River delta and several floodplain reaches upstream support population densities in excess of 500 people per km². The Mekong flood season is effectively a single peaked hydrograph, which begins to rise in May, peaks in September and recedes through October and November. Livelihoods dependent on this established annual pattern are exposed

to a 'double-tailed' hazard if flows are markedly below or above 'normal' levels (Mekong River Commission, 2007). Typhoon incursions from the South China Sea, often occurring in September and October when the discharge levels are already high, are the most likely causes of significant hydrograph peaks superimposed on the annual flood cycle. In 2000, floods associated with typhoons displaced 45 million people in Southeast Asia (Douben and Ratnayake, 2006). Through its data archive, the Mekong River Commission is able to refine estimation of flood event characteristics and probability distributions for the timing and magnitude of minimum, maximum and transition flows. Such data provide a solid basis for quantitative estimates of physical vulnerability, but must be bridged with consideration of socio-economic vulnerability and quality-of-life issues. Three environmental concerns in the Mekong are the uncertainties associated with climate change, land cover transformation and deliberate channel interventions on the dynamics of water and sediment flow. Of particular concern is the impact of dam construction in the upper Mekong in the Chinese province of Yunnan, accompanied by a scheme to increase the navigability of the main channel by removing numerous bedrock reefs. Downstream in Thailand and Laos, there is anecdotal evidence from communities dependent on the river for their livelihoods that intra-season flow levels have become more variable and that bank erosion has increased (Figures 21.4 and 21.5). Establishing the geomorphological evidence to link cause and effect in meaningful ways to promote sustainable livelihoods is an extremely difficult task. Nevertheless, modelling can provide guidance on likely impacts downstream. Kummu and Sarkkula (2008) have identified the potential impact of increased dry season flow and reduced flood season flow on the sensitive ecosystems of Tonle Sap, the shallow lake in Cambodia that

FIGURE 21.4. Bank erosion impacting fishing villages on the Lao (left bank) side of the Mekong River, near Chang Saen, Thailand.

FIGURE 21.5. Bank erosion impacting fishing villages on the Thai (right bank) side of the Mekong River, near Chang Saen, Thailand.

experiences reverse flow between dry and wet seasons. The ecosystem productivity, on which many livelihoods depend, may be compromised by a combination of increased inundation and loss of gallery forest. In the delta, flood protection infrastructure has some negative impacts on crop and aquaculture activity (Hoa *et al.*, 2008). Similarly, the question of bank erosion hazard can be tackled through remote sensing surveys (Kummu *et al.*, 2008).

Large rivers are not the only concern. There are numerous incidents each year of floods and debris flows affecting communities in upland basins, particularly in Indonesia. The media and authorities typically link these hazards to upstream deforestation, though it appears that few scientific studies have established specific causal relationships. Geomorphology can contribute to hazard reduction in two ways. Firstly, experimental research on flow and sediment dynamics under undisturbed and disturbed tropical forest has enabled the development of sustainable logging guidelines (Thang and Chappell, 2005). Secondly, it is apparent that downstream modification of river channels and increasing occupation of floodplains increases human vulnerability to flood hazard (Douglas, 2004). Holistic approaches to integrated catchment management are required to ensure that measures to reduce hazard in one part of the basin are not compromised by decisions made elsewhere. In an integrated framework for sustainable management of areas subject to hazard, geomorphology can act as a template upon which land management decisions are made. For example, funds available through carbon trading schemes, which incentivize afforestation projects in tropical regions, could be targeted to reduce soil erosion and dampen hydrograph response.

The final example of a sustainability-led approach to hazard management in Southeast Asia comes from Singapore. Singapore has successfully reduced the impact of flood hazard through a massive infrastructure plan, which constructed a dense network of drains and concrete-lined canals to evacuate discharge from frequent high-intensity storms. This engineering approach reduced the area susceptible to flooding from over 3,000 ha in 1970 to just 155 ha in 2004. However, few could argue that the concrete structures are sympathetic to ecological or aesthetic concerns. A change in policy direction to increase the amount of catchment area used to harness domestic water supplies dictates the need to rethink the function of the canals so that more of the storm runoff can be retained for water resource requirements. In turn, this provides an opportunity to address issues of urban flood hazard management and river rehabilitation. Demonstration projects to remove some of the structural controls and to landscape

ecologically productive aquatic environments are currently on the drawing board (Higgitt, 2008).

21.6 Conclusions

It is logical that the aims and objectives of sustainable development and of natural hazard management converge. History demonstrates that while geomorphology, science and engineering have contributed to improved understanding of the dynamics of environmental processes that pose harm to society, the financial and human cost of extreme events has continued to escalate. The move towards a sustainability framework seeks to integrate the contribution of science and social science towards the management of hazards, the maintenance of environmental quality, improved living conditions and issues of equity. Geomorphology is one of the knowledge domains contributing to this interdisciplinary arena. Innovative techniques, particularly those involving handling of geospatial data, numerical modelling and development of data archives will provide further insights to understanding event characteristics. Communicating the results and their implications to other scientists, policy makers, stakeholders and the general public is essential (Brierley, 2009). While many goals of fostering sustainability (Table 21.1) may seem far removed from the bread and butter of geomorphology, it is useful for geomorphologists to be aware of the current debates, the progress made by social science in developing methods to characterize vulnerability and resilience, the influence of sustainability thinking on global research and policy agendas and the opportunities that may arise for new applications of geomorphology as the paradigms of hazards research shift towards the sustainability agenda.

References

Adger, W. N. (2006). Vulnerability. *Global Environmental Change*, **16** 268–281.

American Geological Institute (1984). *Glossary of Geology*. Falls Church, VA: AGI.

Andres, L. and Strappazzon, G. (2007). Natural hazard management and sustainable development: a questionable link. *Revue de Géographie Alpine*, **95**, 2, 29–50.

Baker, V. R. (1977). Stream-channel response to floods, with examples from central Texas. *Geological Society of America Bulletin*, **88**, 1057–1071.

Bankoff, G. (2002). *Cultures of Disaster: Society and Natural Hazards in the Philippines*. London: Routledge.

Bankoff, G. (2004). The historical geography of disaster: 'vulnerability' and 'local knowledge' in western discourse. In G. Bankoff, G. Frerks, D. Hilhorst and T. Hilhorst (eds.),

Mapping Vulnerability: Disasters, Development and People. London: Earthscan, pp. 25–36.

Beer, T. (2004). Geophysical risk, vulnerability and sustainability. In R. S. J. Sparks and C. J. Hawkesworth (eds.), *State of the Planet: Frontiers and Challenges in Geophysics.* Washington, D.C.: American Geophysical Union, pp. 375–385.

Berkes, F. (2007). Understanding uncertainty and reducing vulnerability: lessons from resilience thinking. *Natural Hazards,* **41**, 283–295.

Bird, M. I., Cowie, S., Ong, J. E. *et al.* (2007). Indian Ocean tsunamis: environmental and socio-economic impacts in Langkawi, Malaysia. *Geographical Journal,* **173**, 103–117.

Blaikie, P., Cannon, T., Davis, I. and Wisner, B. (1994). *At Risk: Natural Hazards, People's Vulnerability and Disasters.* London: Routledge.

Brierley, G. J. (2009). Communicating geomorphology. *Journal of Geography in Higher Education,* **33**, 3–17.

Burton, I. and Kates, R. W. (1964). The perception of natural hazards in resource management. *Natural Resources Journal,* **3**, 412–441.

Burton, I. R., Kates, R. W. and White, G. F. (1993). *The Environment as Hazard,* 2nd edition. Oxford: Oxford University Press.

Campagna, M. (ed.) (2006). *GIS for Sustainable Development.* London: Taylor and Francis.

Carpenter, S., Walker, B., Anderies, J. M. and Abel, N. (2001). From metaphor to measurement: resilience of what to what? *Ecosystems,* **4**, 765–781.

Cashman, K. V. and Cronin, S. J. (2008). Welcoming a monster to the world: myths, oral tradition and modern societal response to volcanic disasters. *Journal of Volcanology and Geothermal Research,* **176**, 407–418.

Chorley, R. J. (1978). Bases for theory in geomorphology. In C. Embleton, D. Brunsden and D. K. C. Jones (eds.), *Geomorphology: Present Problems and Future Prospects,* Oxford: Oxford University Press, pp. 1–13.

Cutter, S. L. (1996). Vulnerability to environmental hazards. *Progress in Human Geography,* **20**, 529–539.

Cutter, S. L., Boruff, B. J. and Shirley, W. L. (2003). Social vulnerability to environmental hazards. *Sociological Quarterly,* **84**, 242–261.

Douben, N. and Ratnayake, R. M. W. (2006). Characteristic data on river floods and flooding: facts and figures. In J. van Alpen, E. van Beek and M. Taal (eds.), *Floods, from Defence to Management.* London: Taylor & Francis, pp. 19–35.

Douglas, I. (2004). People induced geophysical risks and sustainability. In R. S. J. Sparks and C. J. Hawkesworth (eds.), *State of the Planet: Frontiers and Challenges in Geophysics.* Washington D.C.: American Geophysical Union, pp. 387–397.

Gillespie, T. W., Chu, J., Frankenberg, E. and Thomas, D. (2007). Assessment and prediction of natural hazards from satellite imagery. *Progress in Physical Geography,* **31**, 459–470.

Halvorson, S. J. and Hamilton, J. P. (2007). Vulnerability and the erosion of seismic culture in mountainous Central Asia. *Mountain Research and Development,* **27**, 322–330.

Haque, C. F. and Burton, I. (2005). Adaptation strategies for hazards and vulnerability mitigation: an international perspective. *Mitigation and Adaptation Strategies for Global Change,* **10**, 335–353.

Higgitt, D. L. (2008). Catchment management: from engineering to 'imagineering'. *Innovation: The Magazine of Science and Technology,* **8**, 38–40.

Hillman, M. and Brierley, G. J. (2008). Restoring uncertainty: translating science into management practice. In G. J Brierley and K. A. Fryirs (eds.), *River Futures: An Integrative Scientific Approach to River Repair.* Washington, D.C.: Island Press, pp. 255–272.

Hoa L. T. V, Shigeko H., Nhan, N. H. and Cong, T. T. (2008). Infrastructure effects on floods in the Mekong Delta, Vietnam. *Hydrological Processes,* **22**, 1359–1372.

Horton, B. P., Bird, M. I., Birkland, T. *et al.* (2008). Environmental and socioeconomic dynamics of the Indian Ocean tsunami in Penang, Malaysia. *Singapore Journal of Tropical Geography,* **29**, 307–324.

IUGG GeoRisk (2002). *The Budapest Manifesto on Risk Science and Sustainability.* IUGG Commission on Geophysical Risk and Sustainability. Available at http://www.iugg-georisk.org/ (accessed November 2008).

Jackson, J. (2008). The May 2008 Sichuan earthquake: a herald of things to come. *Geology Today,* **24**, 178–181.

Janssen, M. A. and Ostrom, E. (2006). Resilience, vulnerability and adaptation: a cross-cutting theme of the International Human Dimensions Programme of Global Environmental Change. *Global Environmental Change,* **16**, 237–239.

Kates, R. W. (2001). Queries on the human use of the Earth. *Annual Review of Energy and the Environment,* **26**, 1–26.

Klein, R. J. T., Nicholls, R. J. and Thomalla, F. (2003). Resilience to natural hazards: how useful is this concept? *Environmental Hazards,* **5**, 35–45.

Kreps, G. A. (2001). Sociology of disaster. In N. J. Smelser and P. B. Bates (eds.), *International Encyclopedia of Social and Behavioural Sciences.* Amsterdam: Elsevier.

Kummu, M. and Sarkkula, J. (2008). Impact of the Mekong River flow alteration on the Tonle Sap flood pulse. *Ambio,* **37**, 185–192.

Kummu, M., Lu, X. X., Rasphone, A., Sarkkula, J. and Koponen, J. (2008). Riverbank changes along the Mekong River: remote sensing detection in the Vientiane-Nong Khai area. *Quaternary International,* **116**, 100–112.

Lamond, J. E. and Proverbs, D. G. (2008). Flood insurance in the UK: a survey of the experience of floodplain residents. In D. G. Proverbs, C. A. Brebbia and E. Penning-Rowsell (eds.), *Flood Recovery, Innovation and Response.* Southampton: WIT Press, pp. 325–334.

May, D. J., Burley, R. J., Eriksen, N. J. *et al.* (1996). *Environmental Management and Governance: Intergovernmental Approaches to Hazards and Sustainability.* London: Routledge.

Mekong River Commission (2007). *Annual Flood Report 2006.* Available at http://www.mrcmekong.org/flood_report/2006/ (accessed January 2008).

Mileti, D. S. (1999). *Disasters by Design: A Reassessment of Natural Hazards in the United States*. Washington, D.C.: Joseph Henry Press.

Munich Re (2006). *Natural Catastrophes 2006: Analyses, Assessments, Positions*, Topics Geo. http://www.munichre.com/publications/302-05217 en.pdf.

National Research Council (2006). *Facing Hazards and Disasters: Understanding Human Dimensions*. Washington, D.C.: The National Academies Press.

Pelling, M. (2006). Measuring urban vulnerability to natural disaster risk: benchmarks for sustainability. *Open House International*, **31**, 125–132.

Schneider, R. O. (2002). Hazard mitigation and sustainable community development. *Disaster Prevention and Management*, **11**, 141–147.

Slaymaker, O. and Spencer, T. (1998). *Physical Geography and Global Environmental Change*. Harlow: Addison Wesley Longman.

Smith, K. (1992). *Environmental Hazards: Assessing Risk and Reducing Disaster*. London: Routledge.

Thang, H. C. and Chappell, N. J. (2005). Minimising the hydrological impact of forest harvesting in Malaysia's rain forests. In M. Bonell and L. A. Bruijnzeel (eds.), *Forests, Water and People in the Humid Tropics*. Cambridge: Cambridge University Press, pp. 852–865.

Treby, E. J., Clark, M. J. and Priest, S. J. (2006). Confronting flood risk: implications for insurance and risk transfer. *Journal of Environmental Management*, **81**, 351–359.

Turner, B. L. II, Kasperson, R. E., Matson, P. E. *et al.* (2003). A framework for vulnerability analysis in sustainability science. *Proceedings of the National Academy of Sciences USA*, **100**, 8074–8079.

Valentine, G. A. (2003). Towards integrated natural hazard reduction in urban areas. In G. Heiken, R. Fakundiny and J. Sutter (eds.), *Earth Science in the City: A Reader*. Washington, D.C.: American Geophysical Union, pp. 63–73.

WCED (1987). *Our Common Future*. New York: World Commission on Environment and Development.

White, G. F. (1945). *Human Adjustments to Floods*. Research Paper No. 29, University of Chicago, Department of Geography.

White, G. F. and Haas, E. (1975). *Assessment of Research on Natural Hazards*. Cambridge, MA: MIT Press.

White, G. F., Platt, R. H. and O'Riordan, T. (1997). Classics in human geography revisited: commentary on 'Human adjustment to floods'. *Progress in Human Geography*, **21**, 423–429.

Wisner, B., Blaikie, P., Cannon, T. and Davis, I. (2004). *At Risk: Natural Hazards, People's Vulnerability and Disasters*, 2nd edition. London: Routledge.

Wolman, M. G. and Gerson, R. (1978). Relative scales of time and effectiveness of climate in watershed geomorphology. *Earth Surface Processes*, **3**, 189–208.

Wolman, M. G. and Miller, W. P. (1960). Magnitude and frequency of forces in geomorphic processes. *Journal of Geology*, **68**, 54–74.

22   Geomorphology and disaster prevention

Irasema Alcántara-Ayala

22.1 Geomorphological hazards

The occurrence of a threatening condition derived from a natural phenomenon in a defined space and time can be considered as a natural hazard (Alcántara-Ayala, 2002). It is important to understand, however, that the damage produced by such hazards can extend beyond the exact moment when they occur. In other words, they could have a significant impact in the long term. The notion of natural hazards is very frequently associated with geological, geophysical and hydrometeorological processes; nonetheless, as they are mostly significant constituents of the Earth's surface dynamics, they should be also viewed and analyzed from a geomorphological viewpoint.

Physical phenomena, such as volcanic activity, seismicity, flooding and landsliding, turn into hazards when they pose a danger to landscapes, both cultural and natural. Cultural landscapes are shaped from a natural setting by a cultural group, where culture is the agent, the natural area is the medium, and the cultural landscape the result (Sauer, 1925). While geomorphology aims at the understanding and appreciation of landforms and natural landscapes (Bauer, 2004), from a practical point of view it also recognizes and understands the processes and landforms that are related to dangerous conditions.

Different conceptual terms have been addressed in order to define and comprehend the complexity of natural landscape evolution: endogenic–exogenic forces; destructive–constructive action; erosional–depositional forms; stress–strength relationships, and polygenesis and inheritance (Bauer, 2004). By virtue of their nature, geomorphological hazards can be understood by means of all those concepts. Concerning the forces involved, geomorphological hazards can be classified into three main categories: endogenous, exogenous, and climate and land-use change induced. Neotectonics and volcanic activity are endogenous processes, while floods, karst collapses, snow avalanches, erosion, sedimentation, landslides, tsunamis, coastal hazards, and others, are among the exogenous processes. Furthermore, also linked to climate and land-use change are erosional processes, such as desertification, land degradation, salinization and floods (Slaymaker, 1996). Landforms such as volcanoes are fashioned by constructive action, whereas, by contrast, destructive action is expressed by denudation processes responsible for weathering and associated hazards. What is most common, however, is the coexistence of landforms and processes involving both destructive and constructive action in addition to erosional and depositional forms; a clear example is flooding. In addition, mass movement processes are the perfect case to illustrate stress–strength relationships as the action of gravity stresses the landscape system and strength decreases due to interactions among hillslope-forming materials.

Notwithstanding the development of a vast quantity of investigations concerning hazards from various geomorphological approaches, relatively few definitions of geomorphological hazards have been established (Slaymaker, 1996). Embleton et al. (1989) defined them as those events or processes, natural or human-induced, producing a change in Earth surface features that was detrimental to humans and their activities. On the other hand, for Panizza (1996), a geomorphological hazard is the probability that a certain phenomenon of geomorphological instability and of a given magnitude may occur in a certain territory in a given period of time. Gares et al. (1994) characterized them as the group of threats to human resources resulting from the instability of the Earth's surface features.

Even though it was 15 years ago that Rosenfeld (1994) pointed out that geomorphologists were getting

Geomorphological Hazards and Disaster Prevention, eds. Irasema Alcántara-Ayala and Andrew S. Goudie. Published by Cambridge University Press. © Cambridge University Press 2010.

progressively more and more engaged with hazards and recognized that the increase in population density had led to the occurrence of disasters due to human occupance of hazard-prone zones that had previously been avoided, authorities at the international and national levels were not then sufficiently aware of the future likely scenarios linked to global climate change, and consequently to the need for establishing and implementing adequate hazard and risk assessment strategies. The outcome was disasters such as hurricane Mitch in 1998, the Southeast Asia tsunami in 2004, and hurricane Katrina in 2005, among countless others.

The evaluation of potential geomorphological hazards includes the identification of physical parameters such as magnitude, frequency, duration, areal extent, speed of onset, spatial dispersion and temporal interval (Burton et al., 1978). Additionally, and based on White's paradigm (White, 1974), Gares et al. (1994) suggested five essentials for hazard evaluation: '(1) the dynamics of the physical processes; (2) the prediction of the occurrence; (3) the determination of the spatial and temporal characteristics; (4) an understanding of the impact of physical characteristics on people's perception; and (5) knowledge about how the physical aspects can be used to formulate adjustments to the event'.

Although quite clearly within the international literature terms such as natural, geological, geophysical and hydrometeorological hazards are widely used, the concept of geomorphological hazard is still rather unknown. Geomorphological hazards can be defined as the ingredients, in terms of landforms and processes, through which the likely impact of Earth's surface dynamics on natural and cultural landscapes is expressed in time and space. While magnitude, frequency, duration, areal extent, speed of onset, spatial dispersion and temporal interval are determined by the nature of the interaction among processes and landforms, the spatial and temporal dimensions of the impact on humans are derived from the character of the exposed cultural landscape. Hence, just as processes shape landforms, so the cultural landscape is fashioned by individuals and risk is configured by the combination of both nature and humanity.

Under the same conceptual umbrella, progress in geomorphology has undoubtedly been linked to the rising interest in environmental issues and particularly to the global understanding of climate and land-use change, vulnerability, hazard assessment, risk management and disaster prevention. This has led to a new direction in modern geomorphology. Coined by Haff (2002), 'neogeomorphology', this is the study of anthropically driven Earth surface processes. Consequently, the panorama of interest of geomorphologists has been broadening consciously towards a wider appreciation of the complexity of the bidirectional dynamics and interactions among the Earth's surface processes and landforms and anthropogenic activities.

22.2 Disasters: the international framework

Scientific developments during the 1950s and 1960s pointed to the natural character of disasters taking place on defenceless societies (Fritz, 1961, 1968; Barkun, 1974). Soon after, the work carried out by White, Kates and Burton (Kates, 1962, 1971; Burton and Kates, 1964; Burton et al., 1968, 1993; White, 1973) and later on by Susman, O'Keefe and Wisner (1983) drew attention to the social and economic characteristics of the regions where hazards occurred. Nonetheless, frequently taken as a synonym for disasters, natural hazards were envisaged as the only components that needed to be understood in order to prevent disasters.

By the end of the twentieth century, initiatives such as the United Nations International Decade for Natural Disaster Reduction (IDNDR) still did not have an accurate comprehension of disasters, as targets were directed towards 'the reduction of the loss of life, property damage and social and economic disruption caused by natural disasters such as earthquakes, windstorms, tsunamis, floods, landslides, volcanic eruptions, wildfires, grasshopper and locust infestations, drought and desertification and other calamities of natural origin'. In fact, the blame was still placed on nature.

Specifically the goals of the decade were depicted as follows:

(a) To improve the capacity of each country to mitigate the effects of natural disasters expeditiously and effectively, paying special attention to assisting developing countries in the assessment of disaster damage potential and in the establishment of early warning systems and disaster-resistant structures when and where needed;

(b) To devise appropriate guidelines and strategies for applying existing scientific and technical knowledge, taking into account the cultural and economic diversity among nations;

(c) To foster scientific and engineering endeavours aimed at closing critical gaps in knowledge in order to reduce loss of life and property;

(d) To disseminate existing and new technical information related to measures for the assessment, prediction and mitigation of natural disasters;

(e) To develop measures for the assessment, prediction, prevention and mitigation of natural disasters through programmes of technical assistance and technology transfer,

TABLE 22.1. *Yearly number of disasters associated with natural hazards during the IDNDR (1990–1999)*

	1990	1991	1992	1993	1994	1995	1996	1997	1998	1999	*Total*
Earthquakes	43	29	25	20	25	26	13	23	30	33	**267**
Floods	60	77	59	84	88	95	91	95	94	122	**865**
Mass movement processes	7	12	12	26	11	16	24	13	22	18	**161**
Storms	138	66	76	108	81	81	77	79	88	106	**900**
Volcanoes	2	10	5	6	6	5	5	4	4	5	**52**
Total	**250**	**194**	**177**	**244**	**211**	**223**	**210**	**214**	**238**	**284**	**2245**

Elaborated with the information provided by the EM-DAT database

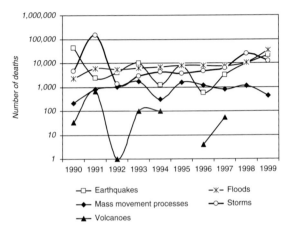

FIGURE 22.1. Number of deaths associated with natural hazards during the IDNDR (1990–1999). (Elaborated with the information provided by the EM-DAT database.)

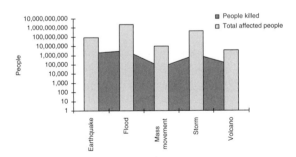

FIGURE 22.2. People killed and affected globally as a result of disasters associated with earthquakes, floods, mass movement processes, storms and volcanoes from 1900 to 2000. (Elaborated with the information provided by the EM-DAT database.)

hazards; however, there was one piece of the jigsaw missing: society = vulnerability.

Paradoxically, during the IDNDR, disasters were quite severe around the globe (Figures 22.1, 22.2 and Table 22.1). Just in terms of events associated with earthquakes, floods, mass movement processes, storms and volcanoes, 2,245 disasters occurred and human losses reached 440,411. On average, more than 220 disasters took place every year. Additionally, the largest numbers of disasters and victims, 284 and 71,394 respectively, were ironically registered in 1999, the last year of the decade (based on the information provided by EM-DAT database).

On the basis of documents such as Agenda 21 and the Rio Declaration on Environment and Development, and as part of a mid-decade assessment within the framework of the IDNDR, the Yokohama Strategy and Plan of Action for a Safer World was established in 1994. It became appreciated that the impact of *natural disasters* in terms of human and economic losses had risen and that society in general had become more vulnerable to *natural disasters*. Disaster prevention, mitigation, preparedness and relief were considered as the four contributing elements to the implementation of sustainable development policies, whereas the information, knowledge and some of the technology necessary to reduce the effects of natural disasters were seen as resources that should be freely available. Likewise, community involvement and active participation were contemplated as ways to gain greater insight into the individual and collective perception of development and risk.

Lack of convergence between goals and outcomes was the source of a further extension of the IDNDR: the establishment of the International Strategy for Disaster Reduction (ISDR). The key role of the ISDR was expressed in terms of its mission as: 'Building disaster resilient communities by promoting increased awareness of the importance of disaster reduction as an integral component of sustainable development, with the goal of reducing human, social, economic and environmental losses due to

demonstration projects, and education and training, tailored to specific disasters and locations, and to evaluate the effectiveness of those programmes.

In reality, the aim was merely to develop measures and strategies to increase the scientific understanding of natural hazards and related technological and environmental disasters'. This task has encompassed the adoption of different resolutions on the topics of natural disasters and vulnerability, and international cooperation to reduce the impact of the El Niño phenomenon.

Nourished by the preceding Kyoto Protocol, the Millennium Declaration and the Johannesburg Plan of Implementation of the World Summit on Sustainable Development (2002), among others, and as specific gaps and challenges were identified in various spheres (governance; risk identification, assessment, monitoring and early warning; knowledge management and education; reducing underlying risk factors; and preparedness for effective response and recovery), during the 2005 World Conference on Disaster Reduction, the Hyogo Framework for Action 2005–2015: Building the Resilience of Nations and Communities to Disasters was launched.

The Hyogo Framework (2005) aims at:

(a) The more effective integration of disaster risk considerations into sustainable development policies, planning and programming at all levels, with a special emphasis on disaster prevention, mitigation, preparedness and vulnerability reduction;

(b) The development and strengthening of institutions, mechanisms and capacities at all levels, in particular at the community level, that can systematically contribute to building resilience to hazards;

(c) The systematic incorporation of risk reduction approaches into the design and implementation of emergency preparedness, response and recovery programmes in the reconstruction of affected communities.

The resulting work, derived from international initiatives, has been lively and varied, but in the end somewhat shapeless. Strategies for vulnerability reduction and scientific sharing remain the biggest challenges.

22.3 1900–2000: beyond a century of disasters

As geomorphologists are involved with understanding the dynamics of the Earth's surface, processes and landforms, it is clear that earthquakes, floods, mass movement processes, storms and volcanic activity are among the hazards most closely linked to the discipline. Under such a scheme, a review of the disasters linked to the occurrence of those particular types of hazards is presented in this section.

From 1900 to 2000, a total of 5,902 disasters associated with natural hazards affected more than 2,800 million people, killed more than 6.5 million and produced economic damage of US$ 906 billion (based on information provided by the OFDA-CRED database). Asia was by far the most affected region. The aftermath of 2,587 disasters included 5.6 million people killed, 2,682 million affected and a total damage of approximately US$ 430 billion. These figures represent 43.83% of the total number of disasters, 87.01% of deaths, 93.82% of affected people and 47.43% of global economic damage (Table 22.2 and Figure 22.3).

In terms of number of disasters and people killed, Oceania was the least affected region with 337 events and 9,243 casualties. Regarding total affected people, the smallest impact, 5,822,547 inhabitants, was located in North America, whereas the least economic damage, slightly less than 13 billion US$, was registered in Africa.

Derived from floods there were 3,311,762 people killed (50.75% of the total), followed by 1,860,049 during earthquakes (28.50%), 1,204,118 casualties associated with storms (18.45%), 95,749 due to volcanic activity (1.47%), and 53,771 as a result of mass movement processes (0.82%). As far as affected people are concerned, 80.01% of the global figure was associated with floods. It corresponded to more than 2.2 billion

TABLE 22.2. *Yearly number of deaths associated with natural hazards during the IDNDR*

	1990	1991	1992	1993	1994	1995	1996	1997	1998	1999	*Total*
Earthquakes	42,853	2,454	4,033	10,088	1,242	7,739	576	3,159	9,573	21,869	**103,586**
Floods	2,251	5,852	5,315	6,150	6,771	7,962	8,041	7,685	10,653	34,807	**95,487**
Mass movement processes	214	814	1,035	1,749	307	1,571	1,155	801	1,141	445	**9,232**
Storms	4,641	146,297	1,342	2,965	4,239	3,763	4,581	6,150	24,935	12,274	**211,187**
Volcanoes	33	683	1	99	101	0	4	53	0	0	**974**
Total	**49,992**	**156,100**	**11,726**	**21,051**	**12,660**	**21,035**	**14,357**	**17,848**	**46,302**	**69,395**	**420,466**

Elaborated with the information provided by the EM-DAT database

TABLE 22.3. *Number of disasters and their impact associated with natural hazards from 1900 to 2000*

	Number of disasters	People killed	Injured	Affected	Homeless	Total affected	Total damage (US$ thousands)
Oceania	337	9,243	4552	6,274,177	396,561	6,675,290	13,264,357
Asia	2,587	5,678,303	2,704,755	2,562,417,779	117,544,554	2,682,667,088	429,807,023
Central America and the Caribbean	599	218,880	212,328	29,947,309	4,497,333	34,656,970	42,688,802
Africa	524	40,616	75,591	40,066,853	6,233,714	46,376,158	12,728,213
Europe	742	309,653	66,217	20,276,844	3,188,054	23,531,115	174,784,672
South America	531	236,772	299,333	54,730,626	4,470,969	59,500,928	28,930,251
North America	582	31,982	18,685	5,409,989	393,873	5,822,547	203,910,450
Total	**5,902**	**6,525,449**	**3,381,461**	**2,719,123,577**	**136,725,058**	**2,859,230,096**	**906,113,768**

Elaborated with the information provided by the EM-DAT database

(a)

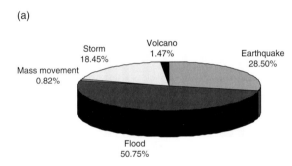

(b)

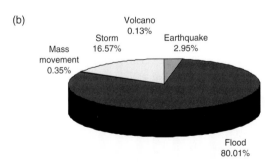

FIGURE 22.3. Percentage of people killed (a) and affected (b) as a result of disasters associated with earthquakes, floods, mass movement processes, storms and volcanoes from 1900 to 2000. (Elaborated with the information provided by the EM-DAT database.)

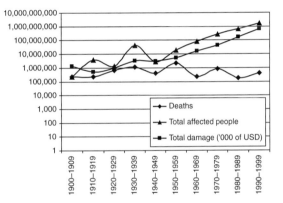

FIGURE 22.4. Decadal impact generated by disasters associated with earthquakes, floods, mass movement processes, storms and volcanoes during the twentieth century. (Elaborated with the information provided by the EM-DAT database.)

Disasters happen almost on a daily basis. As their frequency increases, losses are sky-rocketing, but the impact varies considerably in space and time. Even though during the 1950s there was the highest number of decadal casualties of the twentieth century (2,125,117 deaths), it was the 1990s when the highest number of people were affected and the largest economic damage took place (Figures 22.4 and 22.5).

Examples of the significant impact of disasters associated with hazards such as earthquakes, floods, mass movement processes, storms and volcanoes are illustrated in Table 22.4, in which the top 10 most important disasters associated with each of the mentioned hazards are presented according to the highest numbers of people killed.

inhabitants, whereas 473 million, the equivalent to 16.57% of the total were affected by storms, more than 84 million by earthquakes (2.95%), almost 10 million by landslides (0.35%) and 3.7 million by volcanoes (0.13%) (Figure 22.3).

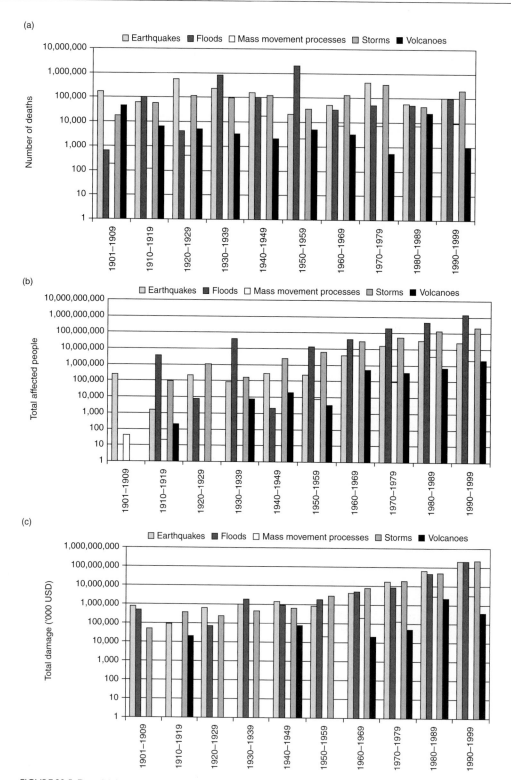

FIGURE 22.5. Decadal distribution of people killed (a) affected (b) and total economical damage (c) as a result of disasters associated with earthquakes, floods, mass movement processes, storms and volcanoes in the world from 1900 to 1999. (Elaborated with information provided by the EM-DAT database.)

TABLE 22.4. *Most important disasters associated with natural hazards in terms of people killed for the period 1900 to 2008.*

Volcanoes			Earthquakes		
Country	*Date*	*Killed*	*Country*	*Date*	*Killed*
Martinique	5/8/1902	30,000	China P Rep	7/27/1976	242,000
Colombia	11/13/1985	21,800	China P Rep	5/22/1927	200,000
Guatemala	10/24/1902	6,000	China P Rep	12/16/1920	180,000
Indonesia	3/23/1905	5,500	Indonesia (Tsunami)	12/26/2004	165,708
Indonesia	5/1/1919	5,000	Japan	9/1/1923	143,000
Guatemala	4/12/1905	5,000	Soviet Union	10/5/1948	110,000
Papua New Guinea	1/15/1951	3,000	China P Rep	5/12/2008	87,476
Cameroon	8/24/1986	1,746	Italy	12/28/1908	75,000
Indonesia	1/3/1963	1,584	Pakistan	10/8/2005	73,338
St Vincent and The Grenadines	5/7/1902	1,565	China P Rep	12/26/1932	70,000

Floods			Storms		
Country	*Date*	*Killed*	*Country*	*Date*	*Killed*
China P Rep	7/1/1931	3,700,000	Bangladesh	11/12/1970	300,000
China P Rep	7/1/1959	2,000,000	Bangladesh	4/29/1991	138,866
China P Rep	7/1/1939	500,000	Myanmar	5/2/2008	138,366
China P Rep	4/18/1905	142,000	China P Rep	7/27/1922	100,000
China P Rep	3/25/1905	100,000	Bangladesh	10/1/1942	61,000
China P Rep	7/1/1949	57,000	India	4/18/1905	60,000
Guatemala	10/1/1949	40,000	China P Rep	Aug-1912	50,000
China P Rep	Aug-1954	30,000	India	10/14/1942	40,000
Venezuela (Flash flood)	12/15/1999	30,000	Bangladesh	5/11/1965	36,000
Bangladesh	7/1/1974	28,700	Bangladesh	5/28/1963	22,000

Mass movement		
Country	*Date*	*Killed*
Soviet Union	5/2/1905	12,000
Peru	Dec-1941	5,000
Honduras	9/20/1973	2,800
Italy	10/9/1963	1,917
Philippines	2/17/2006	1,126
India	10/1/1968	1,000
Colombia	9/27/1987	640
Peru	3/18/1971	600
China P Rep	3/23/1934	500
India	9/18/1948	500

Elaborated with the information provided by the EM-DAT database

22.4 Geomorphology: a brief account of contributing research, methodologies and techniques

Within geomorphology, one of the earliest and most distinctive contributions to disaster prevention was related to the understanding of the dynamics of the Earth, including the relationships among landforms and processes.

As science and technology have developed, so the methodologies and techniques used in geomorphology have significantly progressed. They include such procedures as: (a) geomorphometry, (b) geomorphological mapping,

(c) detailed survey, (d) Global Positioning Systems, (e) dendrogeochronology, (f) dendrogeomorphology, (g) lichenometry, (h) slope instability assessments, (i) quantitative assessment, (j) Geographical Information Systems, (k) Remote sensing (satellite and radar imagery), (l) LiDAR (Light Detection and Ranging), (m) processes modelling, (n) simulations, (o) instrumentation and (p) monitoring.

While some earth scientists seek to make their studies more precise in terms of the understanding of specific mechanisms, geomorphologists are concerned with the interactions of a wide variety of factors controlling the Earth's surface dynamics, including the role of anthropogenic activities. Consequently, the quantity and quality of disaster prevention related investigations undertaken by geomorphologists have improved markedly in recent years.

Geomorphological approaches have catalyzed and very often facilitated applied disaster investigations, and the progress made has been illustrated in many publications. An extensive review of the different investigations developed by geomorphologists in terms of disaster mitigation and prevention is beyond the scope of this chapter. However, the role of geomorphologists in hazard understanding and assessment, vulnerability analysis, risk management, sustainable development and disaster prevention has been expressed not only by the daily achievements of our discipline, but through all the contributions in this volume, which include processes and applications.

As we have seen in this volume, studies devoted to processes usually include: morphotectonics and seismic hazards; volcanic hazards; mountain hazards; snow avalanches; landslide hazards, sedimentary budgets and climatic change; floods and high-magnitude floods; coastal hazards, weathering hazards, hazards associated with karst, dune migration; soil erosion; and desertification and land degradation in arid and semi-arid regions. Applications of geomorphology to risk assessment and management involve: GIS for the assessment of risk; hazard assessment for risk analysis and risk management; vulnerability analysis in geomorphic risk assessment; geomorphological hazards and global climate change; and geomorphic hazards and sustainable development.

22.5 Conclusions: the future agenda

In contrast to the efforts made during the IDNDR, at the beginning of the twenty-first century disaster reduction underwent another wave of environmental and public policy relevance, driven by the global impact of several disasters, particularly of the South Asia tsunami and hurricane Katrina in 2004 and 2005, respectively.

Geomorphology contributed not only to the understanding of these types of events but also to the generation and application of theoretical and methodological currents within disaster science and the evident links with social scientists. It is very important to acknowledge, however, that thanks to the efforts made by those social scientists – geographers among them – the real meaning of disaster risk is now beginning to be mapped worldwide.

With the benefit of hindsight it is easy to see why a geomorphological approach focuses attention on the complex interactions of both the natural and cultural landscapes, giving attention to the impact generated by anthropogenic activities. This recognizes the central role of climate change in future hazard scenarios in addition to the spatial contrasts of societies with high vulnerability and low resilience.

Future perspectives on disaster occurrence can be derived from the climatic scenarios produced by the IPCC (2008a, b). By 2100, climate change is expected to cause an increase of 1.1–6.4 °C in the global average surface air temperature, and a rise in sea level of between 18 and 59 cm. What is more, it is probable that heatwaves and heavy precipitation events will take place more frequently. More intense precipitation would take place at higher latitudes, and the activity of tropical cyclones and hurricanes might become more intense and frequent at lower latitudes.

Although the extent to which future likely climate change scenarios are understood and accepted varies among scientists, new pathways developed by geomorphologists will lead to greater comprehension of the anthropogenic impact on the environment, and particularly on the landscape. Some consequences of global warming (Table 22.5) that will impact upon hazards include: an increase in the percentage of precipitation that falls as rain rather than snow; increasing snowfall in high latitudes; a larger spread, frequency and intensity of cyclones; inundation of lower areas; accelerated coast recession and many others (Goudie, 2004).

The inescapably political character of the impact generated by disasters should be seen as a marriage of convenience between academia and public policy implementation. At least part of the problem can be traced to the establishment of a set of agendas, strategies, plans and projects to bring together reduction and mitigation actions in order to prevent, confront and recover from disasters. Now that the range of reliable scientific techniques has been extended to understand hazards, risk management needs to be directed towards reducing vulnerability, minimizing social impact, ensuring the best possible security conditions, and decreasing material losses within community-based and trans-disciplinary

TABLE 22.5 *Some geomorphological consequences of global warming (Goudie, 2004).*

Hydrologic
Increased evapotranspiration loss
Increased percentage of precipitation as rainfall at expense of winter snowfall
Increased precipitation as snowfall in very high latitudes
Possible increased risk of cyclones (greater spread, frequency and intensity)
Changes in state of peatbogs and wetlands
Less vegetational use of water because of increased CO_2 effect on stomatal closure

Vegetational controls
Major changes in latitudinal extent of biomes
Reduction in boreal forest, increase in grassland, etc.
Major changes in altitudinal distribution of vegetation types (*c.* 500 m for 3 °C)
Growth enhancement by CO_2 fertilization

Cryospheric
Permafrost, decay, thermokarst, increased thickness of active layer, instability of slopes, river banks and shorelines
Changes in glacier and ice-sheet rates of ablation and accumulation
Sea-ice melting

Coastal
Inundation of low-lying areas (including wetlands, deltas, reefs, lagoons, etc.)
Accelerated coast recession (particularly of sandy beaches)
Changes in rate of reef growth
Spread of mangrove swamp

Aeolian
Increased dust storm activity and dune movement in areas of moisture deficit

Soil erosion
Changes in response to changes in land use, fires, natural vegetation cover, rainfall, erosivity, etc.
Changes resulting from soil erodibility modification (e.g. sodium and organic contents)

Subsidence
Desiccation of clays under conditions of summer drought

academic research frameworks (Alcántara-Ayala, 2004; Alcántara-Ayala *et al.*, 2004). Moreover, it should be clear that disasters are not simply natural, but are the result of a social configuration in space and time (Westgate and O'Keefe, 1976; Maskrey, 1993; Quarentelli, 1998).

Considering the complexity of disasters, development and implementation of prevention strategies in the future could only possibly be achieved by undertaking a trans-disciplinary and holistic approach reinforced predominantly with solid and down-to-earth actions to reduce vulnerability; the latter being indeed the challenge of the present, derived from the lessons of the past, in the light of wisdom for the future (Alcántara-Ayala, 2007).

Finally, from the contemporary mass of remarks allied to disaster prevention the following words will stand out for ever:

More effective prevention strategies would save not only tens of billions of dollars, but save tens of thousands of lives. Funds currently spent on intervention and relief could be devoted to enhancing equitable and sustainable development instead, which would further reduce the risk for war and disaster. Building a culture of prevention is not easy. While the costs of prevention have to be paid in the present, its benefits lie in a distant future. Moreover, the benefits are not tangible; they are the disasters that did NOT happen. (Kofi Annan)

Acknowledgements

Disaster information was kindly provided by the OFDA/CRED database (EM-DAT).

Special thanks are due to Regina Below and Jean Michel Scheuren from Université Catholique de Louvain.

References

Alcántara-Ayala, I. (2002). Geomorphology, natural hazards, vulnerability and prevention of natural disasters in developing countries. *Geomorphology*, **47**, 107–124.

Alcántara-Ayala, I. (2004). Flowing mountains in Mexico: incorporating local knowledge and initiatives to confront disaster and promote prevention. *Mountain Research and Development*, **24**(1), 10–13.

Alcántara-Ayala, I. (2007). International Geographical Union, E-Newsletter 9, July. Report on the ICSU Young Scientists Conference, Lindau, Germany, 4–6 April, 2007.

Alcántara-Ayala, I., Lopez-Mendoza, M., Melgarejo-Palafox, G., Borja-Baeza, R. C. and Acevo-Zarate, R. (2004). Natural hazards and risk communication strategies among indigenous communities: Shedding light on accessibility in Mexico's mountains. *Mountain Research and Development*, **24**(4), 298–302.

Barkun, N. (1974). *Disaster and the Millenium*. New Haven, CT: Yale University Press.

Bauer, B. (2004). Geomorphology. In A. S. Goudie (ed.), *Encyclopedia of Geomorphology*. London: Routledge and International Association of Geomorphologists, pp. 428–434.

Burton I. and Kates, R. W. (1964). The perception of natural hazards in resource management. *Natural Resources Journal*, **3**, 412–441.

Burton, I., Kates, R. W. and White, G. F. (1968). *The Human Ecology of Extreme Geophysical Events*. Department of Geography, Natural Hazards Research Working Paper No. 1, University of Toronto.

Burton, I., Kates, R. W. and White, G. F. (1978). *The Environment as Hazard*. New York: Oxford University Press.

Burton, I., Kates, R. and White, G. (1993). *The Environment as Hazard*, 2nd edition. New York: Guilford Press.

Embleton, C., Federici, P. R. and Rodolfi, G. (eds.) (1989). *Geomorphological Hazards*. Supplementi di Geografia Fisica e Dinamica Quaternaria, http://www.dst.unipi.it/gfdq/sup_2.html.

Fritz, C. E. (1961). Disasters. In R. K. Merton and R. A. Nisbet (eds.), *Contemporary Social Problems*. New York: Harcourt, pp. 651–694.

Fritz, C. E. (1968). Disasters. In *International Encyclopedia of the Social Sciences*. New York: The Macmillan Company & The Free Press, pp. 202–207.

Gares, P. A., Sherman, D. J. and Nordstrom, K. F. (1994). Geomorphology and natural hazards. *Geomorphology*, **10**, 1–18.

Goudie, A. S. (2004). Global warming. In A. S. Goudie (ed.), *Encyclopedia of Geomorphology*. London: Routledge and International Association of Geomorphologists, pp. 479–485.

Haff, P. K. (2002). Neogeomorphology. *American Geophysical Union EOS Transactions*, **83**(29), 310–317.

Hyogo Framework (2005). *Hyogo Framework for Action 2005–2015: Building the Resilience of Nations and Communities to Disasters*. http://www.unisdr.org/eng/hfa/docs/Hyogo-framework-for-action-english.pdf.

IPCC (2008a). *Fourth Assessment Report, Working Group I, Summary for Policymakers*. http://195.70.10.65/pdf/assessment-report/ar4/wg1/ar4-wg1-spm.pdf.

IPCC (2008b). *Fourth Assessment Report, Working Group II Report*. http://195.70.10.65/ipccreports/ar4-wg2.htm.

Kates, R. W. (1962). *Hazard and Choice, Perception in Flood Plain Management*. Research Paper No. 78. Department of Geography, University of Chicago.

Kates, R. W. (1971). Natural hazards in human ecological perspective: hypotheses and models. *Economic Geography*, **47**, 428–451.

Maskrey, A. (1993). *Los Desastres no son Naturales*. Santa Fé de Bogotá, Colombia: Tercer Mundo Editores.

Panizza, M. (1996). *Environmental Geomorphology*. Oxford: Elsevier.

Quarantelli, E. L. (1996). Basic themes derived from survey findings on human behavior in the Mexico City earthquake. *International Sociology*, **11**(4), 481–499.

Rosenfeld, C. L. (1994). The geomorphological dimensions of natural disasters, *Geomorphology*, **10**, 27–36.

Sauer, C. O. (1925). The morphology of landscape. *University of California Publications in Geography*, **2**(2), 19–53.

Slaymaker, O. (1996). Introduction. In O. Slaymaker (ed.), *Geomorphic Hazards*. Chichester: Wiley, pp. 1–7.

Susman, P., O'Keefe, P. and Wisner, B. (1983). Global disasters: a radical interpretation. In K. Hewitt (ed.), *Interpretations of Calamity*. Boston: Allen & Unwin, pp. 264–283.

Westgate, K. N. and O'Keefe, P. (1976). *Some Definitions of Disaster*. Disaster Research Unit Occasional Paper No. 4, Department of Geography, University of Bradford.

White, G. F. (1973). Natural hazards research. In R. J. Chorley (ed.), *Directions in Geography*. London: Methuen, pp. 193–216.

White, G. F. (ed.) (1974). *Natural Hazards: Local, National, Global*. Oxford: Oxford University Press.

Yokohama Strategy and Plan of Action for a Safer World (1994). *Guidelines for Natural Disaster Prevention, Preparedness and Mitigation*, World Conference on Natural Disaster Reduction, Yokohama, Japan, 23–27 May, http://www.unisdr.org/eng/about_isdr/bd-yokohama-strat-eng.htm.

23 Geomorphology and the international agenda: concluding remarks

Irasema Alcántara-Ayala

Natural hazards are a part of life. But hazards only become disasters when people's lives and livelihoods are swept away. The vulnerability of communities is growing due to human activities that lead to increased poverty, greater urban density, environmental degradation and climate change.

> UN Secretary General, Kofi Annan, 8 October 2003,
> International Day for Disaster Reduction

Empirical investigations of geomorphological hazards have existed since humankind wondered about the nature of the Universe. The ancient Greeks, for example, paid attention to phenomena such as earthquakes and volcanoes. During the Renaissance, scientists such as Leonardo da Vinci contributed to the early scientific development of topics related to geomorphology. Later on, at the beginning of the nineteenth century as specialization took place, instrumentation and measurement technology improved and diverse schools of thought appeared on the scene (Bauer, 2004). Contributions became concentrated on the complexity of nature.

Accordingly, the duality between nature (hazards) and society (vulnerability) can not be regarded as new. In the nineteenth century, Humboldt and Ritter were already aware of the necessity of addressing the planet from both natural and human global perspectives. As Ritter remarked, 'The earth and its inhabitants stand in the closest mutual relations, and one element cannot be seen in all its phases without the others.' It seems inexorable therefore that understanding risk involves a full comprehension of the interactions among natural and cultural landscapes.

In the beginning, the term disaster was practically understood as a calamity caused by nature; as a consequence, most of the scientific investigations focused on disasters were concentrated on the understanding of natural hazards. Notwithstanding the scientific progress, the impact of disasters continued to grow during recent decades (see Chapters 21 and 22). As a response, social scientists made a major contribution in terms of pointing out the great significance of vulnerability within the risk equation. For geomorphologists however, the challenges to be faced have gone far beyond demanding; thus, natural geomorphological hazards need to be understood at the same time as those derived directly or indirectly from anthropogenic activities, in addition to integrating the comprehension of vulnerability factors, which coupled, govern the levels of risk and consequently, disasters occurrence.

Besides contributing to the comprehension of geomorphological hazards, geomorphologists are progressing very fast in their understanding of anthropically driven Earth surface processes; this has given birth to the term 'neo-geomorphology' (Haff, 2002). However, although geomorphologists are among the scientists who have been naturally involved in the study of hazards and risks, perhaps at this stage one of the obvious indicators to geomorphology's utter lack of visibility in disaster prevention is their limited presence in international strategies and programmes.

Conversely, as geomorphology enters the twenty-first century, new directions have started to be established. As suggested by Slaymaker (see Chapter 4), for instance, concepts such as panarchy have much to offer to our discipline and to the field of disasters. The dimension of panarchy can be defined in terms of understanding the structure in which systems – both natural and human – are interconnected in constant adaptive cycles of growth, accumulation, restructuring, and renewal at varying spatial and temporal scales. Hence, 'each adaptive system, at its own spatial scale, evolves towards a critical condition leading either to collapse (hazard) or to self-reorganization (adaptation or mitigation in socio-economic context)' (Slaymaker, this volume).

At this point, and within the sphere of Earth sciences, it is clear that many strands of the dynamics of natural hazards

Geomorphological Hazards and Disaster Prevention, eds. Irasema Alcantara-Ayala and Andrew S. Goudie. Published by Cambridge University Press. © Cambridge University Press 2010.

have been analyzed and properly understood by scientists, yet, there still remains the need to comprehend and reduce vulnerability. Above all, we must never forget that it is vulnerability that determines the magnitude of the impact of hazards on societies (Alcántara-Ayala, 2002), and therefore it can be considered as the major, dynamic and complex driving force of disasters. Accordingly, disasters are not natural; although increasing human and economic losses from disasters are continuously reported on a yearly and global basis, they result from the fact that societies are becoming more and more vulnerable in space and time.

Particularly, the spatial-temporal dimensions of geomorphological hazards are related to the assessment of physical parameters such as magnitude, frequency, duration, areal extent, speed of onset, spatial dispersion and temporal interval. While their magnitude and frequency can be assessed in terms of the phenomena *per se*, impact on societies is derived from the speed and extent of the impact of those hazards on the exposed vulnerable communities. Therefore, risk management should necessarily involve a full understanding of the interactions among hazards, vulnerability and resilience; in other words, besides natural events, it should take into account weaknesses and capabilities of human groups.

While hazards are addressed from a scientific perspective – in this particular case, from a geomorphological point of view – decision makers should create feedback conditions associated with scientific developments to benefit societies (Figure 23.1). It should be also clear that there is a need for the development and implementation of risk management strategies in accordance with social realities and based on sound approaches, involving all actors: the scientific community, stakeholders, decision makers, authorities and exposed societies. In this fashion, risk management should ensure conditions for reducing vulnerability, lessening the social and economic impact of disasters, guaranteeing the best possible security environment, and decreasing direct and indirect losses based on community approaches (Alcántara-Ayala, 2004; Alcántara-Ayala *et al.*, 2004), trans-disciplinary academic frameworks and supported by the understanding of the trilogy hazards–vulnerability–resilience (Figure 23.2), as suggested in the Atomium model (Alcántara-Ayala, 2009).

As this volume was divided into two parts, it was possible to analyze both processes and processes and applications of geomorphology to risk assessment and management, including vulnerability. In terms of processes, the key points made by the contributions were concentrated on three main aspects: (i) level of uncertainty with reference to hazard evaluation and prediction (Gares *et al.*, 1994, see Chapters 3, 5, 8, 13, 17, 18, 19, 20, 21); (ii) methodological approaches

FIGURE 23.1. Elements of risk management for disaster prevention: decision making should be based on hazard understanding and take into account resilience of human groups in order to reduce vulnerability.

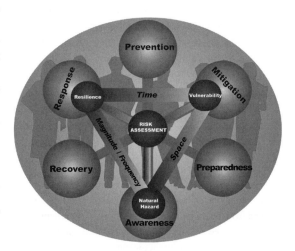

FIGURE 23.2. The Atomium risk management model. (Source: Alcántara-Ayala (2009).)

for understanding and assessing geomorphological hazards, such as geomorphometry, geomorphological mapping, detailed survey, lichenometry (Chapter 3), dendrogeochronology, dendrogeomorphology, palaeohydraulic estimations, palaeohydrologic records (Chapters 3, 8, 10), palaeoclimate and palaeoenvironmental reconstructions (Chapters 4, 6, 8, 10, 15, 20), palaeolimnology (Chapter 18), modelling and simulations; and (iii) the role of new technology, including Global Positioning Systems (GPS), Geographical Information Systems (GIS) and Web-GIS (Chapter 17), remote sensing, Light Detection and Ranging (LiDAR), instrumentation and monitoring.

As far as applications of geomorphology to risk assessment and management are concerned (Rosenfeld, 1994), there is an urgent need to understand the extent to which humans influence (see Chapter 18) and transform geomorphological systems (Goudie, 2006) either directly (through land-cover and land-use changes) or indirectly (through climate change). Additionally, there is no doubt that, on the basis of multi- and trans-disciplinary approaches, more holistic and sensitive social and scientific progress is required.

It is clear that although further scientific developments on natural hazards can realistically be achieved by geomorphologists and other Earth scientists, a reduction in vulnerability and hence of risk will continue to be highly desirable. Indeed, as global economies sink, lack of financial rationale, in addition to political commitment and awareness, will play an adverse role in development. As patterns of climate conditions are changing (IPCC, 2008a and b), and population is growing, vulnerability is undoubtedly increasing whilst resilience is decreasing. Consequently, the demands of international agendas for sustainable development and disaster prevention would become more difficult to achieve. Science is moving forwards, but humankind has become the largest obstacle for guaranteeing its own progress.

Finally, within the dimensions of geomorphology, the challenges that need to be achieved in order to move into the international agenda and to fulfil the desire of preventing disasters, have to be directed towards solving a complex puzzle, a conundrum that can be called integrated and multidimensional assessment and management of risks. The latter must be integrated by several indispensable pieces: adequate interpretation of results and uncertainties derived from evaluations of hazards; full comprehension of dynamic vulnerabilities; understanding and reasonable use of scales – for both hazards and vulnerability issues – in space and time; land use planning; differentiation between forecasting and prevention; establishment of non-static mitigation measures; inclusion of risk perception; identification of acceptable risks; implementation of risk communication initiatives; formation of human resources and professionals in the field; targeting real sustainable development and co-responsibility of actors and actions.

Acknowledgements

Special thanks are due to Miss Laura Diana López-Ascencio from the Institute of Geography, UNAM, who kindly assisted in editorial work of the book.

References

Alcántara-Ayala, I. (2002). Geomorphology, natural hazards, vulnerability and prevention of natural disasters in developing countries. *Geomorphology*, **47**, 107–124.

Alcántara-Ayala, I. (2004). Flowing mountains in Mexico: incorporating local knowledge and initiatives to confront disaster and promote prevention. *Mountain Research and Development*, **24**(1), 10–13.

Alcántara-Ayala, I. (2009). Geomorphosite management in areas sensitive to natural hazards. In E. Reynard, P. Coratza, and G. Regolini (eds.), *Geomorphosites: Assessment, Mapping and Management*. Munich: Pfeil Verlag, pp. 163–173.

Alcántara-Ayala, I., López-Mendoza, M., Melgarejo-Palafox, G., Borja-Baeza, R. C., Acevo-Zarate, R. (2004). Natural hazards and risk communication strategies among indigenous communities: shedding light on accessibility in Mexico's mountains. *Mountain Research and Development*, **24**(4), 298–302.

Bauer, B. (2004). Geomorphology. In A. S. Goudie (ed.), *Encyclopedia of Geomorphology*. London: Routledge and International Association of Geomorphologists, pp. 428–434.

Gares, P. A., Sherman, D. J. and Nordstrom, K. F. (1994). Geomorphology and natural hazards. *Geomorphology*, **10**, 1–18.

Goudie, A. S. (2006). *The Human Impact on the Natural Environment*, 6th edition. Oxford: Blackwell Publishing.

IPCC (2008a). *Fourth Assessment Report, Working Group I, Summary for Policymakers*. http://195.70.10.65/pdf/assessment-report/ar4/wg1/ar4-wg1-spm.pdf.

IPCC (2008b). *Fourth Assessment Report, Working Group II Report*. http://195.70.10.65/ipccreports/ar4-wg2.htm.

Rosenfeld, C. L. (1994). The geomorphological dimensions of natural disasters. *Geomorphology*, **10**, 27–36.

Index